Ecological Processes in Coastal and Marine Systems

MARINE SCIENCE

Coordinating Editor: Ronald J. Gibbs, *University of Delaware*

A Continuation Order Plan is available for this series. A continuation order will bring delivery of each new volume immediately upon publication. Volumes are billed only upon actual shipment. For further information please contact the publisher.

Ecological Processes in Coastal and Marine Systems

Edited by
Robert J. Livingston

Florida State University
Tallahassee, Florida

PLENUM PRESS • NEW YORK AND LONDON

Library of Congress Cataloging in Publication Data

Main entry under title:

Ecological processes in coastal and marine systems.

(Marine science; v. 10)
"Proceedings of a conference . . . conducted by the Florida State University Grad-
ate Research Council, the Department of Biological Science (Florida State Uni-
versity) and the Center for Professional Development and Public Service, and held
at the Florida State University, Tallahassee, April 13–15, 1978."
Includes index.
1. Coastal ecology—Congresses. 2. Marine ecology—Congresses. I. Livingston,
Robert J. II. Florida. State University, Tallahassee. Graduate Research Council.
III. Florida, State University, Tallahassee. Dept. of Biological Science. IV. Florida.
State University, Tallahassee. Center for Professional Development and Public Service.
QH541.5.C65.E26   574.5'26   79-21388
ISBN 0-306-40318-8

Proceedings of a conference on Ecological Processes in Coastal and Marine Systems
conducted by the Florida State University Graduate Research Council, the Department of Biological Science (Florida State University), and the Center for Professional
Development and Public Service and held at the Florida State University, Tallahassee,
April 13-15, 1978.

© 1979 Plenum Press, New York
A Division of Plenum Publishing Corporation
227 West 17th Street, New York, N.Y. 10011

Preface

This volume is based on the proceedings of a conference held at Florida State University in April, 1978. This conference was supported by the Florida State University Graduate Research Council, the Department of Biological Science (F.S.U.), and the Center for Professional Development and Public Service. Particular recognition should be made of the efforts of Dr. Anne Thistle in the organization of the conference and the completion of this book. Julia K. White and Sheila Marrero produced the typescript.

The principal objective of the conference was to assemble a group of marine scientists from diverse disciplines to discuss the state of marine ecology with particular attention to new research directions based on previous studies. Emphasis was placed on the integration of different research approaches and on the application of established procedures to various environmental problems. An effort was made to eliminate traditional disciplinary boundaries which often hinder our understanding of marine systems. There was generally wide latitude for review and speculation concerning such topics as physico-chemical processes, productivity and trophic interactions, population distribution and community structure, and natural or anthropogenic disturbance phenomena. Throughout, the usual miniaturization of the scope of discussion was subordinate to a frank appraisal of the present status of marine research.

Although many introductory ecological texts stress the so-called ecosystem approach, individual marine research projects seldom encompass this broad course. There is, in fact, a real need for system-wide studies at both the theoretical and applied levels. Strict separation of mutually exclusive disciplines is characteristic, and such compartmentalization tends to obscure natural interactions. This is not to say that all marine research should be based on the systems approach. However, there is presently a gap in the research effort which reflects a lack of attention to interdisciplinary relationships which, after all, should form the basis for the establishment of ecological principles. Spatial/temporal variability is often ignored or underestimated while cause and effect relationships are based on bold assumptions and tenuous

data. In spite of recent recognition of the importance of hypothesis testing in the environmental sciences, there has been a
lamentable lack of objective appraisal of the assumptions underlying
research approaches and modelling efforts. Many "long-term" studies
span only one- to two-year periods -- closer to the duration of a
typical masters thesis than to the actual temporal structure of a
particular system. The tautology becomes complete when hypothetical
formulations are modelled, thereby enabling mere speculation to
become ecological law through repetition rather than proof.

One important issue regarding the testing of any given hypothesis should be a clear definition of its underlying assumptions.
Ideally, they should be based on empirical knowledge. This tenet,
again, is currently out of favor with a broad spectrum of investigators. The extension of research paradigms to areas unrelated
to those in which they were originally derived is lamentably
commonplace in ecological research today. The tacit assumptions
associated with such old chestnuts as the rocky intertidal concept
of competition and predation, the role of marshes in coastal
productivity, and the rigid compartmentalization in many trophic
schemes need re-examination. The relationship between coastal
(inshore) and offshore components remains virtually unknown for
most marine systems. This has not prevented workers in these
fields from being extremely active in the preparation of predictive
models, both for the purpose of describing natural processes and
for development of impact criteria for an array of pollutants.
Overall, the marine sciences remain fragmented and without much
discernible purpose, from either the standpoint of redefining past
generalizations or that of proposing new research goals.

The marine environment, in its inaccessibility to routine
observation, remains difficult to study and refractory to generalization. There is an established framework of ecological
observation, but it remains incomplete and largely bereft of consistent (apparent) organization. The interactive relationships
of various diverse biological processes underlie the basic
complexity of many different marine systems. Yet, current research
programs increasingly emphasize the obscure in what has become a
highly inefficient random walk to significant findings. The
observational and experimental approaches in ecology are often
treated as mutually exclusive or even opposing methods of analysis
rather than as the complementary and interdependent methods they
are. Experimental ecology is often carried out without regard to
or acknowledgement of ambient physical variability and the natural
history of component species. Hence, the effects of predation are
researched without the slightest idea of the timing and structure
of the trophic interactions of a given system. On the other hand,
descriptive studies are frequently directed at causal phenomena
without acknowledgement of factors which can be explained, in a
functional sense, through experimental procedures. Hence, the

integration of experimental and field approaches to marine ecology
remains undeveloped and underestimated.

When systems models stress simplicity as a substitute for
accuracy the result is the celebration of accommodation and the
tacit denial of the complexity of natural systems. To derive impact
criteria from perturbation or stress models without basic knowledge
of system performance is akin to ascribing functional features to
a mythical organism. The identification of optimal system parameters based on the assumption of some higher-level state of
equilibrium is an essential step toward the final acceptance of
such circular reasoning. The continual avoidance of the often
tedious but necessary background work concerning variability at
many levels of biological hierarchies simply postpones substantive
advances in our understanding of marine ecosystems. Breaking up
essential research goals into costly half-steps by individual
researchers not only renders this approach expensive, but leads
to the perpetual development of hypotheses which are not worthy
of testing.

It is hoped that this series of papers will contribute to the
airing of some new questions.

                              Robert J. Livingston
                              Tallahassee, Florida, 1979

Contents

I. Introduction

ECOLOGY: A SCIENCE AND A RELIGION

Paul K. Dayton

Scripps Institution of Oceanography

INTRODUCTION

Is ecology a healthy, vibrant science? Whereas a rigorous definition of science is difficult, it is nonetheless clear that there are symptoms of ill health in ecology. The lay enthusiasm and concern for ecology stem more from the public's failure to understand nature, and to influence society to relate wisely to nature, than they do from any notable achievements of ecologists. Indeed, there are few, if any, universally accepted and acclaimed breakthroughs in ecology, and, for that matter, few are likely because the field is conspicuously splintered into subdisciplines with serious loss of communication. For example, there are dichotomies between pure and applied branches of ecology which, although the two often pursue almost identical questions, involve different workers from different schools who have different values and seemingly speak different languages. In the same sense, there are serious dichotomies between theoretical and empirical ecologists which have led to a mutual contempt that is counterproductive to the proper pursuit of knowledge. In addition to the loss of communication between the theorists and empiricists, we find that the field is splintered into fairly rigidly defined "camps" or disciplines such as those working with descriptive problems, systematics, population regulation, ecosystem-system analysis, energetics, theory, or experimental ecology. The lack of communication and mutual respect between these groups is distressing.

Another symptom of a serious problem is that after at least 70 years, during which ecology has been considered a respectable scientific discipline, we are usually unable to offer substantial positive contributions to the many societal problems

3

confronting us. Thus, I may not be capable of a rigorous definition of a healthy science -- indeed, the philosophers themselves
seem to have little common ground and even subdivide *themselves*
and each other into subscripts (see *Feyerabend*, 1970; *Kuhn*, 1970c;
Lakatos, 1970)--but there are abundant indications that ecology
lacks traits of other, apparently more successful sciences (*Platt*,
1964). At the same time, I emphasize that I am not arguing that
any of the problems discussed in this essay are unique to ecology:
on the contrary, our problems are omnipresent in all scientific
disciplines, and they are already well recognized in the literature. The purpose of this essay is to focus attention on some
of the issues with the hope that, once they are clearly identified,
they can be resolved.

I believe there are three general explanations for some of
our problems: (1) the practical difficulties encountered in
describing our complex natural world; (2) the serious problems of
scientific methods apparent in the approach of many ecologists to
science; and (3) a general lack of realistic creativity that goes
beyond the problems of methods.

PRACTICAL PROBLEMS AND INADEQUATE QUESTIONS

Certainly all of us are aware of the enormous possibility of
nature and agree that an infinite number of facts might be
collected. Furthermore, we are familiar with researchers who
collect descriptive data and then rely on hindsight or, more
commonly, on computers to give their "research" meaning. There
is often a pathetic appeal to mathematicians and physicists to
help us find meaning in these data collected without questions.
Accordingly, we are familiar with a large and growing literature
dealing with the inner meanings of summary statistics and sampling programs. Other disciplines have long recognized the flaws
in the statement by John Stuart Mill that science progresses by
the continuous collection of data which will eventually show
important patterns; but we find in ecology a continuing popu-
larity of the unquestioning collection of data and the reliance
upon summary statistics which few understand but which have camps
of disciples (*Hurlbert*, 1978). While adequate descriptive
ecological data are important, most of us would agree with *Eccles*
(1970) that "scientific literature is overwhelmed by mere
reportage of observations that are published merely as observations without organic relationship to precisely formulated
hypotheses. Such observations are scientifically meaningless.
They are boring and soon to be forgotten."

Ecologists who know this seem to suffer from an insecurity
caused by *ex post facto* questioning and from an inability to frame
pertinent questions around which they can arrange the essential

data.   Thus we have a history of fads or issues that serve as the
"questions" or as reasons for "puzzle solving" (*Kuhn*, 1970a) or
"busy work" that may serve as research.   Certainly "good" questions of general interest can be, and often are, very popular;
it is not obvious how to differentiate a popular but worthwhile
question from an influential "fad."   Similarly, a once-important
question can develop into unproductive "busy work."   Nevertheless,
with hindsight we can make such differentiations and can see that
we have been influenced by many fads that have served as "questions"
for ecological research.   These run the gamut from a fascination
with pure typological descriptions of nature (including defini-
tions of biomes, biotic provinces, life zones, and various
associations) to community development and succession; from food
pyramids and energy flow to population regulation that is density-
independent or dependent; from the diversity-stability debacle to
systems analysis, and then to simplified models alleged to help
us comprehend evolutionary mechanisms.   In each case, ecologists
rarely grapple with the issue of what type of explanations they
seek (see *Weaver*, 1964); instead, they rally around brilliant
leaders who often acquire messianic stature and whose opinions
are rarely challenged by their disciples.   In hindsight, we
could say that these large questions or approaches acquire
paradigm status [I use *Kuhn's* (1970a) term in a utilitarian sense,
in full sympathy with the problems *Shapere* (1971) and others have
with an exact definition] since counter examples are discarded as
being "noise" and the larger overview is rarely challenged.

METHODS OF SCIENCE

I have been greatly influenced by John *Platt's* (1964) observation that some scientific disciplines move forward rapidly,
while others remain stagnant, with the practitioners of the latter
defeatedly wringing their hands and complaining about the complexities and difficulties of their fields.   Platt argued that
the problem is simply one of rigorous scientific methodology and
that practitioners in the vigorous, rapidly moving fields pose
their questions as alternate, testable hypotheses with sufficient
precision that some alternatives are rejected.   It is apparent
that ecologists have had difficulty moving ahead in this fashion.
Certainly the values of induction and proper experimentation were
clearly and eloquently demonstrated as early as the middle of the
13th Century by Roger Bacon (1214-1292; see *Barnes*, 1965).   Louis
Pasteur and other scientists also used these techniques with
striking efficiency.   Further, ecologists are not alone in
struggling with highly complex problems, as most scientific
disciplines are inherently complex.   In fact, there was nothing
particularly new in Platt's paper; most of the ideas were
effectively expressed in the late 1800's by *Chamberlain* (reprinted
in 1965).   Clearly the importance of critically testing and

negating hypotheses is acknowledged by modern philosophers of science. Consider the following quotations:

"The method of science is bold conjectures and ingenious and severe attempts to refute them" - *Popper*, 1963.

"...only if you can tell me how your theory might be refuted, or falsified, can we accept your claim that your theory has the character of an empirical theory... as long as we cannot describe what a possible refutation of a theory would be like, that theory is outside empirical science" - *Popper*, 1963.

To be sure, Popper never claims that irrefutable theories are false or even meaningless. The critical demarcation of scientific statements is that such hypotheses be *capable* of being critically tested.

With the above ideals in mind, it is interesting to consider the behavior and conceptual overviews of numerous ecologists today. Many have difficulty isolating important mechanistic questions, and others continue to collect descriptive data and pray for a serendipitous appearance of some form of dendrogram that will force a question into their collection of data. But even then, the descriptive results are idiosyncratic and lack convincing general interest. And even when there are provocative mechanistic questions, ecologists have difficulty posing them as testable hypotheses and often ignore obvious alternate explanations. These problems are becoming evident, and many ecologists are unable to meet the challenge and therefore rely on the so-called "hard" sciences, such as chemistry and physics, for intellectual guidance. This appeal to physicists and engineers seems to be a continuing characteristic; as *Cohen* (1970) declares, "Physics-envy is the curse of ecology."

It is interesting to note that although we look to physics as a model of hard science, its appeal to ecology results from the successes of Newtonian physics. Ecologists are largely unaware that the Newtonian physics they envy is in a "zone of middle dimensions" and that, in both the large and small dimensions, the Newtonian physics becomes highly probabilistic with many theological parallels discussed by *Capra* (1975). Thus, honest physicists have reason to be rather modest; but unfortunately, when ecologists seek guidance, physicists are happy to lead us to the promised land through a constant production of deterministic, usually linear models. Because the models seem to offer instant relevance, many ecologists piously follow this exodus from an understanding of nature. Certainly much the same can be said of some engineers and system analysts who ignore the

heavily stochastic reality of our world and force nature into
electrical circuitry or elaborate mathematical models, apparently
in the belief that an understanding of nature can be found in a
computer terminal. In both cases, the many exceptions and natural
variations are treated simply as bothersome noise which can be
either ignored with lofty pronouncements about the search for
generality rather than precision (*Levins*, 1968), or suppressed
with elaborate mathematics or ingenious wiring of the circuits,
such that the natural variations disappear (H. T. *Odum*, 1971;
Patten, 1971, 1975; *Nihoul*, 1975; E. P. *Odum*, 1977). What is
lost is empirical reality, a critical distinction between science
and theology.

For several reasons, the efforts to force mathematical
dendrograms or engineering determinism onto ecological systems
tend to cloud the search for understanding:

1) they actually hide true biological relationships;
2) they cannot acknowledge the existence of key species
   that have disproportionately important roles in the
   communities;
3) they are static and ignore probabilistic functions
   which are basic parameters of ecological systems;
4) they are often not general;
5) they have such large scale that they utterly deny
   critical tests.

As *Simberloff* (1979) has pointed out, "What physicists view
as noise is music to the ecologist; the individuality of populations and communities is their most striking, intrinsic, and
inspiring characteristic." He points out that there are several
types of indeterminacy that make ecology unique. First, there
is a basic noise of genetic variability that cannot be determin-
istically modeled yet is the core of evolution. This noise is a
fundamental characteristic of ecology and evolution. Secondly,
populations are governed by many non-linear but realistic systems
of difference equations, whose sum effects are difficult or
intractable mathematically and add such "chaos" to the system as
to make it appear random (*May*, 1974; *Oster*, 1975; *May and Oster*,
1976). The third and final type of ecological indeterminacy is
engendered by the enormous numbers of entities even in simple
ecological systems. Furthermore, these entities themselves
interact in complex and subtle ways such that deterministic
models are usually hopelessly inadequate (see *Hedgpeth*, 1977).

Unfortunately, the guidance of some physicists and engineers
leads us outside the realm of Popperian science because many of
the models, flow diagrams, and elaborate mathematic models are
based on untested and often untestable assumptions and are
almost always immune to critical testing. Indeed, the immunity

is often so carefully built into most of the models that counter-examples are readily discounted or ignored. Thus, our messiahs continue their prodigious production of models with little or no interest in critical tests. As a result, many of these models can be cited as the "questions" about which the ecologists piously collect data. Consider, for example, the bursts of papers and theses following *Gause's* (1934) work and *Hardin's* (1960) restatement of the "competitive exclusion principle." *MacArthur's* (1955) diversity-stability model, and especially his broken stick model (*MacArthur*, 1957; 1960), produced an enormous literature despite the fact that they were both irrelevant to the issues they addressed. (Interestingly, the inapplicability of the broken stick model was demonstrated by *Hairston* in 1959 before the model became so popular, suggesting that the loyalty of its believers overcame their critical acuity.) Other models that have acquired faithful followers can be found in *MacArthur and Levins* (1967), *MacArthur and Wilson* (1967) and *Levins* (1968). Many researchers have devoted substantial portions of their careers to trying to answer such "questions." The theological parallels seem obvious.

SCIENTIFIC METHODS AND CREATIVITY

The above is a consideration of ecology as a traditional scientific methodology. In recent years, controlled experimentation has become accepted as an important, powerful, scientific tool in ecology (*Paine*, 1977). This is an important lesson, but I am concerned because often, especially in marine benthic studies, cages are installed and other manipulations are performed without the formulation of precise questions or hypotheses. Therefore, the experiments are sometimes carried out without adequate control of the artifacts attendant to field manipulations. One reason why experiments tend to be inadequately controlled is that they are set up, consciously or subconsciously, to *verify* rather than to test critical hypotheses.

It is obvious that improperly controlled experiments can be seriously misleading. But much more serious is the lack of creative questions in much ecological work. Indeed, I believe that the influence of learned and socially reinforced world views or paradigms and the lack of original questioning are the major causes of the diverging and confusing arguments in the philosophic literature (*Kuhn*, 1970a; *Lakatos and Musgrave*, 1970; *Shapere*, 1971; *Holton*, 1973; *Merton*, 1975). These arguments often seem to be semantic nigglings about the true meaning of "normal science," "paradigms," and the methods by which science actually proceeds (see especially *Feyerabend*, 1975). While much of this is semantic, I believe there is an important difference between "normal science" -- which is based on the proper use of established scientific methods as defined by *Chamberlain*, 1965; *Platt*, 1964; *Popper*, 1963; and many others -- and the rare occurrence of true "Scientific Revolutions" (see *Kuhn*, 1970b). *Capra* (1975), *Pirsig* (1974), *Maslow* (1976) and probably others have argued that our exaltation of scientific methods clouds the critical issue of genuine creativity. That is, the traditional scientific method is itself simply a tool which, at best, produces simple hindsight. It provides a method for testing and eliminating various hypotheses and thus uncovers truth in an accurate but somewhat backwards manner. *Maslow* (1976) even suggests that "science is a technique .. whereby even unintelligent people can be useful in the advance of knowledge." It is obvious that the methodology itself ultimately depends on the preexistence of the hypotheses to test. There is simply no way for proper scientific method to generate bright new hypotheses; without new questions or hypotheses from elsewhere, it is empty. Thus, traditional knowledge is only the collective memory of old hypotheses that have been tested. No general method exists for looking ahead and deciding in what direction science should go. Thus there are an infinite number of facts for us to consider, but it is difficult and critical to observe and to recognize the "right facts" with which we construct new conjectures and hypotheses (cf. *Feyerabend*, 1975).

Given that scientific techniques and methods are sterile tools dependent upon the availability of relevant hypotheses, the actual cutting edge of creativity is in an arena that is rarely emphasized in most of the literature. Creativity consists of an almost subliminal choice of what facts are observed and synthesized into innovative hypotheses. Such syntheses and speculation are critical because these choices determine the path of future hypotheses upon which the tool of Popperian methodology functions. I believe that in ecology this synthesis depends both upon (1) an appreciation of the fact that a basic scientific objective is to develop general principles and (2) the "feel" of competent naturalists regarding proper questions to ask nature. This emphasizes the value of speculation to effective science. Note that this discussion superficially parallels *Feyerabend's* (1975) anarchisms, but differs in the critical sense that constructive and creative hypotheses are spun from the highly organized and structured "real world," perhaps best read by those naturalists with the ability to consider these implications in a generalized theoretical framework.

The largely semantic differences between Kuhn and Popper (*Lakatos and Musgrave*, 1970; see also *Westman*, 1978) seem to lie in this extremely important but conceptually difficult region of science. Thus, in my own interpretation, the bold conjectures or leaps that Popper refers to should be, in fact, the syntheses of these new and original hypotheses upon which the classic scientific method depends. The paradigms and themata of Kuhn and Holton are larger psychological, social, and historical constraints within which we make our bold conjectures. To use my own intertidal research as an example, I believe that I used reasonably good Popperian methods, following the advice of Platt and Chamberlain. But any creativity I may have achieved was developed from an appreciation or understanding of some of the natural history of the phenomena that structure the intertidal community. However, to some extent I cheated in developing the hypotheses because several of the null hypotheses which I negated were weak. That is, I already had a reasonably accurate picture of nature, and in many cases I already knew that the null hypothesis which I tested would be negated. Thus, I was working within a larger paradigm which assumed that the intertidal community is structured around identifiable biological interactions which can be observed on the shoreline -- especially competition, predation and, in some cases, mutualistic relationships. In some of my experiments, I attempted to "prove" or "verify" these assumptions by falsifying weak alternatives. Thus, one effect of my thesis research was to entrench rather than to topple this larger paradigm. While counter-examples were uncovered and pet hypotheses were falsified, most of the hypotheses I first generated were spun directly from the larger paradigm, and "unobservables" (see *Eddington*, 1958), such as planktonic processes, were ignored. I believe that this is generally true in science and is, in fact, the essence of Kuhn's argument. Thus Kuhn's "normal science" includes the creative conceptual leaps of Popper, but it is still housed within the larger overview or blinders of the paradigm.

REFUTATION OF A WORLD VIEW?

How do we engineer a scientific revolution? Is it possible, deliberately, to shake up and topple an established paradigm? Kuhn presents historical descriptions of such events but does not provide a formula for engineering a scientific revolution in the present or in the future. Pirsig's discussion of *mu* provided me with an approach, actually inherent in the work of Kuhn, Popper, and Holton, for dealing with nature in such a way that we can actually test, modify and even destroy the larger paradigms within which the more precise scientific methodology is constrained. I suggest that we use traditional scientific methods as outlined by Popper, Platt, Eccles, etc., with a much more rigorous application of Pirsig's *mu* concept. That is, the traditional approaches that I have always used have been designed to force Mother Nature to give me *yes* or *no* answers to my questions. But Pirsig, and undoubtedly others, have argued that logically there is a third alternative, which Pirsig describes as *mu*, a Japanese term for "no thing." *Mu* simply means that the

context of the question is bad and that it is wrong to ask nature
that particular *yes* or *no* question. *Mu* instructs the questioner
to "unask the question." It is obvious that *mu* exists in the
"real" world of science and there is a (possibly apocryphal)
story that Einstein remarked of some particularly irrelevant
research that it was "not even wrong."

Many of my own hypotheses were carefully designed to force
yes or *no* answers from nature when, in fact, nature may have been
crying out *mu* because, for example, my time scale may have been
inadequate. If such is the case, the *mu* answer is very important
because it says that the framework of my question is wrong and
that I should therefore reevaluate the entire intellectual context of the question. This is an extremely important concept
that is basic to Kuhn's philosophy and one which I was unlikely
to understand in my stubborn adherence to Popperian methods as
interpreted to me by Platt. An extension of this trap might be
to generalize from situations which in different seasons or years
could yield very different patterns. Thus the complexity of
ecology emphasizes the importance of paying careful attention to
possible *mu* answers.

How, then, do we recognize a *mu* answer from nature? To a
certain extent, I believe that an honest and competent scientist,
following traditional scientific methods but at the same time
listening carefully for *mu* answers, will intuitively recognize
them. But one must be responsive to this alternative and not
insist on wringing *yes* or *no* answers from nature. On a larger
scale, however, much of the "noise" and natural variation that so
bewitch ecologists working within their highly structured paradigms might be a *mu* answer. Thus a consequence of such
Platonically idealistic outlooks is the intellectual dishonesty
that ignores or suppresses this "noise" so that it becomes
impossible to challenge the larger paradigms within which we
work. For example, in my past research I focused on *in situ*
benthic processes and defined my questions so as to preclude
"noise" from issues such as larval availability or relatively
rare events whose consequences may structure the community for
many years. Such intellectual blinders are restrictive.

SOCIOECOLOGY

I want to shift gears briefly at this point and consider
ecological research as a sociological phenomenon. This is
relevant because the paradigms that direct our research are
themselves a historical-sociological-psychological phenomenon,
and it is worthwhile to consider some of the social pressures
that maintain them. Furthermore, as Kuhn repeatedly argues,
there is resistance to a scientific revolution in which a

paradigm is overthrown, and this resistance comes from the
"puzzle solvers" themselves who are anxious to maintain their
puzzles and, in effect, their self-importance.

Robert K. *Merton* has discussed the mechanisms of rewards in
the communication systems of science (1968, 1973; but see *Glaser*,
1964). He focuses his essay on the established leaders of
science, especially the historical notables and the Nobel
laureates, and describes many psychosociological mechanisms by
which the rich get richer and the less famous are actively
neglected. As a result, established scientists often reach a
particular threshold where their fame becomes a self-fulfilling
prophesy characterized by linear achievement and logarithmic
reward. This phenomenon affects not only the individual but also
the laboratory or institution with which the scientist is
associated, because people flock to work with him, bringing
additional fame and financial reward. Merton argues that this
is a common facet of modern science and designates it the
"Matthew effect," a term derived from the Gospel according to
St. Matthew: "For unto every one that hath shall be given, and
he shall have abundance; but from him that hath not shall be
taken away even that which he hath" (Matthew, XXV, 29).

I believe that the "Matthew effect" is alive and well in
ecology. Consider the intellectual influences and the mystique
of our leaders, such as Odum, Watt, Hutchinson and MacArthur.
In some cases, the mystique has extended to their students, some
of whom may have contributed little toward earning their own
"Matthew effect" but who occasionally have cultivated their
associations with the messiah. For example, I have frequently
heard ecologists explain that "the problem with so-and-so" is
that he or she has no interest in or understanding of evolution.
The implication is that they themselves do; the context of the
conversation usually implies that their keen evolutionary
insights were honed on the clay of the feet of their idol.

The resistance to new ideas, so clearly discussed by Thomas
Kuhn, is also a common scientific event; but it does not seem to
be quite so common in ecology, simply because there are not many
new exciting ideas to resist. But a corollary to this is the
tenacious refusal of many scientists to reject ideas or to
challenge the paradigms (*Barber*, 1961; *Barber and Hirsch*, 1962;
Stent, 1972). For example, one often hears theoretical ecologists
lament that their natural-historian critics are "stamp collectors,"
who are unable to analyze the complex problems of ecology and
evolution, and who have impeded progress in ecology for years.
In addition, it is often opined that the era of natural history
as "real science" ended long ago. Often this opinion is followed
by a generous concession that natural historians and taxonomists
are necessary functionaries who make no contributions to real

science. On the other hand, all of us have heard counter-examples
of natural historians who, wallowing in the unsynthesized and
unsynthesizable facts that they collect, fulminate against the
theoreticians who "know nothing of the real world." (See the
introduction to *Fretwell's* (1972) book for a discussion of how
these attitudes truncate the proper development of science.)

A parallel, but perhaps a more insidious, problem is the
recent trend among anonymous reviewers of grant proposals and
manuscripts to ostracize other scientists for "not being *real*
ecologists." I have read this precise expression in reviews
of four manuscripts and five proposals, and have heard it
expressed in many more conversations. This elitism is corrosive
to the growth of knowledge, as it censors the ideas and counter-
examples that are essential to a healthy science. Unfortunately,
this attitude actually influences the publication of papers and
awarding of grants. *Van Valen and Pitelka's* (1974) note hints
at a behind-the-scenes power-play regarding a manuscript submitted to a leading journal, as well as other examples of
intellectual censorship. Another current example relates to an
important and influential hypothesis in marine ecology, which is
criticized by two authors in an unpublished paper. Rather than
offer corollaries and alleged predictions of the hypothesis, the
authors consider and negate each of the assumptions upon which the
hypothesis was based. They then erect and defend a falsifiable
hypothesis in its place. Unfortunately, despite reviews which
suggested modifications rather than rejection, the complete
manuscript may never be published because one of our most
"open-minded" journals rejected it as being trivial. An editor
wrote one of the authors that in the paper they "...confuse the
establishment of a correlation with the falsifyability [**sic**] of a
model. It does not follow that because a model is not falsify-
able that a correlation does not nevertheless exist." This
attitude certainly contradicts the Popperian argument that it is
the falsifiability of the statement, rather than its accuracy,
that demarcates science from non-science.

In summary, we seem to suffer a strong resistance to the
rejection of pet theories. These problems are particularly
destructive when they limit or even prevent the publication of
new ideas and interpretations which should be discussed and
argued in a rational manner within the scientific community at
large. While similar examples may be common, or even be integral
parts of science as practiced in the "real world," they certainly
vitiate the search for knowledge.

THE ROAD TO GLORY

While regrettable, these examples probably are not unusual
and would not surprise Merton, as they are described in his writ-

ings. After reading Merton's papers and observing the behavior
of our fellow ecologists, we can see that many are consciously or
subconsciously working to enhance their own "Matthew effect."
For example, some ecologists cultivate their charisma and spend
considerable time on the "seminar circuit," probably cognizant
of the fact that citation indices are heavily influenced by
gossip and by personal familiarity with somebody's research.
Another aspect of the "Matthew effect" can be discerned in the
publication of reviews and their subsequent citations in which
the author of the review often is mistakenly credited with
everything that he reviews. This is not to argue against reviews,
because good interpretive reviews are extremely helpful; but some
are mere distortions and rediscoveries of the reviewed papers or
are lengthy rehashes of the reviewer's own publications. In such
cases it seems that the authors might be searching for or enhanc-
ing their own "Matthew effect." Again, I do not argue that any
of this is unique to ecology; indeed, as *May* (1973) alleges in
the preface of his book, it may be that ecology is relatively free
of some of the destructive behavior that may characterize other
disciplines. Nonetheless, I argue that our field has a sociology
similar to that of any other science and that it is helpful to
be aware of some of the ramifications.

Watkins (1970) argues that Kuhn developed a strong parallel
between science and theology; *Kuhn* (1970c) does not deny it, and
inasmuch as we practice such strong allegiance to our paradigms
and demigods, I believe that this is true. But, in a way, this
is an integral part of our culture, as we are all influenced by
the small nomadic bands of people who eked out a living from their
harsh desert habitat thousands of years ago. These people
differed from others of their era because they actually tran-
scribed their oral history, and fragments of the transcripts have
been passed down to us as the Old Testament. Like modern
ecologists, the early poets struggle to understand the diverse
and wonderful natural patterns they encountered. They learned
that in order to coexist with nature, they had to obey many
natural rules, and they therefore developed practical and valuable codes which were probably essential to a harmonious society
in such a rigorous environment. These codes, well tested through
time, became the written laws. While the codes were practical,
the authors were a curious people who also created subjective
explanations for the origins and workings of their world. These
explanations were recorded in Genesis and are still consistent
with observations.

It seems to me that most ecologists (or, perhaps, all
scientists) are striving to offer alternate hypotheses to Genesis.
But unfortunately many of our influential hypotheses are associated
with corroboratory correlations little better than those supporting
the Genesis model; in fact, there are strong parallels in the

critical behavior patterns of the respective adherents. Until we shake the shackles of non-scientific methods, rigid paradigms, and the psychological and social crutches and alliances so prevalent in ecology, the theological parallels will continue to be strongly associated with our work.

ACKNOWLEDGMENTS

I dedicate this essay to John Isaacs, whose unfettered mind suffers no paradigms. I am especially grateful to George N. Somero for introducing me to many of these topics, and to George and the students in our Philosophy of Science seminars who have discussed some of the references with me. John Oliver, Tim Gerrodette, Linda Moore, Don Strong, Loren Haury and George Somero have made constructive criticisms of the manuscript, for which I am grateful.

REFERENCES

Barber, B., 1961. Resistance by scientists to scientific discovery, *Science* <u>134</u>: 596-602.

Barber, B. and W. Hirsch, 1962. *The Sociology of Science*, 602 pp., The Free Press, New York.

Barnes, H. E., 1965. *An Intellectual and Cultural History of the Western World*, 1381 pp., Dover Publications, New York.

Capra, F., 1975. *The Tao of Physics*, 330 pp., Shambhala Publications, Berkeley, California.

Chamberlain, T. C., 1965. The method of multiple working hypotheses, *Science 148:* 754-759.

Cohen, J. E., 1970. Review of *An Introduction to Mathematical Ecology* (E. C. Pielou, 1969. 286 pp., Wiley-Interscience, New York). *Amer. Sci. 58:* 699.

Eccles, J. C., 1970. *Facing Reality*, 210 pp., Springer-Verlag, New York.

Eddington, A., 1958. *The Philosophy of Physical Science*, 230 pp., Univ. of Michigan Press, Ann Arbor.

Feyerabend, P., 1970. Consolations for the specialist, In: *Criticism and the Growth of Knowledge*, edited by I. Lakatos and A. Musgrave, 197-230, Cambridge University Press, Aberdeen.

Feyerabend, P., 1975. *Against Method*, 339 pp., Humanities Press, Atlantic Highlands, N. J.

Fretwell, S. D., 1972. *Populations in a Seasonal Environment*, 217 pp., Princeton University Press, Princeton, N. J.

Gause, G. J., 1934. *The Struggle for Existence*, 163 pp., Williams and Wilkins, Baltimore.

Glaser, B. G., 1964. Comparative failure in science, *Science 143:* 1012–1014.

Hairston, N. G., 1959. Species abundance and community organization, *Ecology 40:* 404–416.

Hardin, G., 1960. The competitive exclusion principle, *Science 131:* 1292–1297.

Hedgpeth, J. W., 1977. Models and muddles, some philosophic observations. *Helgoländer wiss. Meeresunters 30:* 92–104.

Holton, G., 1973. *Thematic Origins of Scientific Thought: Kepler to Einstein*, 495 pp., Harvard University Press, Cambridge, Mass.

Hurlbert, S. H., 1978. A paean to H, *Ecology 59:* 442.

Kuhn, T. S., 1970a. *The Structure of Scientific Revolutions*, 172 pp., University of Chicago Press, Chicago, Illinois.

Kuhn, T. S., 1970b. Logic of discovery or psychology of research, In: *Criticism and the Growth of Knowledge*, edited by I. Lakatos and A. Musgrave, 1–23, Cambridge University Press, Aberdeen.

Kuhn, T. S., 1970c. Reflections on my critics, In: *Criticism and the Growth of Knowledge*, edited by I. Lakatos and A. Musgrave, 231–278, Cambridge University Press, Aberdeen.

Lakatos, I., 1970. Methodology of scientific research programs, In: *Criticism and the Growth of Knowledge*, edited by I. Lakatos and A. Musgrave, 91–96, Cambridge University Press, Aberdeen.

Lakatos, I. and A. Musgrave, 1970. *Criticism and the Growth of Knowledge*, 282 pp., Cambridge University Press, Aberdeen.

Levins, R., 1968. *Evolution in Changing Environments*, 120 pp., Princeton University Press, Princeton, N. J.

MacArthur, R. H., 1955. Fluctuations of animal populations and a
    measure of community stability, *Ecology 36:* 533-536.

MacArthur, R. H., 1957. On the relative abundance of bird
    species, *Proc. Nat. Acad. Sci. 43:* 293-295.

MacArthur, R. H., 1960. On the relative abundance of species,
    Amer. Naturalist 94: 25-36.

MacArthur, R. H. and R. Levins, 1967. The limiting similarity,
    convergence, and divergence of coexisting species, *Amer.
    Naturalist 101:* 377-385.

MacArthur, R. H. and E. O. Wilson, 1967. *The Theory of Island
    Biogeography,* 203 pp., Princeton University Press,
    Princeton, N. J.

Maslow, A. H., 1976. *The Farther Reaches of Human Nature,*
    407 pp., Penguin Books.

May, R. M., 1973. *Stability and Complexity in Model Ecosystems,*
    235 pp., Princeton University Press, Princeton, N. J.

May, R. M., 1974. Biological populations with nonoverlapping
    generations: stable points, stable cycles, and chaos,
    Science 186: 645-647.

May, R. M. and G. F. Oster, 1976. Bifurcations and dynamic
    complexity in simple ecological models, *Amer. Naturalist
    110:* 573-599.

Merton, R. K., 1968. The Matthew effect in science, *Science
    159:* 56-63.

Merton, R. K., 1973. *The Sociology of Science,* 605 pp.,
    University of Chicago Press, Chicago.

Merton, R. K., 1975. Thematic analysis in science: notes on
    Holton's concept. *Science 188:* 335-338.

Nihoul, J. C. J. (ed.), 1975. *Modelling of Marine Ecosystems,*
    272 pp., Elsevier, Amsterdam.

Odum, E. P., 1977. The emergence of ecology as a new integra-
    tive discipline, *Science 195:* 1289-1293.

Odum, H. T., 1971. *Environment, Power and Society,* 331 pp.,
    Wiley-Interscience, New York.

Paine, R. T., 1977. Controlled manipulations in the marine inter-
    tidal zone and their contributions to ecology theory, In:
    The Changing Scenes in Natural Sciences, 1776-1976, 245-270,
    Academy of Natural Sciences, Special Publication 12.

Patten, B. C. (ed.), 1971. *Systems Analysis and Simulation in
    Ecology. Vol. I.* 610 pp., Academic Press, New York.

Patten, B. C., 1975. Ecosystem linearization: an evolutionary
    design problem. *Amer. Naturalist 109:* 529-539.

Pirsig, R. M., 1974. *Zen and the Art of Motorcycle Maintenance,*
    412 pp., William Morrow & Co., New York.

Platt, J. R., 1964. Strong inference, *Science 146:* 347-353.

Popper, K. R., 1963. *Conjectures and Refutations: The Growth
    of Scientific Knowledge,* 412 pp., Harper and Row, New York.

Shapere, D., 1971. The paradigm concept, *Science 172:* 706-709.

Simberloff, D. S., 1979. A succession of paradigms in ecology:
    idealism to materialism and probabilism. *Synthese,* in press.

Stent, Gunther S., 1972. Prematurity and uniqueness in scientific
    discovery. *Scientific American 227:* 84-93.

Van Valen, L. and F. A. Pitelka, 1974. Commentary--intellectual
    censorship in ecology, *Ecology 55:* 925-926.

Watkins, J., 1970. Against normal science. In: *Criticism and
    the Growth of Knowledge,* edited by I. Lakatos and
    A. Musgrave, 25-37, Cambridge University Press, Aberdeen.

Weaver, W., 1964. Scientific explanation. *Science 143:* 1297-
    1300.

Westman, R. S., 1978. The Kuhnian perspective. *Science 201:*
    437-438.

II. Primary Production and Export Processes

PRODUCTION AND EXPORT OF SEA GRASSES FROM A TROPICAL BAY

Joseph C. Zieman[1], Gordon W. Thayer[2],
Michael B. Robblee[3], and Rita T. Zieman[1]

[1]*University of Virginia,* [2]*Southeast Fisheries Center,*
Beaufort Laboratory, [3]*University of Virginia, and*
West Indies Laboratory, Fairleigh Dickinson University

ABSTRACT

*Seagrass meadows have been shown to be highly productive and
of great value to nearshore marine regions. In addition to the
utilization of their production by both direct grazing and
detrital food chains, considerable amounts of seagrass are
transported offshore, often to great distances, where it may
serve as food for both surface feeding and benthic feeding
organisms.*

*Seagrass leaves are detached by senescence, storms, and the
action of herbivores such as sea urchins and parrotfish. Leaves
of turtle grass (*Thalassia testudinum*) are sometimes seen
drifting on the surface, but usually sink rapidly or remain on
the bottom, whereas* Syringodium filiforme *leaves, with larger
lacunal spaces, nearly always float to the surface.*

In Tague Bay, St. Croix, U.S.V.I., only about 1% of
Thalassia *production is exported, whereas 60-100% of the*
Syringodium *production is carried out of the system.*

INTRODUCTION

During the past decade many studies have demonstrated the
value of the highly productive seagrass beds to the nearshore
marine region (*Wood, Odum, and Zieman,* 1969; *Thayer et al.,* 1975;

Zieman, 1975; *McRoy and Helfferich,* 1977). These and other
studies (*e.g., Ryther,* 1963; *McRoy and McMillan,* 1973) also have
shown that seagrass meadows are among the most productive of
marine ecosystems. The major large consumers of seagrasses are
green turtles, dugongs, manatees, waterfowl, and fishes. Since
the decline of the large herbivores such as green turtles and
manatees as a result of excessive exploitation, the primary
utilization of the seagrasses as a food source has been through
the detrital food chain (*Thayer et al.,* 1975; *Zieman,* 1975;
Thayer et al., in press). In tropical areas such as the U. S.
Virgin Islands, however, direct herbivory also may account for
consumption of a considerable proportion of the seagrass production (*Ogden and Zieman,* 1977). The organic material fixed
by the seagrasses, in addition to being directly utilized and
decomposed within a grass bed, may be transported out of these
meadows by currents. This exported material can thus serve as a
food source for both benthic and surface feeding organisms at
considerable distances from the original source of its formation.

    Although there is only scattered confirmatory evidence,
export of seagrass material appears to be a common phenomenon,
the relative importance of which at present can only be estimated.
Menzies et al. (1967) collected leaves and fragments of tropical
and subtropical turtle grass, *Thalassia testudinum,* off the
North Carolina coast in 3,160 meters of water. Photographs
from the Blake plateau showed densities of up to 48 blades per
photograph. The turtle grass was seen at several localities,
indicating that it may constitute an important food source in
the deep sea, although it was estimated that the nearest source
material was 500 to 1,000 km distant. *Menzies and Rowe* (1969)
later established the quantitative distribution of *Thalassia*
blades to depths greater than 5,000 m. *Roper and Brundage* (1972)
surveyed the Virgin Island Basin and, in over 5,000 photographs
taken at an average depth of 3,900 m, observed seagrass blades
or fragments in nearly every one. Most were identifiable as
Thalassia testudinum or thin, cylindrical leaves of *Syringodium
filiforme.* *Wolff* (1976) collected seagrasses by trawling in
three Caribbean deep sea trenches and found seagrass material in
all trenches studied. Most of the material collected was
Thalassia and Wolff found evidence of consumption of the
seagrass fragments by deep water organisms. *Thayer and Bach*
(unpublished data) have estimated that from 1 to 20% of the
production of the temperate seagrass *Zostera marina* is exported
from the embayment grass beds. *Greenway* (1976) estimated that
9.5% of the weekly production of *Thalassia* drifted away from
Kingston Harbor, Jamaica, because of echinoid grazing. Much of
this material may eventually be transported to the ocean or sink
and be utilized by the benthos of unvegetated areas.

Many authors who have observed seagrass blades in the deep sea and transported great distances from the source of their production have associated this phenomenon with storms and hurricanes (*Moore*, 1963; *Menzies et al.*, 1967; *Menzies and Rowe*, 1969). *Menzies and Rowe* (1969) suggested that the flow of water down the continental slope was largely responsible for the transport of seagrasses to great depths, while *Roper and Brundage* (1972) felt that simple sinking of the grasses was the primary means by which seagrasses reached deeper waters. The studies cited above show that the occurrence of seagrasses at great depths, especially in the Caribbean and on the eastern coast of the United States, is a widespread phenomenon, and because of the scarcity of food in deeper waters, could be a quantitatively important food source. What has not been evaluated is the proportion of seagrass production which is exported and the manner in which it moves out to the deeper waters. The purpose of this paper is to describe preliminary data on the export of seagrass material from a tropical embayment in the U. S. Virgin Islands. This is a continuing study, as is similar work on open-water and embayment temperate *Zostera Marina* beds (*Thayer and Bach*, unpublished data), with the overall objective of further defining the role of these submerged grass bed systems.

AREA AND METHODS

Tague Bay, on the northeastern tip of St. Croix, U. S. Virgin Islands (Fig. 1), is the study site. The segment of the bay used is approximately 500 m wide and 1.2 km in length with a total area of 62 hectares. The long axis of the bay runs east and west and is parallel to the prevailing winds. During this experiment in March, 1978, the winds varied from calm to 8 m/sec, and, except for brief periods during rain squalls, came from $050^\circ - 100^\circ$, with the dominant winds being from $075^\circ - 090^\circ$. As a result of the dominant easterly winds, water movement through the bay is from east to west. Since there is only a small tidal range (.2-.3 m), currents within the bay appear to be primarily wind driven. During the experimental period the average surface current was about 9 cm/sec.

Tague Bay is a shallow embayment with a depth ranging from 3 to 6 m in the study area (Fig. 1). It is bounded by the island of St. Croix on the south and a partially emergent coral reef to the north. The bottom is calcareous sand and mud and is covered with various mixtures of the seagrasses *Thalassia testudinum*, *Syringodium filiforme*, and *Halodule wrightii*, as well as numerous calcareous green algae such as *Penicillus* and *Halimeda*. Also common in this bay are patch reefs and *Callianassa* mounds.

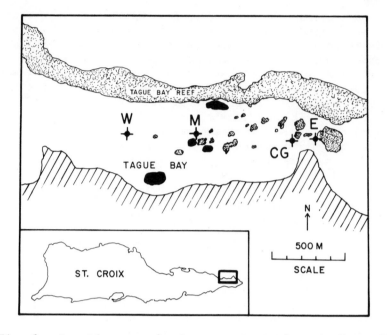

Fig. 1. Location map showing export stations in Tague Bay.
The hatched zones show where the productivity and standing crop
samples are located. Stippled areas in Tague Bay are patch reefs.

     The densities of *Thalassia* and *Syringodium* have been
measured at numerous sites in Tague Bay over the past two years.
Production of *Thalassia* has been measured during the same period
by the staple technique of *Zieman* (1974), while the production
of *Syringodium* has been measured using this technique, but
employing tape rather than staples. Transport of seagrasses
through Tague Bay was measured at the water surface and at
the water-sediment interface. Surface nets, shown in Fig. 2,
have an opening 0.61 x 0.35 m. Each net is about 0.7 m deep
and is constructed of a reinforcing rod frame with 0.5 mm Nitex
mesh liners. Each is supported by 3 styrofoam floats such that
about 70% of the net is below the water surface at all times.
The nets are tied to a surface float with a 3 m harness and
the surface float is anchored to the bottom. The float and
harness arrangement allows the net to swing with the currents,
but does not pull the nets under the surface in rough water.
The bedload nets are of similar design but have a mouth opening
1.3 x 0.3 m and a depth of 1 m; no flotation is used. Mesh
sizes of 1 mm and 0.5 mm were tried for these nets with the 1 mm
size being preferred because it was easier to maneuver underwater.

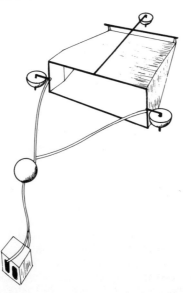

Fig. 2.   Drift nets used in the experiments.   Flotation is
provided by three halved fishing floats.   The float in the middle
serves to keep the net from being pulled under in rough water.

Stations were located at four points in Tague Bay (Fig. 1).
The easternmost station (E) was placed at the extreme eastern end
of the bay, just behind the reef, to measure material coming in
from the Caribbean.   Station M was in the middle portion of the
bay about 700 m west of E, and the westernmost station (W) was
500 m further to the west, giving a distance between the two most
distant stations (E and W) of 1.2 km.   A fourth station was
added at Cotton Garden Point (CG) in order to estimate material
transported into Tague Bay from Boiler Bay, the most eastern bay
on the north shore of St. Croix.   With the exception of the CG
station, all stations had 3 nets spaced about 50 m apart across
the main axis to determine the variability of the transported
material.   Only three bedload nets were used, one each at
stations E, CG, and M.   The nets were fished continually over
a 5-day period from 1600 on March 13 to 1600 on March 18.
Material in the nets was collected periodically throughout the
day and night to attempt to determine any diurnal periodicity
to the movement, but the results are only reported here as
daily values.

At collection the boat was pulled up to each of the surface
nets in succession and the net contents were dumped into a piece
of plastic sheeting and then placed in a marked plastic bag.   The
boat then was anchored to a fixed buoy and suspended particulate

samples were collected at the surface, mid-depth and at 10 cm
above the sediment surface, using 1-liter bottles filled at depth
by a SCUBA diver. The bedload net then was transported to the
boat and emptied. Current velocities were taken at 1-m intervals using a Marsh-McBirney electromagnetic current meter. The
suspended particulate matter samples were returned to the
laboratory, filtered onto Whatman GF-C glass fiber filters,
and frozen. The drift samples were drained and weighed wet.
Each sample then was sorted to species, and dry weights ($105^{O}C$)
and ash-free dry weights ($500^{O}C$) obtained. Only the cumulative
daily dry weights are reported here except as specifically noted.

RESULTS AND DISCUSSION

Because of the predominant unidirectional current flow in
Tague Bay resulting from the dominant easterly winds, we
hypothesized at the outset of this study that the concentration
of seagrass material being exported from this system would
increase in an east to west direction and that this export
would amount to a significant fraction of the total production
of seagrasses within Tague Bay. The major questions posed to
examine this hypothesis were: 1) how much seagrass material
is exported from the system?, 2) what proportion of the standing
crop and production does the exported material represent?, 3) how
is the material exported (*i.e.*, as bedload or surficially)?, and
4) what is the fate of the material removed from Tague Bay? Our
preliminary data provide insight into the first three questions,
and demonstrate the positive contribution that the grass beds
make to offshore regions.

To determine the standing crop and production of seagrasses
in Tague Bay, samples were taken at the areas shown in Fig. 1.
The results of these measurements are shown in Table 1 in terms
of standing crop, production and turnover for *Thalassia* and
Syringodium, the dominant grasses within Tague Bay. As part
of the continuing International Decade of Ocean Exploration's
Seagrass Ecosystem Study Program based at St. Croix, several
hundred measurements of seagrass standing crop and productivity
have been made. The data in Table 1 are typical of our measurements and serve to characterize the environment. *Thalassia
testudinum* dominates the system, representing about 94% of the
total dry weight biomass and 91% of the daily productivity.
Syringodium, representing 6% of the seagrass standing crop and
10% of the total production (Table 1), is particularly notable
in having an extremely fast turnover rate. This species produces a new crop every 18 days or approximately 20 crops of
leaves each year.

TABLE 1. Production and standing crop in study area

	Thalassia	*Syringodium*
STANDING CROP	77 gm/m^2	5.8 gm/m^2
% of total	94%	6%
PRODUCTION	2.7 gm/m^2/d	0.32 gm/m^2/d
% of total	90.5%	9.5%
TURNOVER	3.5%/d	5.8%/d
TURNOVER TIME	24 days	18 days

In general, there was an increase from east to west both in the suspended particulate organic material throughout the water column and in leaf material present in surface waters of Tague Bay. Our preliminary estimates of suspended particulate matter indicated that about 8.5 mg/ℓ organic matter was suspended in water entering Tague Bay over the reef. This was primarily material stirred up by waves breaking over the reef. At cotton Garden and the Middle station values were lower (7-7.4 mg/ℓ) while at the westernmost station particulate density was higher. The high values we observed at this station may have been partly the result of sediment reworking by benthic invertebrates, since this station was located in an area of dense *Callianassa* burrows.

Although the magnitude of increase in drift seagrasses over the beds varied from day to day, there was an increase in seagrass content of surficial waters in terms of the daily and overall averages as the water moved down the bay across the seagrass beds (Fig. 3). Even though the winds were from the east during the entire study, it is readily seen that areas east of the reef contribute little to the floating grass in Tague Bay (Fig. 4). The average increase in concentration (Table 2) was greater between the Middle and West stations than between the East and Middle stations. Using the distances between stations (700 m between E and M, and 500 m between M and W), we computed an average daily contribution to surface export of 0.36 g dry wt/m^2 for the western portion of the bay and 0.18 g dry wt/m^2 for the eastern portion. Thus, during the study period, the contribution of exported leaf material from Tague Bay would

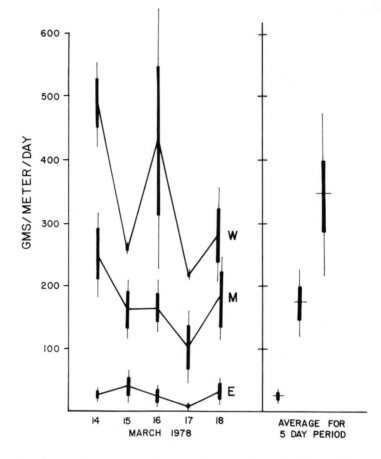

Fig. 3. Seagrass export during period of 14 to 18 March 1978. The left portion shows the amount of material collected in a 1-meter wide section of the bay at the three main stations (E,M,W). The right portion of the figure shows the average values for the 5-day period. Wide vertical lines are 1' standard error of the mean, narrow vertical lines are 1 standard deviation.

amount to 150 kg dry weight each day, based on a study area of 62 hectares (Table 2).

Although *Thalassia* is quantitatively the most abundant seagrass in Tague Bay, the major constituent of the floating material exported from the system was *Syringodium*. Factors involved in this difference appear to be lacunar space and the degree of epiphytism. On a relative basis, *Syringodium* leaves have considerably larger air spaces (lacunae) than *Thalassia*,

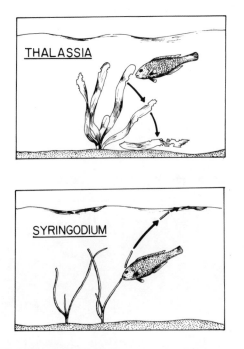

Fig. 4. The effects of grazing on the seagrasses *Thalassia* and *Syringodium* by the bucktooth parrotfish, *Sparisoma radians*.

and the leaves of *Syringodium* always are bouyant. By comparison, the leaves of *Thalassia* are denser and only float when relatively young and green. Senescent leaves of this species not only are less bouyant but also are heavily coated with calcareous epiphytes, a factor increasing their density. An increase in epiphytism during senescence also has been shown in the flat leaf temperate seagrass *Zostera marina* (*Thayer et al.*, 1977).

Mechanisms of detachment of *Thalassia* and *Syringodium* blades and subsequent contributions to surficial or bedload transport appear to differ. The two major mechanisms, other than through natural exfoliation, are the action of grazers, which may chop leaves off at the bottom, and storms which tear up and detach sometimes large quantities of leaves. *Menzies et al.* (1967) mentioned that drifting *Thalassia* frequently was seen following storms, and was often green and apparently still capable of photosynthesizing. The dominant herbivores in the Virgin Islands are the bucktooth parrotfish, *Sparisoma radians*, and the urchins *Tripneustes ventricosus* and *Diadema antillarum*. At St. Croix, parrotfish and urchins graze extensively, directly consuming 10 to 20% of the daily production of *Thalassia* (*J. C. Zieman*,

TABLE 2a. Loss from beds to export.

Stations	W	M
Exported at Surface	.36 gm/m^2/d	.18 gm/m^2/d
Bedload Loss	—	.02 gm/m^2/d

TABLE 2b. Export from Tague Bay (Study Area = 6 x 10^5 m^2).

Surface Export	15 x 10^4 gm/d
% of Total	92.5%
Bedload Export	1.2 x 10^4 gm/d
% of Total	7.5%
TOTAL EXPORT	16.2 x 10^4 gm/d

in prep.). Although feeding on both species, parrotfish apparently are the main grazers of *Syringodium* and the effects of their mode of grazing on the two plant species are quite different (Fig. 4). A parrotfish bite leaves a characteristic half-circular hole similar to that of a paper punch at the edge of a *Thalassia* blade. Grazing is primarily on the older and more epiphytized portions of the leaf, but the blade is seldom detached by this action. The blades usually become encrusted with epiphytes as they become senescent and gradually float lower in the water until they detach and become part of the litter layer. A parrotfish bite on *Syringodium*, however, normally severs completely the 1-1.5 mm diameter cylindrical blade, and the upper portion floats to the surface.

The above observation suggests that most detached blades of *Thalassia*, being denser than *Syringodium*, remain on the bottom and are transported, if at all, as bedload, while *Syringodium* is transported near the surface. In our study, over 95% of the surface drift material was *Syringodium* leaves, while bedload material consisted of *Thalassia* and a variety of drifting algae, but no *Syringodium*. The bedload sampler at station M collected 14.6 gm of dry material per day, indicating a loss to export of 0.02 gm/m^2/day (Table 2). Extrapolated to the bay, this figure would give a contribution of 12 kg/day as *Thalassia* bedload. This fraction was more variable than the surface drift and would increase considerably with increased turbulence of the water column due to high winds and storms. The measured bedload export of *Thalassia* of 0.02 gm/m^2/day accounts for only about 1% of the *Thalassia* production. However, surficial *Syringodium* export of 0.18-0.36 gm/m^2/day accounted for 60 to 100% of the *Syringodium* production.

Our data indicate a major difference in the function of these two seagrasses within the system. *Thalassia* is dense and tends to remain on the bottom of the bay. As a result it is the primary, or in many cases the exclusive, component of the litter layer, and when it is transported from the system it is primarily as bedload. *Syringodium*, however, is transported out on the surface of the water. The quantity of the *Syringodium* which is exported would indicate that the grazing pressure on this seagrass, or at least the number of bites taken from the blades, is quite high, with the majority of the production being exported from the system. It would appear, therefore, that the majority of *Thalassia* production is available for utilization within the system either by herbivores or detritivores and decomposers, whereas *Syringodium* may also serve this function but in areas distant from its original source of production.

ACKNOWLEDGMENT

The authors wish to thank John C. Ogden, Robert F. Dill and the staff of the West Indies Laboratory for their aid in this study. This study was supported by The International Decade of Ocean Exploration grant OCE-77-27051.

REFERENCES

Greenway, M., 1976. The Grazing of *Thalassia testudinum* in Kingston Harbour, Jamaica, *Aquat. Bot. 2:* 117-126.

McRoy, C. P. and C. McMillan, 1973. Production ecology and physiology of seagrasses, Review paper Productivity/ Physiology Working Group, 29 pp., Int. Seagrass Workshop, Leiden, Netherlands.

McRoy, C. P. and C. Helfferich (editors), 1977. *Seagrass Ecosystems: A Scientific Perspective,* 314 pp., M. Dekker, New York.

Menzies, R. J., J. S. Zaneveld, and R. M. Pratt, 1967. Transported turtle grass as a source of organic enrichment of abyssal sediments off North Carolina, *Deep-Sea Res. 14:* 111-112.

Menzies, R. J. and G. T. Rowe, 1969. The distribution and significance of detrital turtle grass *Thalassia testudinum* on the deep sea floor off North Carolina, *Int. Rev. Gesamten. Hydrobiol. 54:* 217-222.

Moore, D. R., 1963. Distribution of the sea grass *Thalassia* in the United States, *Bull. Mar. Sci. Gulf and Carib. 13:* 329-342.

Ogden, J. C. and J. C. Zieman, 1977. Ecological aspects of coral reef – seagrass bed contacts in the Caribbean, *Proc., 3rd Int. Coral Reef Symp.,* 371-382, Rosentiel School Mar. and Atm. Sci., Univ. of Miami.

Roper, C. F. E. and W. L. Brundage, Jr., 1972. Cirrate octopods with associated deep-sea organisms: new biological data based on deep benthic photographs (Cephalopoda), *Smithson. Contrib. Zool.* No. *121:* 1-46.

Ryther, J. H., 1963. Geographic variations in productivity in the sea, In: *The Sea; Ideas and Observations on Progress in the Study of the Seas. Vol. 2,* edited by M. N. Hill, 347-380, John Wiley & Sons, N. Y.

Thayer, G. W., D. A. Wolfe, and R. B. Williams, 1975. The impact of man on seagrass systems, *Amer. Scient. 63:* 288-296.

Thayer, G. W., D. W. Engel, and M. W. LaCroix, 1977. Seasonal distribution and changes in the nutritive quality of living, dead and detrital fractions of *Zostera marina* L., *J. Exp. Mar. Biol. Ecol. 30:* 109-127.

Thayer, G. W., P. L. Parker, M. W. LaCroix, and B. Fry, in press. The stable carbon isotope ratio of some components of an eelgrass, *Zostera marina,* bed, *Oecologia.*

Wolff, T., 1976. Utilization of seagrass in the deep sea, *Aquat. Bot. 2:* 161-174.

Wood, E. J. F., W. E. Odum, and J. C. Zieman, 1969. Influence of seagrasses on the productivity of coastal lagoons. *Laguna Costeras*, UN Simposio Mam. Simp. Intern Lagunas Costeras, 495-502, Nov. 28-29, 1967. Mex. DF.

Zieman, J. C., 1974. Methods for the study of the growth and production of turtle grass, *Thalassia testudinum* Konig, *Aquaculture 4:* 139-143.

Zieman, J. C., 1975. Quantitative and dynamic aspects of the ecology of turtle grass, *Thalassia testudinum*, *Estuarine Res. 1:* 541-562.

INTERACTIONS BETWEEN GEORGIA SALT MARSHES AND COASTAL WATERS:

A CHANGING PARADIGM

Evelyn B. Haines

University of Georgia Marine Institute

Problems worthy of attack
Prove their worth by hitting back

PIET HEIN *(Grooks)*

ABSTRACT

Salt marsh estuaries on the Georgia coast have been sites of classic studies leading to broad generalizations about the functioning of shallow estuarine ecosystems and the exchange of material between salt marshes and coastal waters. Several recent studies have suggested that these fundamental concepts need to be reevaluated. Not only are we uncertain as to the quantities of material fluxing in coastal ecosystems, we are also unsure of even the basic directions and mechanisms of exchange. Along the Georgia coast, the idea that salt marshes enrich coastal waters by the outwelling of organic detritus has been revised. The emerging concepts concerning these systems include: 1) algal-derived organic matter, from phytoplankton and benthic diatom photosynthesis within the estuary and just offshore, forms the bulk of organic seston, 2) vascular plant detritus is largely accumulated and consumed in marsh and estuarine sediments, 3) estuarine food webs are much more complex than has been thought, and 4) the most important roles of salt marshes in estuarine food webs are as refuges

Contribution No. 382 of the University of Georgia Marine Institute

and feeding habitat for young and small animals, and as exporters of protein in the form of fish, crabs, and shrimp to coastal waters.

INTRODUCTION

The salt marsh estuaries of the Georgia coast are sites of important research concerning the structure and function of detritus-based ecosystems (*e.g. Teal*, 1962; *Odum and de la Cruz*, 1967; *Wiegert et al.*, 1975). However, the current ruling theory, or paradigm, of material exchanges between salt marshes and coastal waters developed from these studies is now being seriously challenged (*e.g. Haines*, 1977). Two fundamental ideas concerning this current paradigm (as diagrammed in Fig. 1) that are being reevaluated are 1) that significant plant detritus is transferred from the salt marsh to estuarine waters, and 2) that phytoplankton production represents a negligible contribution to total estuarine production.

Although the present discussion focuses on Georgia estuaries, recent work on other coastal systems having extensive tidal marshes has also questioned the importance of transport of

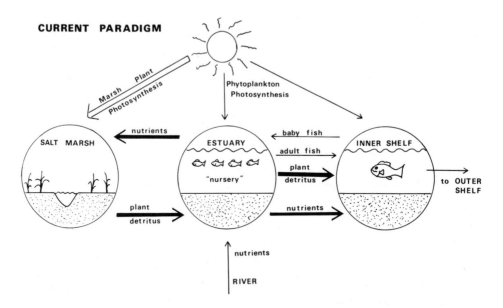

Fig. 1. The current concept of material exchanges between salt marshes, open estuaries, and nearshore shelf waters on the Georgia coast.

vascular plant material from the marshes to the open estuary.
This research includes the import/export studies of *Nadeau* (1972)
for a New Jersey salt marsh, of *Heinle and Flemer* (1976) for low
salinity tidal marshes in the Patuxent Estuary in Maryland, of
Hackney (1977) for a Mississippi *Juncus roemerianus* salt marsh,
and of *Woodwell et al.* (1977) for the Flax Pond salt marsh in
New York. In addition, whole-system studies of Rhode Island
estuaries have suggested that salt marsh plants might be less
significant, and macrophytic algae more significant, as sources
of organic matter in the open estuary than was previously thought
(*Nixon et al.*, 1976; S. *Nixon*, pers. comm.).

Here I present my interpretation of the changing, or
emerging, paradigm concerning interactions between Georgia salt
marshes and estuarine and nearshore shelf waters (as diagrammed
in Fig. 2). The points which I will emphasize are: 1) the
quantitative and qualitative importance of algal production
in the estuary, 2) the dominant role of marsh soils and
estuarine sediments as sites of accumulation, consumption,
and remineralization of organic matter, 3) the complexity of
estuarine food webs, and 4) the role of salt marshes as refuges
and feeding habitat for estuarine animals.

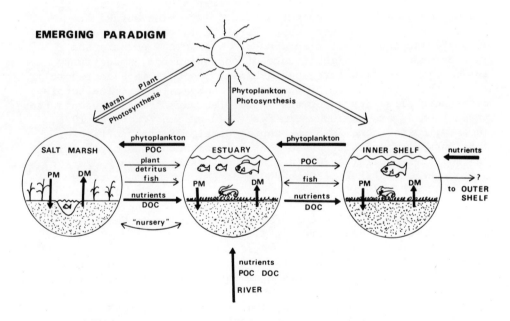

Fig. 2. An interpretation of the emerging ideas about
ecosystem function and material exchanges in Georgia coastal
waters. POC, DOC: Particulate and Dissolved Organic Carbon;
PM,DM: Particulate and Dissolved Materials.

Estuaries along the Georgia coast comprise a 5-8 km wide
band of salt marshes dominated by the cordgrass *Spartina*
alterniflora behind coastal barrier islands. The marshes are
drained by tidal creeks, rivers, and sounds, which compose 1/3
of the total surface area of the estuaries (*Teal*, 1962). The
estuaries receive little freshwater inflow except locally at
the mouths of the Savannah and Altamaha Rivers. The tidal
range is 2-3 meters; at low tide the influence of estuarine
water offshore is confined to a zone within 10-15 km of the
coast (*Pilkey and Frankenberg*, 1964; *Oertel*, 1974).

IMPORTANCE OF ALGAL PRODUCTION

Phytoplankton in Georgia estuaries (which may include some
benthic algae mixed into the water column, see *Roman and Tenore*,
1978) appears to be a significant source of organic matter in
the estuary both quantitatively, in terms of high rates of
carbon fixation, and qualitatively, in terms of rapid turnover
of carbon and ease of utilization by animals.

The concept that phytoplankton production was not very
important in the turbid waters of Georgia estuaries was based
in part on the light/dark bottle experiments of *Ragotzkie* (1959).
Recent studies of water column productivity using uptake of
$^{14}CO_3^-$ have shown, however, that there are in fact high rates
of phytoplankton carbon fixation both within the estuary and
just offshore. The highest rate of phytoplankton productivity
in Georgia coastal waters, 547 g $C/m^2/yr$, occurs in a nearshore
zone within 10 to 15 km of the coast (*Thomas*, 1966). The turbidity and shallow depths of the inshore estuary limits phytoplankton
production to around 300 g $C/m^2/yr$ (*Sellner and Zingmark*, 1976;
Whitney, Haines, and Pomeroy, unpub. data). Farther offshore,
nutrients become limiting, so that phytoplankton photosynthesis
on the inner shelf (to 20 m depth) averages about 285 g $C/m^2/yr$,
and on the outer shelf (20 to 200 m depth) photosynthesis is
about 130 g $C/m^2/yr$ (*Haines and Dunstan*, 1975). The interaction
of nutrient concentration and depth of euphotic zone which
results in this distribution of algal production in Georgia
coastal waters is diagrammed in Fig. 3.

The high phytoplankton production just off the coast is
probably "fueled" by nutrients from three sources: outflow of
river and estuarine water (*Thomas*, 1966), inflow from intrusions
of nutrient-rich subsurface water at the edge of the shelf
(*Dunstan and Atkinson*, 1976), and regeneration of nutrients in
the water and sediments (*Haines*, 1975). This phenomenon could
retard nutrient loss from the estuary, in that dissolved nutrients
flowing out of the estuary can be incorporated into particulate
matter via algal photosynthesis, which can then be carried by

tidal flow back into the estuary and retained by sedimentation
(*e.g. Meade*, 1969) (see Fig. 2).

An independent line of investigation suggesting the
importance of phytoplankton in Georgia estuaries is stable
carbon isotope ratio analysis. There are differences in the
natural abundance ratios of ^{12}C and ^{13}C in plants depending on
the photosynthetic mechanism and source of CO_2 (*Smith and Epstein*,
1971). Once fixed into organic matter during photosynthesis, the
characteristic $^{13}C/^{12}C$ ratio of a particular plant species is not
significantly altered during microbial decomposition or assimilation by animals (*Smith*, 1972; *Minson et al.*, 1975; *Haines and
Montague*, 1976). The stable isotopic composition of seston
in Georgia estuaries implies that this material is derived from
algal production ($\delta^{13}C$* of -21%) rather than from the C_4 grass
Spartina ($\delta^{13}C$ of -13%) or from C_3 vascular land plants

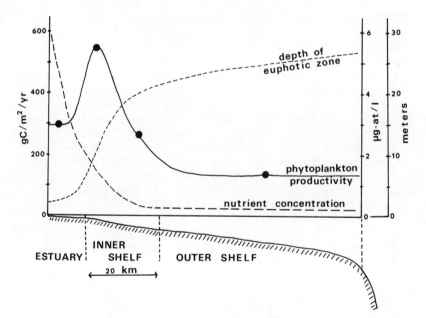

Fig. 3. Relation of phytoplankton production to depth of
euphotic zone and nutrient concentration in Georgia coastal
waters.

$$*\delta^{13}C = (\frac{^{13}C/^{12}C\ sample}{^{13}C/^{12}C\ standard} - 1)\ x\ 1000$$

(δ^{13}C of -27%) (*Haines*, 1977). Even in the mouth of the Altamaha
River, a major southeastern stream, seston and sediment organic
particles have a δ^{13}C of around -27% for seston in freshwater
parts of the river a few km upstream (*Haines and Knowles*,
unpub. ms.).

Besides being a ubiquitous source of food for estuarine
animals, phytoplankton are also much more nutritious than
vascular plant material, which must be colonized by microorganisms before becoming an available food source (*Fenchel*,
1972). Easily digestible naked dinoflagellates and green
flagellates along with pelagic and benthic diatom species are
common in Georgia estuarine waters year-round. The rapid growth
rate and high protein content of the algal cells make them an
important base for the estuarine food web, which has been
previously neglected because of the quantitatively greater
productivity of marsh macrophytes.

THE ROLE OF MARSH SOILS AND ESTUARINE SEDIMENTS

A valid generalization about estuaries is that they tend to
trap water-borne particulate material. Along the east coast, the
rate of net sediment accumulation in estuaries is on the order of
mm per year, following the continuing slow rise in sea level
(*Redfield*, 1967). However, the annual flux of elements to and
from marsh soils and estuarine sediments is orders of magnitude
greater than the net accumulation of materials (*Nixon et al.*,
1975; *Haines et al.*, 1977). Thus benthic deposits in shallow
estuarine systems are important catalytic sites where organic
materials are accumulated, consumed, and remineralized. In
addition, much of the particulate material in the water is
resuspended sediment (*Oviatt and Nixon*, 1975; *Roman and Tenore*,
1978).

The importance of the sediment-water interface in estuarine
processes is obvious, but has not been sufficiently emphasized.
It is here that optimum conditions prevail for the degradation
of organic detritus (*Haines and Hanson*, unpub. ms.), and also
here that many important estuarine animals, including shrimp,
crabs, and fish such as mullet, feed (*Darnell*, 1961; *Odum*, 1970).

Because of the hypothesized role of estuarine sediments as dominant sites for accumulation and utilization of particulate organic
materials, it follows that the important material outflux from salt
marshes may be as dissolved rather than as particulate matter. A
net outflux of dissolved organic matter from Georgia salt marsh
soils of 22 g C/m^2/yr has been shown (*Pomeroy et al.*, 1977).
Marsh soils are also sources of ammonia, silicate, phosphate, and

dissolved organic nitrogen for estuarine waters (*Haines*, 1979; *Pomeroy et al.*, 1972; *Gardner*, 1975).

THE COMPLEXITY OF THE ESTUARINE FOOD WEB

Previous studies in macrophyte–dominated estuaries have assumed that the food web was in large part based on non–living plant material, and that animals in the second trophic level were for the most part detritivores rather than herbivores (*Teal*, 1962; *Odum and Heald*, 1975). This concept has caused misunderstanding in that 1) there has been confusion over whether animals assimilate the dead plant material directly or indirectly via digestion of microbes living on the plant material (*e.g. Adams and Angelovic*, 1970), and 2) estuarine food webs have been oversimplified. In the purest sense, the only "detritivores" are the bacteria, fungi, and perhaps polychaete worms which assimilate the dead plant material directly. Associated with these organisms are protozoa, nematodes, and other meiofaunal consumers of the detritus–degraders. Benthic algae in the sediments and phytoplankton in the water column are intimately associated with the microbial and meiofaunal community. Thus the organisms which are usually considered to be detritivores (*e.g.* fiddler crabs, periwinkle snails, and grass shrimp) are more aptly termed "opportunistic omnivores"; they generally feed upon a mixture of microbes, small multicellular animals, and algae, but are not above consuming any carrion they happen upon. *Christian and Wetzel* (1978) have further elaborated some of the complexities of estuarine detritus food webs.

Stable carbon isotopic ratio analysis of estuarine animals has proven to be a powerful technique in determining the ultimate source of plant carbon for individual food chains. Within the salt marsh, many invertebrates have stable carbon isotope compositions which can be explained as assimilation of both *Spartina* and benthic diatom derived organic matter (*Haines and Montague*, 1979). Planktivorous or filter-feeding fauna such as menhaden and oysters, however, have carbon isotopic compositions similar to that of phytoplankton (*Hughes and Haines*, unpub. data). Organisms which are more generalistic feeders such as mullet, killifish, shrimp, and blue crabs have ratios indicating a mixed diet of *Spartina*, benthic algae, and phytoplankton derived carbon (*Hughes and Haines*, unpub. data). Of course, the stable carbon isotopic composition data says nothing about the number of trophic steps between plant and animal, since there is virtually no change in the isotopic composition of the live plant during microbial decomposition and animal assimilation (*Haines and Montague*, 1979).

SALT MARSH AS REFUGE AND FEEDING HABITAT

In view of the importance of algal production as a food
source and of the complexity of the estuarine food web, the salt
marsh may seem relegated to a minor role as a cause of high
secondary productivity in the estuary. However, the salt marsh
serves a dual role as food source and as physical shelter for
young and small animals in the estuary. At high tide, an abundance of aquatic organisms, including large and small fish,
shrimp, and blue crabs, invades the salt marsh to feed on marsh
organisms and to avoid predators. The true nursery-ground of
the estuary is perhaps not so much the large open water rivers
and sounds as the salt marshes and narrow tidal creeks. A
major export of marsh plant production may thus occur not as
particulate detritus but as living organisms. *Herke* (1971)
has convincingly developed this concept for Louisiana salt
marshes, which are similar to Georgia marshes. More work is
needed to evaluate this idea for other coastal systems.

IMPLICATIONS FOR FUTURE RESEARCH

The present uncertainty concerning the nature of interactions
between Georgia salt marshes and coastal waters should make
estuarine ecologists more cautious about comparisons of coastal
ecosystems. Recent studies have stressed the fact that each
estuarine system examined has a unique set of physical and
biological characteristics, the understanding of which is not
immediately amenable to the formation of theory concerning
estuaries in general. Only after similar, intensive studies of
particular systems have been carried out in a number of sites
will we be able to develop general theory concerning the
ecosystem function of salt marsh estuaries. This research
should include a variety of approaches, such as interdisciplinary
studies (*e.g. Pomeroy et al.*, 1977; *Woodwell et al.*, 1977),
modeling of coastal systems (*e.g. Wiegert et al.*, 1975; *Kremer
and Nixon*, 1978), microcosm experiments (*e.g. Pilson et al.*,
1979), field and laboratory manipulations (*e.g. Sherr and Payne*,
1978; *Fallon and Pfaender*, 1976), and fresh techniques such as
the application of $\delta^{13}C$ analysis discussed in this paper.

ACKNOWLEDGMENTS

This work was supported by NSF Grants No. DES 75-20845 and
No. DEB 77-20359 and by grants from the Sapelo Island Research
Foundation. I thank all my colleagues who have contributed to
the formulation of ideas presented in this paper.

REFERENCES

Adams, S. M. and J. W. Angelovic, 1970. Assimilation of detritus and its associated bacteria by three species of estuarine animals, *Chesapeake Science 11:* 249-254.

Christian, R. R. and R. L. Wetzel, 1978. Interaction between substrate, microbes, and consumers of *Spartina* detritus in estuaries, In: *Estuarine Interactions,* edited by M. Wiley, 93-114, Academic Press, N. Y.

Darnell, R. M., 1961. Trophic spectrum of an estuarine community based on studies of Lake Pontchartrain, Louisiana, *Ecology 42:* 553-568.

Dunstan, W. M. and L. P. Atkinson, 1976. Sources of new nitrogen for the South Atlantic Bight, In: *Estuarine Processes, Vol. 1,* edited by M. Wiley, 69-78, Academic Press, N. Y.

Fallon, R. D. and F. K. Pfaender, 1976. Carbon metabolism in model microbial systems from a temperate salt marsh, *Appl. Environ. Microbiol. 31:* 959-968.

Fenchel, T., 1972. Aspects of decomposer food chains in marine bethos, *Verh. Dtsch. Zool. Ges. 65:* 14-22.

Gardner, L. R., 1975. Runoff from an intertidal marsh during tidal exposure-recession curves and chemical characteristics, *Limnol. Oceanogr. 20:* 81-89.

Hackney, C. T., 1977. *Energy Flux in a Tidal Creek Draining an Irregularly Flooded* Juncus *Marsh,* Ph.D. Dissertation, Mississippi State Univ.

Haines, E. B., 1975. Nutrient inputs to the coastal zone: The Georgia and South Carolina shelf, In: *Estuarine Research, Vol. 1,* edited by L. E. Cronin, 303-324, Academic Press, N. Y.

Haines, E. B., 1977. The origins of detritus in Georgia salt marsh estuaries, *Oikos 29:* 254-260.

Haines, E. B., 1979. Pools of nitrogen in Georgia coastal waters, *Estuaries 2:* 34-39.

Haines, E. B., A. Chalmers, R. Hanson, and B. Sherr, 1977. Nitrogen pools and fluxes in a Georgia salt marsh, In: *Estuarine Processes, Vol. 2,* edited by M. Wiley, 241-254, Academic Press, N. Y.

Haines, E. B. and W. M. Dunstan, 1975. The distribution and
     relation of particulate organic material and primary pro-
     ductivity in the Georgia Bight, 1973-1974, *Est. Coast. Mar.
     Sci. 3:* 431-441.

Haines, E. G. and C. L. Montague, 1979. Food sources of
     estuarine invertebrates analyzed using $^{13}C/^{12}C$ ratios,
     Ecology 60: 48-56.

Heinle, D. and D. A. Flemer, 1976. Flows of material between
     poorly flooded tidal marshes and an estuary, *Mar. Biol.
     35:* 359-373.

Herke, W. H., 1971. *Use of Natural, and Semi-impounded,
     Louisiana Tidal Marshes as Nurseries for Fishes and
     Crustaceans,* Ph.D. Thesis, Louisiana State Univ.

Kremer, J. N. and S. W. Nixon, 1978. *A Coastal Marine Ecosystem,
     Simulation and Analysis, Ecological Studies 24,* 217 pp.,
     Springer-Verlag, N. Y.

Meade, R. H., 1969. Landward transport of bottom sediment in
     estuaries of the Atlantic Coastal Plain, *J. Sed. Pet. 39:*
     222-234.

Minson, D. J., M. M. Ludlow, and J. H. Troughton, 1975. Differ-
     ences in natural carbon isotope ratios of milk and hair
     from cattle grazing tropical and temperate pastures,
     Nature 256: 602.

Nadeau, R. J., 1972. *Primary Production and Export of Plant
     Material in a Saltmarsh Ecosystem,* Ph.D. Dissertation,
     Rutgers University.

Nixon, S. W., C. A. Oviatt, J. Garber, and V. Lee, 1976. Diet
     metabolism and nutrient dynamics in a salt marsh embayment,
     Ecology 57: 740-750.

Nixon, S. W., C. A. Oviatt, and S. S. Hale, 1975. Nitrogen
     regeneration and the metabolism of coastal marine bottom
     communities, In: *The Role of Terrestrial and Aquatic
     Organisms in Decomposition Processes,* edited by J. M.
     Anderson and A. MacFaden, 269-283, Blackwell Scientific
     Publ., London.

Odum, E. P. and A. de la Cruz, 1967. Particulate organic
     detritus in a Georgia salt marsh-estuarine ecosystem,
     In: *Estuaries,* edited by G. H. Lauff, 383-388, Pub. No.
     83, AAAS.

Odum, W. E., 1970. Utilization of the direct grazing and plant detritus food chains by the striped mullet, *Mugil cephalus*, In: *Marine Food Chains*, edited by J. Steele, 222–240, University of California Press, Berkeley.

Odum, W. E. and E. J. Heald, 1975. The detritus based food web of an estuarine mangrove community, In: *Estuarine Research*, *Vol. 1*, edited by L. E. Cronin, 265–286, Academic Press, N. Y.

Oertel, G. F., 1974. Delineation of coastal water masses by suspended sediment characteristics, Oral presentation, SEERS meeting, St. Augustine, Florida, Nov. 1974.

Oviatt, C. A. and S. W. Nixon, 1975. Sediment resuspension and deposition in Narragansett Bay, *Est. Coast, Mar. Sci. 3:* 201–217.

Pilkey, O. and D. Frankenberg, 1964. The relict–recent sediment boundary on the Georgia continental shelf, *Bull. Ga. Acad. Sci. 22:* 37–40.

Pilson, M. E. Q., C. A. Oviatt, G. A. Vargo, and S. L. Vargo, 1979. Replicability of MERL microcosms: initial observations, Symposium, Advances in Marine Environmental Research. U.S.E.P.A. (in press).

Pomeroy, L. R., L. R. Shenton, R. D. H. Jones, and R. J. Reimold, 1972. Nutrient flux in estuaries, In: *Nutrients and Eutrophication*, 274–291, Special ASLO Symposium 1.

Pomeroy, L. R., et al., 1977. Fluxes of organic matter through a salt marsh, In: *Estuarine Processes, Vol. 2*, edited by M. Wiley, 270–279, Academic Press, N. Y.

Ragotzkie, R. A., 1959. Phytoplankton productivity in estuarine waters of Georgia, *Publ. Inst. Mar. Sci. Texas 6:* 146–158.

Redfield, A. C., 1967. Postglacial change in sea level in the western North Atlantic Ocean, *Science 157:* 687–692.

Roman, M. R. and K. R. Tenore, 1978. Tidal resuspension in Buzzards Bay, Massachusetts. I. Seasonal changes in the resuspension of organic carbon and chlorophyll a, *Est. Coast. Mar. Sci. 6:* 37–46.

Sellner, K. G. and R. G. Zingmark, 1976. Interpretations of the ^{14}C method of measuring the total annual production of phytoplankton in a South Carolina estuary, *Bot. Marina 19:* 119–125.

Sherr, B. F. and W. J. Payne, 1978. Effect of the *Spartina alterniflora* root-rhizome system on salt marsh soil denitrifying bacteria, *Appl. Environ. Microbiol. 35:* 724-729.

Smith, B. N., 1972. Natural abundance of the stable isotopes of carbon in biological systems, *Bioscience 22:* 226-231.

Smith, B. N. and S. Epstein, 1971. Two categories of $^{13}C/^{12}C$ ratios for higher plants, *Plant Physiol. 47:* 380-384.

Teal, J. M., 1962. Energy flow in the salt marsh ecosystem of Georgia, *Ecology 43:* 614-624.

Thomas, J. P., 1966. *The Influence of the Altamaha River on Primary Production beyond the Mouth of the River,* M. S. Thesis, Univeristy of Georgia.

Wiegert, R. C., R. R. Christian, J. L. Gallagher, J. R. Hall, R. D. H. Jones, and R. L. Wetzel, 1975. A preliminary ecosystem model of coastal Georgia *Spartina* marsh, In: *Estuarine Research, Vol. 1,* edited by L. E. Cronin, 583-601, Academic Press, N. Y.

Woodwell, G. M., D. E. Whitney, C. A. S. Hall, and R. A. Houghton, 1977. The Flax Pond ecosystem: exchanges of carbon in water between a salt marsh and Long Island Sound, *Limnol. Oceanogr. 22:* 833-838.

ECOLOGICAL CONSIDERATIONS OF DETRITAL AGGREGATES IN THE SALT MARSH

Ben W. Ribelin and Albert W. Collier

Florida State University

ABSTRACT

More than 98% of the detrital material exported from the investigated Gulf Coast salt marsh is made up of amorphous aggregates. These detrital aggregates, averaging 25-50 μm in diameter, are produced by the benthic microflora of the marsh, rather than by microbial decomposition of the dominant vascular plant Juncus roemerianus *as the prevailing view holds. Rising tides lift films of the aggregate material from the dense community of benthic algae that carpet the mud surface of the marsh. Ebbing tides transport the floating films into the tidal creeks where even mild water surface disturbances effect the dispersal of the films, which then sink as detrital aggregates into the water column. Ebb tides occurring during daylight hours carry larger quantities of detrital aggregates than tides that ebb during darkness. Detrital aggregate production follows an annual cycle with high production rates in late summer and low rates in the winter. Vascular plant tissue is decomposed beneath the layer of benthic algae and is retained in the marsh. Previously accepted concepts that stress the role of the decomposers in the production of detritus in tidal marshes are examined.*

INTRODUCTION

Contemporary Theory

Current views on salt marsh ecology emphasize the importance of a decomposer-based food web. Grazing herbivores are thought to ingest only a small fraction of the annual macrophyte crop, while

the remainder falls to the marsh floor upon death of the plant
(*Teal*, 1962). Here, bacteria and fungi rapidly colonize the new
substrate and through their activities convert the high-fiber,
low-nitrogen, dead plant material into variously sized particles
of high nutritional value. Eventually, some of the microbe-laden
particles are transported by tidal action from their point of
origin into estuaries or offshore environs, where they are
available to filter-feeding consumers.

It is generally accepted that consumers receive most of their
nutrition from the attached microorganisms and not directly from
the decomposing plant particles (*Odum*, 1970; *Heald*, 1971; *Fenchel*,
1972; *Mann*, 1972; *Olah*, 1972). Egestion of the stripped detrital
particles returns them to the aquatic environment where they are
recolonized and the process is repeated time after time (*Odum and
Heald*, 1975). Abrasion, maceration by consumers, and the continued
action of decomposers result in smaller and smaller fragments.
At some point during the sequence the most minute of the fractions
are considered to become unidentifiable and are often referred to
simply as amorphous aggregates or organic material of undetermined
origin (*Darnell*, 1958; *Odum and de la Cruz*, 1967; *Odum*, 1970).
Although this description might accurately reflect the
sequence of events in some ecosystems, it is unfortunately accepted
as a general description of the carbon flow pathway in salt marshes.
However, the sequence has not been thoroughly documented for salt
marsh ecosystems and especially for those marshes dominated by
the Black Needlerush, *Juncus roemerianus*.

Deficiencies in Contemporary Theory

A review of the literature on salt marsh detritus reveals no
studies in which the complete breakdown of marsh grasses through
progressively smaller fractions to amorphous aggregates has been
observed. Instead, only portions of the process have been
observed and speculation has filled the remaining gaps.

Colonization of dead plant material and detrital particles
has been frequently studied, both quantitatively and qualitatively,
in many aquatic ecosystems (*Fenchel*, 1970; *Olah*, 1972; *Gosselink
and Kirby*, 1974). These studies typically involve placing known
quantities of dead plant material in nylon bags that are then
anchored in tidal creeks or on the marsh floor. By periodic
analysis of samples from the bags, a large volume of data has
been gathered. These data have revealed that a rapid increase
in nitrogen levels (*Newell*, 1965; *Mann*, 1972), amino acids and
proteins (*de la Cruz and Poe*, 1975), and caloric values (*de la
Cruz and Gabriel*, 1974; *de la Cruz*, 1975) accompanies microbial
colonization and decomposition of the detrital material. Rates
of decomposition for many macrophyte species have also been

obtained from similar nylon bag studies (*Burkholder and Bornside,* 1957; *de la Cruz,* 1973; *Reimold et al.,* 1975; *Kirby and Gosselink,* 1976).

Ingestion of colonized particles by consumers, the removal and assimilation of microorganisms during digestion, and the recolonization of the egested fragments have also been observed (*Newell,* 1965; *Odum and Heald,* 1975; *Welsh,* 1975).

The largest missing link in the decomposition pathway is the connection between the minute but still recognizable plant fiber particles and the unidentifiable amorphous aggregates. This organic material of undetermined origin is frequently mentioned in the literature and has been reported as a major component in the gut contents of a variety of detritivores (*Darnell,* 1958; *Odum,* 1970).

Odum and de la Cruz (1967) reported that "aggregates" or "nano detritus" might be the most important component of particulate material exported from a Georgia *Spartina alterniflora* marsh. They separated suspended particles into three size fractions: "course detritus," retained by netting with 0.239 mm apertures; "fine detritus," passing the 0.239 mm net but retained by netting with 0.064 mm apertures; and "nano detritus," passing the 0.064 mm aperture but retained by a Millipore filter with a 0.45 µm pore size. The course and fine fractions (comprising 1% and 4% of the total particulates, respectively), were identified as fragments of vascular plants, but the nano fraction (comprising 95%) was described as "highly decomposed" and "unrecognizable." Nevertheless, *Odum and de la Cruz* (1967) assumed that the decomposition sequence of leaves of *Spartina alterniflora* follows the pathway of coarse to fine to nano (*i.e.,* aggregate) detritus.

New Data Suggest Need for
Reevaluation of Contemporary Theory

Our preliminary observations of the particulate material exported from a Florida Gulf Coast salt marsh indicated that the percentage of amorphous aggregates or "nano detritus" was even greater than the 95% value found in the Georgia *Spartina* marsh study. Quantitative studies further revealed that the crude fiber composition of the marsh detritus was extremely small, and undetectable in some samples. Fluorescence microscopy of detrital samples stained with acridine orange failed to reveal significant quantities of bacteria in or on the aggregate material, a finding inconsistent with the concept that aggregates are the smallest, most active decomposition products of vascular plants (*Odum and de la Cruz,* 1967; *Gosselink and Kirby,* 1974). Light

microscopy further revealed that the few recognizable vascular
plant fragments were frequently smaller than many of the aggregates. However, no particles that appeared to represent an
intermediate or transitional stage between the plant fragments
and the amorphous aggregates were found. Particles were clearly
either vascular plant fragments or aggregates.

These two forms of suspended particles appeared to be so
radically different that a common source or pathway of production
seemed questionable. An investigation into the source of the
aggregate material, its composition, and the dynamics of its
production was begun.

METHODS

Study Site Description

A small tidal marsh (approximately 16 hectares), located on
Apalachee Bay (30°03' N, 84°22' W) about 4.8 km north of the
town of Panacea, Wakulla County, Florida, and lying completely
within the St. Marks National Wildlife Refuge, was chosen as
the focal point for this investigation. The dominant vascular
plant is *Juncus roemerianus* (Scheele), which covers more than
95% of the total marsh area. Narrow bands of *Spartina alterniflora* (Loisel) occur along many of the tidal creeks. A typical
dendritic network of small tributaries empties into a single
large tidal stream approximately 15 m wide at the mouth (at mean
high water) on Apalachee Bay. This stream is frequently reduced
to less than 2 m in width during extreme low tides. Apalachee
Bay is shallow, and numerous oyster reefs are exposed during
most of the semidiurnal low tides. Daily tidal amplitude
averages 0.73 m. Maximum high tides occur during August and
September (+1.07 m)*, whereas minimum high tides occur in
February and March (+0.79 m).

The marsh is well defined by a small elevated dirt road on
the south, a barrier beach strand and Apalachee Bay on the east,
and a band of salt barrens on the west. The northern boundary is
less distinct, but several small salt barrens indicate higher
elevations separating the study site marsh from the drainage
area of the next tidal stream. No fresh water streams enter the
marsh; the only sources of fresh water are local rainfall and
subsurface seepage.

---

*All tide heights are given as distance above (+) or below (-)
the reference point of mean low water.

Collection and Preservation of Samples

A floating, automatic pumping apparatus was used to collect
500 ml water samples hourly for 24 hours during each of the
monthly sampling periods (March 1976 to October 1977). This
apparatus, positioned near the mouth of the large tidal stream,
collected water samples from a depth of 10 cm below the surface.
Other monthly water samples were collected manually in plastic
bottles from additional stations established throughout the tidal
creek system. Samples were poisoned immediately upon collection
with mercuric chloride (4 ml of 2% aqueous solution per liter
of sample).

Preparation for Microscopy

Suspended particulates to be viewed with light microscopy
or scanning electron microscopy (SEM) were separated from the
water samples by slow settling or by low speed centrifugation,
with no observable differences in recovery of particles averaging 10 to 15 µm in diameter or larger.

Samples prepared for SEM were post-fixed in 2% osmium
tetroxide after initial fixation with 3 ml of 5% formalin per
100 ml of sample. Fixed material was filtered from suspension
onto 1.0 µm Nuclepore filters and rinsed with several ml of
distilled water. Filters were then folded closed in envelope
fashion and fastened with cotton thread. Dehydration with
2,2-dimethoxypropane (*Maser and Trimble*, 1977) was followed
by critical point drying with CO_2 (*Anderson*, 1951).

Dried samples, still on the Nuclepore filters, were attached
to aluminum buttons with doublestick tape and coated with
gold/palladium in a vacuum evaporator.

Samples of decomposing *J. roemerianus* leaf fragments were
fixed and dried as above and then attached to SEM buttons with
colloidal silver paint. Several leaf fragments were broken open
after drying and mounted with the broken surface upward, thereby
exposing the internal leaf structure.

Throughout the study period, mud surface samples were
periodically cut from several representative areas of the marsh
floor, placed in petri dishes and returned to the laboratory.

RESULTS

Observations

Observations made on water samples collected from the tidal
marsh creeks revealed that these creeks carried heavy loads of
extremely small suspended particles. Although some of the larger
particles in the samples settled out within 12 to 24 hours, the
majority of the total particulate weight remained in suspension
after 72 hours.

Microscopy of Detrital Particles

When water mounts of the settled detrital particles from
ebb tide collections were viewed with light microscopy, the
settled material was found to consist largely of very small
amorphous particles, although unattached free-floating pennate
diatoms were common. The amorphous material appeared to be
aggregates of diatom frustule fragments, occasional whole diatoms,
and many extremely small (<1µm) particles (Fig. 1), all bound
together by an adhesive, gelatinous substance that was slightly
yellowish-brown in color. The majority of the identifiable
diatoms and frustule pieces, both inside the aggregates and
free-floating, belonged to benthic genera, with species of
Nitzschia and *Navicula* being the most common. Occasional

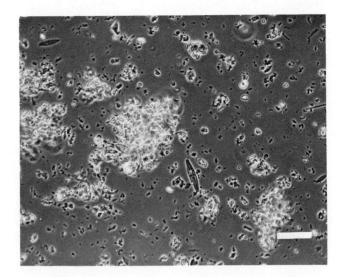

Fig. 1. Light micrograph of detrital aggregates showing
pennate diatoms (both free floating and as inclusions) and
minute inorganic fragments. Scalebar equals 50 µm.

broken frustules of centric diatoms were found in some aggregates
(see Fig. 4). Although common, whole diatoms and frustule fragments appeared to represent a minor constituent of the aggregates.
The remainder of the aggregate material was composed of the
extremely small, irregularly-shaped inclusions and a large
quantity of the amorphous gelatinous substance that bound the
particulate components together. These aggregates ranged from
less than 1 μm to over 100 μm in diameter, with the majority in
the 25 to 50 μm range. Entire slide preparations normally had
only a few particles that could be identified as fragments
derived from vascular plant tissue (Fig. 2). Several cellulose
stains were applied to the settled particles in an attempt to
locate plant fiber material in the aggregates, but only those
few particles which were distinguishable as plant fiber material
before staining exhibited an uptake. Vascular plant fragments
made up less than 1% of the total load of suspended particles.

Particles collected from flood tide waters were generally
quite different in appearance from those in ebb tide samples.
Although aggregates were present in the flood tide samples, they
were much smaller and far fewer in number than those observed in
ebb tide samples. Whole centric diatoms were often present in
the flood tide samples and such representatives as *Chaetocerus*
spp. and *Rhizosolenia* spp. characteristically bore clean and
unbroken spines (Fig. 3a). When samples of the following ebbing
waters were compared to the preceding flood tide samples, the
spine-bearing planktonic forms clearly showed the harsh effects

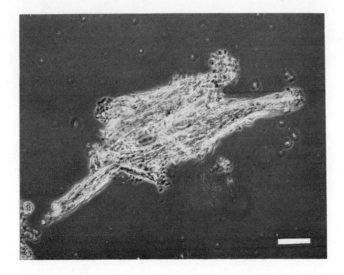

Fig. 2. Light micrograph of decomposing plant fragment.
Scalebar equals 50 μm.

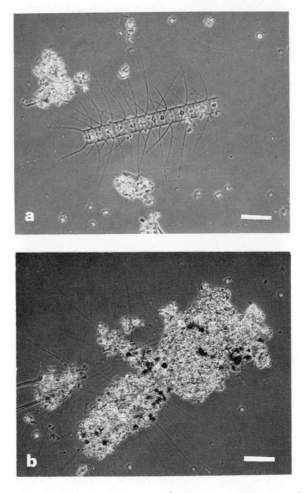

Fig. 3. Comparison of typical *Chaetocerus* chains found in flood tide water samples (a) and in ebb tide samples (b). Scalebars equal 50 μm.

of their journey over the marsh surface. Those spines that remained were usually covered with small aggregates and mucilaginous strands (Fig. 3b).

Fluorescence microscopy of fresh unfixed detrital samples stained with acridine orange failed to reveal significant quantities of bacteria associated with aggregates from ebb or flood tide samples, although bacteria were observed on the surface of those particles that were obviously fragments of vascular plants.

Scanning electron microscopy of the particulate material revealed little more than did phase contrast and fluorescence microscopy. Diatoms were perhaps a bit more clearly distinguished when attached to the surface or protruding from the interior of the aggregates (Fig. 4). SEM failed to reveal a greater number of plant fiber fragments than had been observed with light microscopy. Bacteria likewise were not observed in significant numbers with SEM.

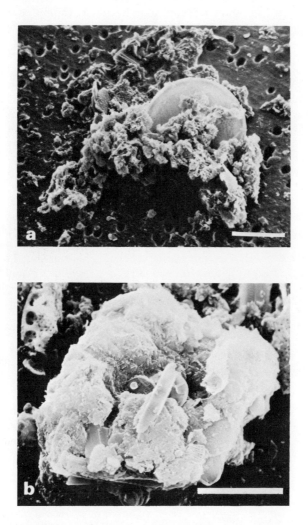

Fig. 4.   Scanning electron micrographs of detrital aggregates with embedded diatom frustules. Scalebar equals 5 μm in (a), 10 μm in (b).

Floating Surface Films and Their
Relationship to Detrital Aggregates

Field observations. A common phenomenon frequently observed
in the marsh provided the key to defining the pathway responsible
for detrital aggregate production. On calm, windless days, large
floating films, often reaching eight to ten meters in length,
appeared on the surface of the tidal creeks during ebbing tides
(Fig. 5). This floating material was brown, oily to the touch,
and easily disrupted, becoming suspended in the water column.
It was further noted that as the water level in the creeks
receded slowly during an ebbing tide, these floating films
continued to appear from the creek banks, spreading slowly
over the surface of the water. Closer observation revealed
that the films were sliding down the banks from higher vegetated
areas. Small pools of water that formed in isolated depressions
in the marsh were often completely covered with the film. As
water continued to drain into the depressions, carrying additional
floating material, the film on the pool surface would thicken,
eventually becoming so thick (1-2 mm), that parallel foldings
would form, creating a corrugated pattern. This same thickening
of the surface film was observed in many small creeks protected
from wind agitation. The slow movement of a boat into one of
these films would cause the same type of corrugating effect.

These films were also observed being lifted from the mud
surface of the marsh by rising tides. As the leading edge of
the incoming water slowly advanced across the marsh, it appeared
to separate the films from the mud surface and continued to lift
them until the tide began to ebb. As the water level then dropped,

Fig. 5. Floating film of detrital aggregate material
typical of those frequently observed in tidal creeks during
ebbing tides.

some of the surface film would be transported into the tidal
streams, where it often combined with other films producing continuously larger sheets. Much of the brown film was not removed,
however, but was redeposited on the mud surface. Slight wind
agitation in the creeks caused the films to disperse and sink
into the water column as they drifted along the creek network
and into the bay. On windless days, large films were often seen
several hundred meters offshore during falling tides.

Upon collection, these floating films dispersed and began
settling into the water column. By the time the samples were
returned to the lab, much of the previously floating material
was present as a thick deposit at the bottom of the collection
bottles.

Microscopy of floating films. Light microscopy revealed
the settled material that originated from the floating films to
be visually indistinguishable from the aggregate material
collected from water column samples, except that the particles
from the surface film were generally larger than typical ebb
tide suspended aggregates. With only slight agitation, these
larger (100–200 µm in diameter) particles broke into particles
identical in appearance to the smaller ones (25–50 µm) from the
water column samples. The majority of diatoms and frustule
fragments in the films were also of benthic species, with
Navicula and *Nitzschia* being the predominating genera. Scattered
remnants of planktonic diatoms, especially species of *Chaetocerus*
and *Cyclotella*, were often embedded inside the aggregates.

Origin of surface films. Studies of the mud surface of the
marsh revealed the source of the floating films and the detrital
aggregates. Samples of this marsh floor mud were found to be
carpeted with a network of filamentous algae (predominately
species of *Vaucheria* and *Ulothrix*) and a dense layer of benthic
diatoms embedded in mucilaginous material that was nearly
identical to the surface films collected in the tidal creeks.
The major difference between this material and that collected
from the water surface was the greater number of diatoms present
in the mud surface material. During summer months, the diatom
carpet became so extensive it completely covered the filamentous
green algae. It was originally believed that the presence of the
filamentous algae might be limited to the winter months, but
probing through the summer diatom layer revealed that the filamentous algae were as dense in summer as in winter. The diatom
layer, however, appeared to be much thinner during the winter
months, leaving much of the filamentous algae exposed.

*Role of benthic diatoms and filamentous algae in surface
film production*. Vertical algal filaments rising a short distance
above the surface of the diatomaceous carpet were densely covered

with diatoms and their associated slimes, thereby producing minute
peaks resembling miniature stalagmites.

When mud surface samples were kept moist but not inundated
and kept in covered dishes under artificial illumination, the
vertical algal filaments grew upward beyond the surrounding
mound of diatoms and mucus-like material. The bright green
filaments would often reach heights of 1-2 mm above the diatom
sheath within a week. After several samples had been kept under
an artificial day/night illumination regime for two weeks, the
dish was filled with sea water. After 24 hours of inundation,
small spike-like formations, visible with a low magnification
dissection microscope, were observed on the new growth of all
the algal filaments. At higher magnification, the "spikes"
were revealed to be pennate diatoms, attached by the tip of the
frustule. Forty eight hours after inundation, numerous other
diatoms were similarly attached along the algal filaments, and,
by the end of 72 hours, the entire new growth of the previously
bare filaments was completely covered with a layer of diatoms
and organic slimes. The slimes, when viewed with light microscopy, were identical to those collected from the natural marsh
surface, and differed from suspended aggregates only in having
a larger living diatom component.

Fate of decaying J. roemerianus *leaves*. A similar overgrowth
was observed when small pieces of decomposing *J. roemerianus*
tissue were placed on the surface of a mud sample under laboratory
conditions. Overgrowth by several filamentous algal strands
occurred within three days when the sample was kept moist. After
inundation of the sample with sea water, the diatom carpet and
associated organic film completely covered the leaf fragments
within one week. Segments of *J. roemerianus* leaves 2 cm in
length and 4 mm in diameter were totally covered by the algal
community in less than two weeks, leaving only a small raised
irregularity on the mud surface.

Overgrowth of these dead leaf fragments under laboratory
conditions helped explain the fate of dead plant tissue in the
natural marsh environment. Field observations on the fate of
dead *J. roemerianus* leaves confirmed the laboratory findings.
Older leaves of *J. roemerianus* plants seldom die and drop to the
marsh floor as single pieces. Instead, dry, dead tissue is
normally found at the tips of all leaves and can extend downward
for up to 25 cm. Small pieces of this dead region break off,
often as a result of mechanical abrasion, and fall to the marsh
floor. Here they are lifted and redeposited by rising and falling
tides for a few days. They are seldom seen floating in the tidal
streams on ebbing tides unless a recent storm has broken off dead
tips from large numbers of plants. Their scarcity in the tidal
streams is probably due to the dense growth of *J. roemerianus*

that serves as a highly effective filter, retaining the dead
fragments in the marsh. If the leaf fragments are left in contact with the mud surface after an ebbing tide, they quickly
become waterlogged and soon collect small amounts of redeposited,
previously-floating films, which loosely cement them to the mud
surface. At this point, the gentle action of rising and falling
tides is no longer able to move the dead macrophyte pieces.
Within two or three days, filaments from the dense filamentous
algal community grow over the dead leaf fragments. This algal
network provides additional attachments that prevent the removal
of dead plant material from the marsh on subsequent ebbing tides.
A thin layer of benthic diatoms and mucilaginous material soon
appears, completely covering the leaf fragment. Most dead
J. roemerianus tissue therefore appears to be rapidly covered
over by the marsh benthic microfloral community and is incorporated into the surface of the marsh floor. This incorporation
occurs before decomposition advances to a stage where small
detrital particles break off and are transported from the marsh
by ebbing tides.

The leaves of J. roemerianus plants growing along creek banks
often experience a somewhat different fate. Dead leaf tips broken
from these plants can fall into the tidal stream and be carried
from the marsh system on the next falling tide. Leaves that break
over at the base of the plant but are not detached, a common
occurrence along creek banks, provide a substrate that is rapidly
colonized by epiphytic green algae. Since the network of mud
surface filamentous algae is seldom as dense on creek banks as
further back in the marsh, the decomposing J. roemerianus leaves
frequently are not anchored to the marsh surface, but may be
raised and lowered freely by tidal action. Samples of this type
of leaf were prepared for SEM viewing. Some leaf sections were
mounted for viewing the covering of epiphytes and others were
broken open after critical point drying and mounted with the
internal structure of the plant exposed.

As seen in Fig. 6a and b, the outer surface of this decomposing leaf is totally covered with filamentous algae. Focusing
down between algal filaments to the leaf surface revealed a layer
of diatom fragments and smaller unidentified fragments (clay
particles?), but no bacteria (Fig. 6c). The inside of the same
leaf is shown in Fig. 6d. Here bacteria and fungi are numerous.
As decomposition progresses, portions of these leaves undoubtedly
break loose and are transported from the marsh. However, the
percentage of plants located along the creek banks and therefore
subjected to this decomposition sequence is very small in comparison with those plants located further back in the marsh, where
they are retained and decomposed under the carpet of filamentous
and diatomaceous algae.

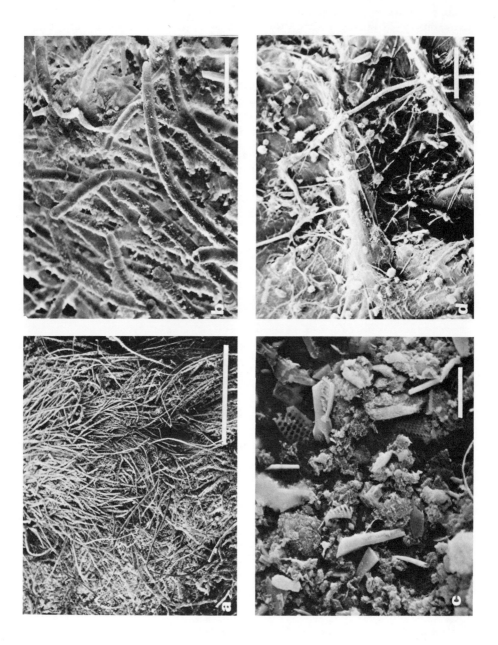

Fig. 6. Scanning electron micrographs of decomposing *Juncus roemerianus* leaf collected from creekbank. (a) dense growth of filamentous algae attached to leaf surface. Scalebar equals 1 mm. (b) higher magnification of (a). Scalebar equals 50 μm. (c) numerous filaments of diatom frustules and smaller fragments (probably clays) cover the surface of the leaf. This view obtained by focusing down between the algal filaments seen in (a) and (b). Note absence of recognizable bacteria. Scalebar equals 10 μm. (d) inside view of same leaf. Bacteria and fungi are abundant. Scalebar equals 5 μm.

DISCUSSION

Role of Decomposers in Salt Marsh
Detritus Production Questioned

The observations that (1) detrital material leaving the
J. roemerianus marsh contains negligible quantities of cellulose,
(2) most *J. roemerianus* leaves are decomposed under a carpet of
benthic algae, and (3) the films produced by the algal carpet
are nearly identical to the suspended aggregate material that
comprises more than 95% of the total particulates in salt marsh
water samples are inconsistent with current concepts of salt
marsh detritus production. The importance of the decomposers
in salt marsh ecology is not questioned, since the tremendous
quantity of nutrient material incorporated in the biomass of
J. roemerianus is almost certainly cycled through this group
of organisms. The role of the decomposers as the actual producers of the particulate material which directly supports a
broad spectrum of coastal marine filter feeding and deposit
feeding organisms, however, is questioned.

Excretion of Organic Material by Algae

Many algae, including blue-green species and diatoms, are
known to liberate considerable quantities of organic material
into the surrounding water. *Hellebust* (1967) reported that
diatom cultures grown in plastic spheres excreted up to 40% of
their photoassimilated carbon and *Fogg* (1962), using radiocarbon,
found that phytoplankton in an English lake liberated from 3 to
90% of the total fixed carbon as soluble organic material. Fogg
also reported that nitrogen-fixing blue-green algae in healthy
cultures liberated substantial amounts of soluble nitrogenous
substances, primarily as peptides. Free amino acids, especially
aspartic acid, glutamic acid, and alanine, were excreted in small
amounts. Substances associated specifically with the nitrogen
fixation process were not found in significant proportions in
the culture filtrate.

Although most studies on algal extracellular products
have focused on dissolved substances, it is well known that many
diatoms excrete mucilaginous materials. The production of extracellular polysaccharides by diatoms in the form of threads,
gelatinous capsules, mucilage pads, stalks and tubes was
discussed by *Darley* (1977). He specifically mentioned capsules
produced by *Navicula pelliculosa* and *Phaeodactylum tricornutum*,
and the mucilage tubes produced by *Amphipleura rutilans*. These
mucilaginous substances were found to be composed of polysaccharides, polyuronides and proteins. Many epiphytic diatoms,
including species of *Cocconeis, Licmophora, Achnanthes, Melosira,*

and *Synedra*, are often attached to their hosts by mucilaginous pads or stalks, and species of *Navicula*, *Nitzschia*, and *Amphora* form masses of cells inside a matrix of gelatinous material (*Main and McIntire*, 1974). All of these genera are well represented in the *J. roemerianus* marsh diatom community. Mucilage is also thought to provide a source of propulsion among certain benthic species that leave a mucilage track behind them (*Harper*, 1977).

The gelatinous material engulfing the benthic algal community of the *J. roemerianus* marsh was found, in this study, to originate from the diatoms. This material provides the adhesive matrix of the aggregate detritus and is removed from the surface of the marsh floor by rising tides.

Diatom films teased directly from the benthic community were almost identical in appearance to the suspended aggregates collected from the water column during ebbing tides. The only visible difference observed in these two types of samples was the greater number of diatoms in materials collected from the benthic films. Diatom migration behavior might be responsible for this occurrence.

Vertical Migration of Diatoms

Harper (1977) provided a literature review of the frequently observed phenomenon of migration and phototaxis in benthic diatoms. *Aleem* (1950) reported that the mud flat diatoms at Whitstable, England migrated diurnally to the surface of the sediments. This diatom community appeared at the surface at low tide if light intensity was sufficient. When low tides occurred near sunset, during the night, or if the day was exceptionally cloudy, the migration did not take place and the diatoms remained below the surface. *Fauré-Fremiet* (1951) likewise observed a phototactic migration of benthic diatoms, including species of *Pleurosigma* and *Hantzschia*, that was related to tidal rhythm. They hypothesized that this movement was a behavioral adaptation that prevented the diatoms from being washed out to sea by the tides.

Floating films and suspensions of aggregate material are removed from the benthic algal carpet in the *J. roemerianus* marsh by ebbing tides. With the onset of a flooding tide, the highly motile diatoms enter the sediments leaving only the mucilaginous film and dead or senescent individuals embedded in the film. Benthic film teased from emersed mud samples collected at low tide, therefore, would be expected to contain large numbers of actively photosynthesizing diatoms.

The explanation for the large member of diatoms in films removed from inundated mud samples that had been held under laboratory conditions for many days is less clear. Since phototaxis and tidal rhymicity seem to be closely interrelated (*Aleem*, 1950), the laboratory community, experiencing no tidal action, might have been responding solely to light, or possibly had ceased all migratory activity.

Desiccation as Possible Inhibitor of Vertical Migration

Another discordance exists between findings reported in the already cited literature on diatom migration and the present observation that diatoms colonized new filamentous algal growth only after the mud samples were inundated. One possible explanation for this occurrence might lie in problems of desiccation. The mud surface layer of mucilaginous material remains moist continuously, so diatoms migrating into this area during periods of illumination would not experience desiccation stress. The vertical algal filaments, rising as high as 2 mm above this mucilaginous film, however, would be considerably drier and might present a suitable substrate for colonization only when immersed. Once the filamentous new growth became colonized and a mucilaginous sheath surrounded it, diatoms embedded inside this covering would receive significant protection from desiccation.

A similar phenomenon might be responsible for the reduced depth of the diatom layer over the marsh floor during winter months. Significantly lower tides occurring during the winter result in a reduced frequency of inundation for the higher elevations of the marsh and also in a longer period of intertidal exposure even for those areas near the tidal creeks. While filamentous algae apparently grow quite well when emersed, the diatoms do not colonize the new substrates produced by the filamentous algae without inundation. As the average height of high tides begins to increase during spring and continues to rise until September, longer inundation periods and greater tidal coverage might allow the colonization of this new substrate by the diatom community that was previously limited by desiccation stress.

Absence of Bacteria in Aggregates Due to Antibacterial Agents?

The nutritional value of detritus has long been considered to be a function of the attached microorganisms. Although many authors have discussed the importance of this microbial community, several investigators have reported finding only very low microbial activity on detrital material collected from a wide range

of habitats. For example, *Wiebe and Pomeroy* (1972) found low
quantities of bacteria associated with particulates collected
from waters of coral reef regions, the Sargasso Sea, subtropical
coastal and continental shelf regions, estuaries and salt marshes.
Only in the interstitial waters of *Spartina alterniflora* rafts
did they observe large numbers of microorganisms. Even in
estuaries, the majority of particles they observed did not
contain recognizable bacteria. *May* (1974) also reported finding
few bacteria on samples of *Spartina alterniflora* which he placed
in the marsh in bags of nylon netting.

Bacteria were likewise not observed in association with the
aggregate detritus collected in this investigation. This absence
of bacteria might be due to inhibitory products in the algal
excretory materials. *Bell et al.* (1974) discussed the general
inhibition of bacteria by rapidly growing diatom cultures. They
reported that generalized bacterial growth occurred only during
bloom senescence, when algal cell lysis resulted in the release
of material favorable to bacterial growth. *Fogg* (1962) also
observed that contamination by bacteria rarely occurred in
rapidly growing algal cultures, and thought that this lack of
contamination might be due to suppression by antibacterial agents
excreted by the algae.

Since most aggregate material appearing in ebb tide waters
had been removed from the marsh surface only hours earlier, any
antibacterial activity in the mucilaginous material would likely
be retained for a short period of time. The existence of antibacterial agents in the excretory products of the diatoms would
certainly limit, if not preclude, the growth of bacteria in the
film of excretory material covering the mud surface. Bacterial
decomposition of vascular plant tissue would therefore be limited
to the subsurface regions below the layer of highest diatom productivity. The presence of bacteria and fungi only in the
interior of decomposing leaves of *J. roemerianus* plants growing
along creek banks might also be due to the action of an antibacterial substance produced by the filamentous algae covering the
outer surface of the leaves. It might even be speculated that
some symbiotic relationships could exist between the algae and
bacteria that colonize the dead leaves of these creek bank plants.
The algal community of the marsh floor is almost certainly the
recipient of minerals as well as vitamins and other organic
products from the decomposer community below the marsh surface.

Origin of Small Quantity of
Vascular Plant Fragments

Since vascular plant decomposition is herein hypothesized to
occur primarily beneath the benthic algal carpet, one might ask

where the small, but nevertheless existent, fraction of identifiable plant fiber detritus originates. *Welsh* (1975) found that
Palaeomonetes pugio accelerated the breakdown of detritus in a
tidal marsh ecosystem by macerating decomposing vascular plant
material into a heterogeneous assortment of particles. The shrimp
plucked away pieces of the cellular material of the detrital
particles, exposing the internal cavities, which were then heavily
invaded by pennate diatoms. *May* (1974) also suggested that a
crustacean, *Cleantis planicauda,* played a major role in the
breakdown of *Spartina alterniflora.* Several species of *Uca* are
common to the *J. roemerianus* marshes of the Gulf Coast, and these
detritivores can be seen in vast numbers working the surface of
the marshes at low tide. It is likely that the occasional fragments of plant fiber detritus observed in ebb tide water samples
result from the feeding activities of the fiddler crabs and other
marsh detritivores, as well as from the erosion of particles from
the decomposing leaves of grass growing along the creek banks.

REFERENCES

Aleem, A. A., 1950. The diatom community inhabiting the mud-flats
    at Whitstable, *New Phytol. 49:* 174-182.

Anderson, T. F., 1951. Techniques for the preservation of three-
    dimensional structure in preparing specimens for the electron
    microscope, *Trans. N. Y. Acad. Sci. 13:* 130-134.

Bell, W. H., J. M. Lang and R. Mitchell, 1974. Selective
    stimulation of marine bacteria by algal extracellular
    products, *Limnol. Oceanog. 19:* 833-839.

Burkholder, P. R. and G. H. Bornside, 1957. Decomposition of
    marsh grass by aerobic marine bacteria, *Bull. Torrey Bot.
    Club 84:* 366-383.

Darley, W. M., 1977. Biochemical composition, In: *Biology of
    Diatoms,* edited by D. Werner, 198-223. Blackwell Scientific
    Publications, Oxford.

Darnell, R. M., 1958. Food habits of fishes and larger inverte-
    brates of Lake Pontchartrain, Louisiana, an estuarine
    community, *Publ. Inst. Mar. Sci. Univ. Tex. 5:* 353-416.

de la Cruz, A. A., 1973. The role of tidal marshes in the
    productivity of coastal waters, *Assoc. S. E. Biol. Bull.
    20:* 147-156.

de la Cruz, A. A., 1975. Proximate nutritive value changes during
    decomposition of salt marsh plants, *Hydrobiol. 47:* 475-480.

de la Cruz, A. A. and B. C. Gabriel, 1974. Caloric, elemental, and nutritive changes in decomposing *Juncus roemerianus* leaves, *Ecology 55:* 882-886.

de la Cruz, A. A. and W. E. Poe, 1975. Amino acids in salt marsh detritus, *Limnol. Oceanog. 20:* 124-127.

Fauré-Fremiet, E., 1951. The tidal rhythm of the diatom *Hantzschia amphioxys, Biol. Bull. 100:* 173-177.

Fenchel, T., 1970. Studies on the decomposition of organic detritus derived from the turtle grass *Thalassia testudinum, Limnol. Oceanog. 15:* 14-20.

Fenchel, T., 1972. Aspects of decomposer food chains in marine benthos, *Verh. Deutsch. Zool. Ges. 65:* 14-22.

Fogg, G. E., 1962. Extracellular products, In: *Physiology and Biochemistry of Algae,* edited by R. A. Lewin, 475-489, Academic Press, New York.

Gosselink, J. G. and C. J. Kirby, 1974. Decomposition of salt marsh grass, *Spartina alterniflora* Loisel, *Limnol. Oceanog. 19:* 825-832.

Harper, M., 1977. Movements, In: *Biology of Diatoms,* edited by D. Werner, 224-249, Blackwell Scientific Publications, Oxford.

Heald, E. J., 1971. The production of organic detritus in a South Florida estuary, *Sea Grant Tech. Bull. 6.*

Hellebust, J. A., 1967. Excretion of organic compounds by cultured and natural populations marine phytoplankton, In: *Estuaries,* edited by G. H. Lauff, 361-366, AAAS Publication 83.

Kirby, C. J. and J. G. Grosselink, 1976. Primary production in a Louisiana Gulf Coast *Spartina alterniflora* marsh, *Ecology, 57:* 1052-1059.

Main, S. P. and C. D. McIntire, 1974. The distribution of epiphytic diatoms in Yaquina Estuary, Oregon, U.S.A., *Bot. Mar. 17:* 88-99.

Mann, K. H., 1972. Macrophyte production and detritus food chains in coastal waters, Mem. Ist. Ital. Idrobiol., *29 Suppl.* 353-383. In: Proc. of the IBP-UNESCO Symposium on detritus and its ecological role in aquatic ecosystems.

Maser, M. D. and J. J. Trimble, III, 1977. Rapid chemical
    dehydration of biologic samples for scanning electron
    microscopy using 2,2-dimethoxypropane, *J. Histochem.
    Cytochem. 25:* 247-251.

May, M. S., III, 1974. Probable agents for the formation of
    detritus from the halophyte *Spartina alterniflora,* In:
    Ecology of Halophytes, edited by R. J. Reimold and
    W. Queen, 429-440, Academic Press, New York and London.

Newell, R., 1965. The role of detritus in the nutrition of two
    marine deposit feeders, the prosobranch *Hybrobia ulvae* and
    the bivalve *Macoma baltica, Proc. Zool. Soc. Lond. 144:*
    25-45.

Odum, W. E., 1970. Utilization of the direct grazing and plant
    detritus food chains by the striped mullet, *Mugil cephalus,*
    In: *Marine Food Chains,* edited by J. H. Steele, 222-240,
    Univ. Calif. Press, Berkeley and Los Angeles.

Odum, W. E. and E. J. Heald, 1975. The detritus-based food web of
    an estuarine mangrove community, In: *Estuarine Research Vol.
    I,* edited by L. E. Cronin, 217-228, Academic Press, New York.

Odum, E. P. and A. A. de la Cruz, 1967. Particulate organic
    detritus in a Georgia salt marsh-estuarine ecosystem, In:
    Estuaries, edited by G. H. Lauff, 383-388, AAAS Publication 83.

Olah, J., 1972. Leaching, colonization, and stabilization during
    detritus formation, Mem. Ist. Ital Idrobiol., *29 Suppl.:* 105-
    127. In: Proc. of the IBP-UNESCO Symposium on detritus and
    its ecological role in aquatic ecosystems.

Reimold, R. J., J. L. Gallagher, R. A. Linthurst, and W. J.
    Pfeiffer, 1975. Detritus production in coastal Georgia
    salt marshes, In: *Estuarine Research Vol. I,* edited by
    L. E. Cronin, Academic Press, New York.

Teal, J. M., 1962. Energy flow in the salt marsh ecosystem of
    Georgia, *Ecol. 43:* 614-624.

Welsh, B. L., 1975. The role of the grass shrimp, *Palaemonetes
    pugio,* in a tidal marsh ecosystem, *Ecology 56:*

Wiebe, W. J. and L. R. Pomeroy, 1972. Microorganisms and their
    association with aggregates and detritus in the sea: a
    microscopic study, Mem. Ist. Ital. Idrobiol. *29 Suppl.:*
    325-352. In: Proc. of the IBP-UNESCO Symposium on detritus
    and its ecological role in aquatic ecosystems.

FACTORS CONTROLLING THE FLUX OF PARTICULATE ORGANIC CARBON FROM ESTUARINE WETLANDS

William E. Odum, John S. Fisher and James C. Pickral

University of Virginia

ABSTRACT

We hypothesize that specific tidal wetland areas may either export or import particulate organic carbon on an annual basis depending upon several geophysical factors. Important among these are: (1) the geomorphology of the wetland drainage basin and (2) the relative magnitudes of the tidal range and freshwater input from upland sources. Determination of the import/export status of individual wetlands further requires (1) continuous long-term measurements to compensate for irregular and infrequent storm events and (2) measurements of particle bedload transport in addition to the usual measurements of suspended and floating particles. Without these components, any study of net organic particle flux will be incomplete.

INTRODUCTION

A classic series of studies of the coastal marshes of Georgia (*Odum*, 1961; *Schelske and Odum*, 1961; *Teal*, 1962; *de la Cruz*, 1965; *Odum and de la Cruz*, 1967) form the basis for the hypothesis that tidal marshes may function as net exporters of particulate organic carbon. This concept was later extended to mangrove swamps (*Heald*, 1969; *Odum*, 1970; *Odum and Heald*, 1975).

More recent investigations from other geographical areas have provided contradictory data. As a result, a controversy has arisen concerning the correctness of the original hypothesis. For example, *Nadeau* (1972), *Boon* (1973), *Heinle and Flemer* (1976), *Woodwell et al.* (1977), *Hackney* (1977) and *Haines* (1977) have been unable

to measure net export from tidal wetlands and in some cases have
found an apparent net import. On the other hand, *Day et al.*
(1973), *Nixon and Oviatt* (1973), *Axelrod* (1974), and *Valiella et
al.* (1978) have found a net export of particulate organic carbon.

In the present paper we argue that the controversy in its
present form is premature for two reasons. First, we suggest
that wetlands may function as either net importers or net exporters
of particulate organic carbon depending upon (1) the geomorphology
of the wetland drainage basin, (2) the tidal amplitude, and (3)
the magnitude of freshwater input to the drainage basin. Second,
all published import/export studies have failed to account
adequately for two important factors: (1) the pulsed (or intermittent) nature of particle transport acting in response to
irregular storm events and (2) the movement of particulate loads
on or near the bed of estuarine creeks, rivers, and embayments.

If these factors are considered, then the marsh import/
export controversy can be replaced by analyses on a marsh by
marsh or estuary by estuary basis. With such an approach we suspect that some wetland areas will be identified as having an
annual input of organic carbon while others will have an output.
Moreover, in many cases it should be possible to predict net flux
on the basis of geomorphological and hydrological factors alone.

A Semantic Point About "Export"

One problem which has further confused the import/export controversy is the varied use of the term "export." To some
researchers "export" is simply the transfer of wetlands-produced
organic carbon to nearby water bodies such as tidal creek systems.
To others the same term applies to long-range movement of particulate carbon out of estuaries into coastal inshore waters. For
the purposes of the present paper, we will consider "export" to
refer to the physical transport of particulate organic carbon
from discrete wetland systems into adjacent, largely aquatic,
systems. The former includes the wetland basin (marsh, mangrove
swamp, etc.) with its closely associated drainage system (small
creeks and embayments which drain almost completely at low tide)
while the latter comprises larger creeks, rivers, and embayments
which function as drainage systems for more than one wetland
basin. The critical question is whether energy fixed at a semi-
aquatic site is made available to aquatic secondary producers.

GEOMORPHOLOGICAL CONSTRAINTS ON PARTICLE FLUX

The geomorphology of a wetland drainage basin clearly has the potential to affect net particle flux. Although no complete field studies have been made in basins with contrasting geomorphological characteristics, we can draw some preliminary conclusions based on existing knowledge of inorganic particle dynamics (*Postma*, 1967; *Steers*, 1967; *Hayes*, 1975). A schematic representation of two extreme basin types and an intermediate case are given in Fig. 1. At one extreme, the wetland basin is connected to a larger body of water by a narrow, shallow channel. The result is a basin which acts as a trap for organic material produced in and around the basin and also for organic particles transported by the incoming tide. Unless flushed by a strong pulse of water from upland sources, this type of basin will probably have little or no net export of particulate carbon. The Flax Pond wetland system of *Woodwell et al.* (1977) is of this type; not surprisingly they found a net import of particulate carbon **on** an annual basis.

At the other extreme is a V-shaped basin which gradually deepens and widens toward the mouth. This type of basin offers little geomorphological impediment to particle export and, in most cases, should be characterized by a net export of particulate carbon.

Many estuaries and wetland basins have geomorphological characteristics intermediate between the two examples which have been discussed. For these cases the relative strengths of the tidal and freshwater inputs along with other hydrologic characteristics of the basin will control the annual net flux of particulate organic carbon.

PARTICLE FLUX AS RELATED TO FRESHWATER

INPUT AND TIDAL AMPLITUDE

To carry our multiple factor hypothesis one step further, organic particle flux in wetland basins is linked closely, not only to geomorphological characteristics, but also to the dynamic forces which control the flow of water through the basin such as astronomical tides, wind tides, and freshwater influx. Although data for organic particle transport are lacking, knowledge of inorganic particle movement allows certain general statements.

For one extreme situation we hypothesize that basins with a relatively slight freshwater input and only a moderate tidal fluctuation will tend to have little net export and may be characterized by a net import of organic carbon. On the incoming

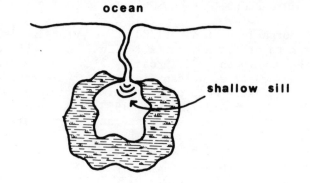

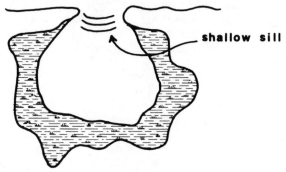

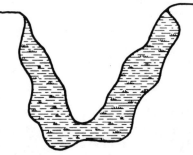

Fig. 1. Three hypothetical types of estuarine basins: (A) a basin with a narrow entrance guarded by a shallow sill, (B) a basin with a wider mouth but still possessing a shallow sill, and (C) a V-shaped basin which gradually deepens and widens toward the mouth. All three basins are ringed by intertidal marshes with associated tidal river and creek networks.

tide oceanic carbon (phytoplankton, pieces of kelp and macroalgae) will be transported into the basin. Much of this material will settle to the bottom during the change of tides and be retained within the basin. This is the case because the out-going tide, lacking the added impetus of a large volume of inflowing fresh water, may be unable to generate sufficient current velocity for extensive particle transport, particularly of the larger size fractions. In addition, carbon produced within the basin will tend to remain within the system.

At the other extreme is the estuarine basin with a modest or large tidal amplitude and a large freshwater input. In this case the water flow on the outgoing tide may be great enough to transport both recently imported carbon particles and particles produced within the basin; the result is an annual net export of particulate organic carbon. This situation could arise in any location with a major freshwater stream or even appreciable groundwater or surface runoff. In both cases we have ignored stratification which will, of course, complicate the examples. For example, a strongly stratified estuary may tend to import oceanic carbon by means of the salt wedge circulation while fine terrestrial material may be exported in the outflowing surface layer.

Since freshwater input may vary seasonally, the status of an estuarine basin in terms of import and export may vary at different times of the year. For example, the Everglades estuary studied by *Heald* (1969) and *Odum* (1970) apparently functions as a net export system during the rainy season but probably has little or no export during the dry season.

For estuarine basins which are intermediate between these two extreme examples, determination of net particle flux may be complex and vary from day to day, season to season, and year to year depending upon variations in precipitation and wind tides. Unfortunately, many estuaries fall into this intermediate category. For these cases the determination of export/import status will require careful long-term monitoring.

THE INTERMITTENT NATURE OF PARTICLE FLUX

Experience from studies of inorganic particle movement at many freshwater stream and river sites shows clearly that most of the net downstream transport of large particles occurs during periodic, often infrequent, extreme events such as heavy rain storms. Considering the geomorphology and hydrologic character- istics of coastal wetland basins, there is no reason to believe that the situation will be any different. There is, in fact, a small amount of evidence to support the hypothesis of irregular

transport of large size fractions. *Pickral and Odum* (1977)
found a net transport of several tons of marsh plant detritus out
of a tidal creek after a heavy rain shower. E. *Kuenzler* (pers.
comm.) has observed that extensive transport of large detritus
particles occurs from North Carolina *Juncus* marshes during extreme
coastal storms which may occur with a periodicity of two to five
years. Finally, E. J. *Heald* and W. E. *Odum* monitored a dramatic
change in suspended organic loadings in the North River estuarine
basin of the Everglades estuary during and after a rain shower at
the beginning of the wet season. This unpublished observation
occurred during a moderate rain shower (two centimeters of rain in
40 minutes) which fell during slack low tide. Within 30 minutes
of the initiation of rainfall the suspended load increased from
15 to 120 mg/liter. This increase was apparently caused by the
input of particles of mangrove detritus scoured from the marsh and
swamp surface and washed into the tidal creeks.

From these few examples and other, unpublished observations,
it appears that storm-induced particle transport occurs during at
least two different meteorological events. The first consists of
a heavy localized shower which starts at or near low tide and
erodes the marsh surface, carrying organic particles into tidal
streams via runoff. The second situation consists of a large
storm with attendant elevated tides, intense rainfall and,
possibly, strong wave action. This second type of storm event is
capable of flushing particulate matter out of extensive wetland
areas, even above the high tide line, but may occur only at
intervals of several years or more.

Future attempts to obtain realistic flux rates from wetland
areas should include measurements during both types of storm
events. The localized shower is almost impossible to predict and,
therefore, requires continuous monitoring to detect its effect.
The impact of the major storm, because it occurs so infrequently,
requires long-term monitoring over periods of years. The acqui-
sition of truly representative particle flux estimates remains a
complicated and expensive task.

TRANSPORT CHARACTERISTICS OF ORGANIC PARTICLES

As previously mentioned, transport studies of organic
particulates are extremely scarce. Fortunately, an extensive
literature exists concerning the transport characteristics of
inorganic particles (e.g. *Leopold et al.*, 1964; *Postma*, 1967;
Meade, 1972; *Gibbs*, 1974). Moreover, in a recent paper (*Fisher
et al.*, 1979) we have used a Shield's Diagram (shear stress
plotted against particle Reynold's number) to show that organic
particles behave in a predictable manner similar to that of
inorganic particles subjected to the same flow conditions.

This conclusion, at least tentatively, validates hypotheses concerning organic particle movement based on inorganic particle observations. The following hypotheses are based on this assumption.

Alternating Flux

Most organic particles do not simply enter or exit from a wetlands drainage basin, but tend to flux back and forth with the tide. This tendency does not preclude a net movement in one direction, but it does mean that an individual particle may be transported back and forth a number of times before permanently entering or leaving a drainage system. This conclusion certainly applies to very fine material which remains in suspension more or less permanently. For larger particles lying on the streambed, transport velocities may not be reached except during extreme tides or after heavy rainfall. In this case transport may occur quickly and irreversibly in one direction.

The alternating flux of suspended particles can be compensated for in monitoring studies by measurements taken over a relatively long period of time. Such monitoring has been included in certain recent studies (*e.g. Valiela et al.*, 1978).

The Importance of Bedload Transport

Depending upon the shape and density of a particle and the flow affecting the particle, there are three ways in which a particle can be transported in water: (1) floating (if the particle density is less than one), (2) suspended (if the density is greater than one and the turbulence is sufficiently high to keep the particle in suspension), or (3) as bedload rolling along the bottom (if the density is greater than one, but the turbulence is not high enough to keep the particle in suspension). Most investigators working with the flux of organic particulates from wetlands have considered the first two pathways, but have ignored bedload transport. We feel that in many situations bedload transport may be important.

This hypothesis concerning the importance of bedload transport was not arrived at in a casual manner, but is derived from a logical series of deductions which can be stated in the following manner:

(1) As particle size increases the current velocity necessary to transport the particle also increases. This relationship is well established for inorganic particles (Fig. 2) and we

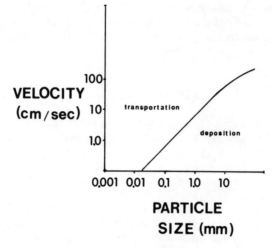

Fig. 2. A generalized representation of the relationship between velocity and inorganic particle size. This relationship is only approximate and may be altered greatly by increased turbulent motion of the current.

have shown that it applies to organic detritus particles as well (*Fisher et al.*, 1979).

(2) Water velocity decreases near the stream bed (Fig. 3), although the exact velocity/depth relationship may vary in response to a number of factors.

(3) The fall velocity or sinking rate of most types of saturated organic particles is relatively high. Several examples for marsh grass detritus are shown in Table 1. Although larger particles and pieces of stem have the highest fall velocities, all of the particles in this table have the potential to settle to the bottom within a few mintues after tidal velocity stops or is reduced.

(4) Combining the previous three statements, we conclude that organic detritus particle concentrations should generally increase with depth and reach the highest values near the bottom. This pattern may not apply to situations in which very fine organic particles (less than 0.1 mm) are subjected to velocities greater than 30-40 cm/sec or to extreme turbulence.

(5) Because of the relatively high density of saturated organic detritus particles and the low water velocity near the stream bed, it is likely that movement of this material consists

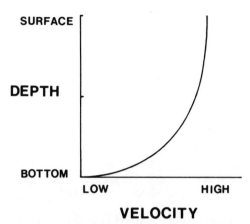

Fig. 3. A typical relationship between velocity and depth in a tidal stream or river. This relationship may be greatly altered by changes in channel geomorphology.

TABLE 1. The estimated fall velocities of several types of organic particles commonly found in estuarine wetland systems. Fall velocities were measured in a settling tube; particle size was determined by wet sieving. Particles are ground *Spartina alterniflora* which had been soaked in water for two weeks.

PARTICLE TYPE	SIZE (mm)	FALL VELOCITY (cmm/sec)
stem	0.5 - 1.0	1.0
stem	1.0 - 2.0	2.5
stem	2.0 - 4.0	3.8
leaf	0.02 - 0.05	0.05 - 0.1
leaf	0.07 - 0.09	0.2 - 0.3
leaf	0.1 - 0.15	0.4
leaf	0.25 - 0.5	0.6
leaf	1.0 - 2.0	0.7
leaf	2.0 - 4.0	0.9

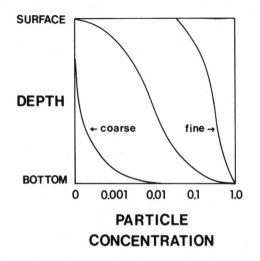

PARTICLE
CONCENTRATION

Fig. 4. The general relationship between inorganic particle concentration and depth. There are a range of possible curves depending upon the size of the particle; three of these curves are shown for three sizes of particles. This series of curves can also be altered by factors such as the degree of turbulent motion and stream geomorphology.

of rolling along or near the bottom as bedload transport. We have observed this phenomenon in a laboratory flume.

To summarize, many organic detritus particles have the potential of sinking to the zone of reduced velocity near the stream bed. Unless the velocity is high or accompanied by excessive turbulence, movement of particles larger than about 0.1-0.2 mm should consist of rolling or bouncing along as bedload transport.

CONCLUSIONS

Based on considerations of geomorphology and transport pro- perties of organic particles, we hypothesize that coastal wetland basins may either import or export particulate organic carbon. Confirmation or rejection of this hypothesis will depend upon long-term field measurements of particle carbon flux in specific wetland basins. These measurements should include all three modes of transport (floating, suspended, and bedload) and must be carried out in all types of weather conditions including extreme storm events.

REFERENCES

Axelrod, D. M, 1974. *Nutrient Flux Through the Salt Marsh Ecosystem*, Ph.D. Thesis, College of William and Mary.

Boon, J. D., 1973. *Sediment Transport Processes in a Salt Marsh Drainage System*, Ph.D. Thesis, College of William and Mary.

de la, Cruz, A. A., 1965. *A Study of Particulate Organic Detritus in a Georgia Salt Marsh Estuarine Ecosystem*, Ph.D. Thesis, University of Georgia.

Day, J. W., Jr., W. G. Smith, and P. R. Wagner, 1973. Community structure and carbon budget of a salt marsh and shallow bay estuarine system in Louisiana, L.S.U. Center Wetland Resources Publ. LSU-SG-72-04.

Fisher, J. S., J. C. Pickral, and W. E. Odum, 1979. Organic detritus particles: initiation of motion criteria, *Lim. Oceanogr. 24*: 529-532.

Gibbs, R. J., 1974. *Suspended Solids in Water*, Plenum Press, New York.

Hackney, C. T., 1977. *Energy Flux in a Tidal Creek Draining an Irregularly Flooded Juncus marsh*, Ph.D. Thesis, Mississippi State University.

Haines, E. B., 1977. The origins of detritus in Georgia salt marsh estuaries, *Oikos 29*: 254-260.

Hayes, M. O., 1975. Morphology of sand accumulation in estuaries: an introduction to the symposium, *Estuarine Research Vol. II*, Academic Press, New York.

Heald, E. J., 1969. *The Production of Detritus in a South Florida Estuary*. Ph.D. Thesis, Univ. of Miami.

Heinle, D. R. and D. A. Flemer, 1976. Flows of materials between poorly flooded tidal marshes and an estuary, *Mar. Bio. 35*: 359-373.

Leopold, L. B., M. G. Wolman, and J. P. Miller, 1964. *Fluvial Processes in Geomorphology*, W. H. Freeman and Co., San Francisco.

Meade, R. H., 1972. Transport and deposition in estuaries, In: *Environmental Framework of Coastal Plain Estuaries, Geol. Soc. Am. Mem. 133*: 91-120.

Nadeau, R. J., 1972. *Primary Production and Export of Plant Material in a Saltmarsh Ecosystem.* Ph.D. Thesis, Rutgers University.

Nixon, S. W. and C. A. Oviatt, 1973. Ecology of a New England salt marsh, *Ecol. Monogr. 43*: 463-498.

Odum, E. P., 1961. The role of tidal marshes in estuarine production, *The Conservationist 15*: 12-15.

Odum, E. P. and A. A. de la Cruz, 1967. Particulate organic detritus in a Georgia salt marsh-estuarine ecosystem. In: *Estuaries*, edited by G. H. Lauff, 383-388, #83. AAAS Pub.

Odum, W. E., 1970. *Pathways of Energy Flow in a South Florida Estuary*, Ph.D. Thesis, Univ. of Miami.

Odum, W. E. and E. J. Heald, 1975. Mangrove forests and aquatic productivity, Chapter Five in: *An Introduction to Land-Water Interactions*, 129-136, Springer-Verlag Ecological Study Series.

Pickral, J. C. and W. E. Odum, 1977. Benthic detritus in a saltmarsh tidal creek, In: *Estuarine Processes, Vol. II*, edited by M. Wiley, 280-292, Academic Press, N. Y.

Postma, H., 1967. Sediment transport and sedimentation in the estuarine environment, In: *Estuaries*, edited by G. H. Lauff, 158-179, AAAS Publ. #83.

Schelski, C. L. and E. P. Odum, 1961. Mechanisms maintaining high productivity in Georgia estuaries, *Proc. Gulf and Carib. Fish. Inst. 14*: 75-80.

Steers, J. A., 1967. Geomorphology and coastal processes, In: *Estuaries*, edited by G. H. Lauff, 100-107, AAAS Publ. #83.

Teal, J. M., 1962. Energy flow in the salt marsh ecosystem of Georgia, *Ecology 43*: 614-625.

Valiela, I., J. M. Teal, S. Volkmann, D. Shafer, and E. J. Carpenter, 1978. Nutrient and particulate fluxes in a salt marsh ecosystem: tidal exchanges and inputs by precipitation and ground water, *Lim. Oceanogr. 23*: 798-812.

Woodwell, G. M., C. A. Hall, and R. A. Houghton, 1977. The Flax Pond ecosystem study: exchanges of carbon in water between a salt marsh and Long Island Sound, *Limn. Oceanogr. 22*: 833-838.

III. Energy Transfer and Trophic Relationships

EFFECTS OF SURFACE COMPOSITION, WATER COLUMN CHEMISTRY, AND TIME OF EXPOSURE ON THE COMPOSITION OF THE DETRITAL MICROFLORA AND ASSOCIATED MACROFAUNA IN APALACHICOLA BAY, FLORIDA

David C. White, Robert J. Livingston, Ronald J. Bobbie

and Janet S. Nickels

Florida State University

ABSTRACT

In experiments where measures of the detrital microfloral biomass, morphology, and activity were compared to the mass, numbers, and diversity of the associated macrofauna on different surfaces (artificial and natural plant detritus) incubated in baskets at the same station in a river-dominated estuary, the changes in microbial mass, morphology, and activity were not correlated with changes in the macrofaunal population attracted to baskets. In experiments where the different surfaces were incubated at two different stations, the gross measures of microbial biomass (lipid phosphate, poly-β-hydroxybutyrate), nutritional history, and respiratory activity were correlated with the particular substrate used, whereas the macrofaunal population was significantly correlated with the water chemistry but not with the gross measures of the detrital microflora.

However, when fine structure of the detrital microbial population was examined by comparison among its components of the proportions of the lipid fatty acids, highly significant correlations between the presence of particular bacterial components of the microflora and the numbers, biomass, and species richness of the detritus-associated macrofauna were evident. Clearly, subtle differences in the population structure of the detrital microflora are associated with the mass and structure of the macrofaunal detrital food web.

INTRODUCTION

Importance of the Detrital Input

The importance of detritus as the basic food source in
shallow estuarine environments was recognized early by *Petersen*
(1913, 1918), *Jensen* (1919), and *Ekman* (1947). Later workers have
correlated high secondary benthic (*Bader*, 1954; *Wieser*, 1960;
Schelske and Odum, 1962; *Wigley and McIntyre*, 1964; *Williams and
Thomas*, 1967; *Rowe and Menzies*, 1969) and fish production (*Darnell*,
1967; *Odum*, 1970; *Jeffries*, 1972, 1975) with areas receiving high
detrital inputs. *Livingston* (1978) has documented the relationship of cyclic detrital input with the very high productivity of
the Apalachicola Bay system.

Detrital "Conditioning"

So-called apparent plants--plants that are large, numerous,
and long-lived--survive herbivory by devoting a large proportion
of their resources to chemical defenses (*Feeney*, 1976). This
strategy results in detrital residue which contains polymers with
resistant structures, such as lignin, which are deficient in
nitrogen and often lack essential components such as aromatic
amino acids, as well as secondary toxic metabolities, such as
tannin, saponin, or alkaloids, which act as metabolic inhibitors
(*Fenchel and Jorgensen*, 1977). Eukaryotes generally lack the
enzymes capable of degrading these resistant polymers, and often
have essential nutrient requirements not met by direct catabolism
of the nitrogen-deficient detritus. The detrital microflora,
however, combines an extraordinary capacity to concentrate
nutrients from the water column, powerful lytic enzyme systems,
the capacity for anaerobic growth, nitrogen fixation, and other
processes to form several populations with diverse metabolic
capabilities. These populations are capable of direct use of
resistent detritus, and in turn provide a nutritious food resource
for the eukaryotic predators that form the next step in the
detrital food web.

The Microbial Problem

Despite the obvious importance of detrital microflora, its
study entails problems for the macroecologist because of several
characteristics of the microbial world: first, structure does
not reflect function, microbes in the same functional group come
in all shapes--spheres, capsules, twisted spirals, filaments.
Second, the species concept is almost without meaning in the
microbial world and bacterial taxonomy has recently been

complicated by the discovery of the widespread distribution of plasmids, in which important, often critical, information sequences can be readily exchanged between species. Lastly, the classical methods of public health microbiology are simply not adequate for studies of the complex microflora of natural environments. Plate counts in artificial media are of necessity highly selective and destroy the complex interactions that are critical to ecological understanding. The morphologic complexity of the detrital microbial world is illustrated in Fig. 1. Viable counts are often several orders of magnitude lower than direct counts or non-selective measures of microbial biomass (*King and White*, 1977; *White et al.*, 1979a).

Measures of the Detrital Microflora

Measurements of a number of biochemical characteristics of microbes can be combined to give a reasonable estimate of the mass and composition of the detrital microflora (*White et al.*, 1979a). Measures of mass include the extractible ATP, the muramic acid content, and the extractible lipid phosphate. Under specific conditions, rates of DNA synthesis, respiratory activity and the activities of esterases such as alkaline phosphatase, phospho-diesterase, sulfatase, β-D-glucosidase, β-D-galactosidase, and α-D-mannosidase are correlated with other measures of microbial biomass (*White et al.*, 1979a, 1979b; *King and White*, 1977; *White et al.*, 1977).

Morphologic features of the population structure can be detected by examination of scanning electron micrographs and by analysis of the lipid components extracted from the microflora (*White et al.*, 1979b; *King et al.*, 1977). Among the lipid components, the fatty acid composition has provided a highly useful tool in the analysis of progressively deeper sedimentary horizons in estuarine mud samples (*White et al.*, 1979b). Preliminary experiments have established that chances in the microbial popula-tion structure on inert surfaces brought about by modifying the chemical composition of the water column cause remarkably well-correlated changes in the results of scanning electron microscopy, the muramic acid to ATP ratios, the extractible steroid levels, and the proportions of the lipid fatty acids (*White et al.*, unpublished data).

With these various techniques the effects of the interaction of the structure and the activity of the microflora with the larger eukaryotic components of the food web can be examined. This paper addresses the interaction between the detrital microflora and the associated animal populations.

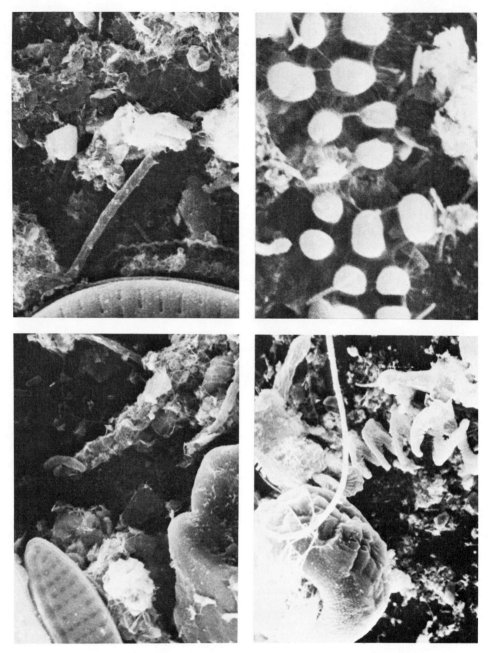

Fig. 1. Scanning electron micrographs of the detrital micro-flora on the artificial surface incubated in Apalachicola Bay in the summer of 1978. Magnification of the upper right and left and lower left panels was about 6500 X. The lower right hand panel was magnified 130 X. (Reduced 40% for purposes of reproduction.)

METHODS, STUDY SITES AND MATERIALS

Study Sites

Station placement was based on previous habitat determinations
(*Livingston*, 1978). Station 5A, in a shallow (1-2 meter) portion
of East Bay, is characterized by relatively low salinity and is
subject to direct effects of storm-water runoff from local rainfall in Tate's Hell Swamp. It is a series of mud flats and
scattered, fringing grass beds (*Ruppia maritima*, *Vallisneria
americana*, *Gracilaria* spp.). Station 3, dominated by Apalachicola
River flow, is a shallow (1-2 meter) mud flat with low salinities
and sparse *Ruppia* beds. In both areas, detrital input includes
leaf and wood matter and benthic macrophyte (*Ruppia*, *Vallisneria*)
leaves. These stations represent major habitat types in the
Apalachicola estuary.

Physico-Chemical Functions

Water samples for physico-chemical analysis (surface and
bottom) were taken at stations 3 and 5A (Fig. 2) at two-week
intervals (corresponding to biological collections) from 17 August
through 26 October, 1977. Temperature and dissolved oxygen were
measured with a YSI dissolved oxygen meter and salinity was
determined using a temperature-compensated refractometer
calibrated periodically with standard sea water. The pH was
sampled with a field meter, and light penetration was estimated
using a standard Secchi disk. Laboratory samples, taken with a
1-liter Kemmerer bottle, were analyzed for turbidity (Hach model
2100A turbidimeter) and water color (American Public Health
Association Platinum-Cobalt standard test). Apalachicola River
flow data (at Blountstown, Florida) were provided by the U. S.
Army Corps of Engineers (Mobile, Alabama), and local rainfall
information came from the East Bay Forestry Station (Florida).

Biological Collections

Litter baskets, constructed of plastic-coated hardware cloth
(6.5 mm mesh) shaped into cubes (30.5 mm/side) with hinged tops,
were fitted at the sides and bottom with an inner fiber-glass
screen liner (2 mm^2 mesh). Each basket (2 replicates/sample) was
filled with one of the following: mixed leaf litter taken from
the banks of the Apalachicola River, blades of *V. americana* taken
from fringing grass beds in upper East Bay, or artificial (plastic)
leaves shaped like the mixed deciduous leaf litter. Baskets were
tied to wooden frames anchored by stakes driven into the sediments.
At two-week intervals, four samples of each substrate were placed

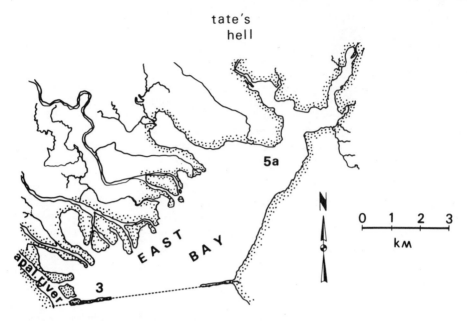

Fig. 2. Map of East Bay in the Apalachicola estuary
showing sampling stations near the Apalachicola River drainage
(3) and the Tate's Hell Swamp drainage (5A).

in the bay. They were retrieved at two- and four-week intervals
(2 baskets for each substrate at each sampling period) from 17
August through 26 October, 1977. The contents of each basket were
rinsed in a bucket and swirled to remove the organisms from the
mesh liner. The water in the bucket was then strained through a
297 μ sieve, and the retained organisms washed into a jar and
preserved in 10% buffered formalin.

In the laboratory, the macrobiota (fishes and invertebrates)
were sorted to species, counted, measured (blue crab carapace
width; fish standard length) and weighed (ash-free dry weight).
All data (physico-chemical, microbial, macrobiological) were
compiled into a permanent file for analysis using an interactive
computer processing system (*Livingston et al.*, 1976). Further
determinations, based on numbers of individuals (N) and species
(S), were carried out, including relative dominance or D_1 (top
species numbers x total numbers $^{-1}$), Margalef richness or D
(Margalef, 1958), the Shannon diversity index or H" (*Shannon and
Weaver*, 1963), and the Hurlbert formulation or $E(S_N)$ (*Hurlbert*,
1971).

Substrate samples were transported in aerated suspensions
of detrital material to the laboratory for microbial analysis

(2-3 hours) and immediately processed. Assays were performed for extractible ATP measures with luciferin-luciferase (*Morrison et al.*, 1977), for muramic acid by colorimetric analysis after chromatographic purification (*King and White*, 1977), for oxygen utilization by polarography (*Morrison et al.*, 1977; *Bobbie et al.*, 1978) for esterase activities by spectrophotometric assay of released p-nitrophenol (*Morrison et al.*, 1977; *Bobbie et al.*, 1978), for extractible lipid phosphate by colorimetric analyses (*White et al.*, 1977; *King et al.*, 1977), and for poly-β-hydroxy-butyrate (PHB) by extraction and spectrometric analysis (*Herron et al.*, 1978). The scanning electron micrographs were prepared as described by *Morrison et al.* (1977).

Fatty acid methyl esters were prepared from the extracted lipids and separated on polar and non-polar columns by gas liquid chromatography (*White et al.*, 1979b). Components were identified by their mobility on the two stationary phases compared to authentic standards before and after hydrogenation with different catalysts (*White and Cox*, 1967). The proportions of the fatty acids were calculated from the response on the recorder generated by methyl palmitate. The reproducibility using this methodology measured in 8 aliquots of the same methyl ester preparation was as follows for the fatty acids: mean value 15-carbon branched ester, ± 8.3%; 16-carbon monoenoic ester, ± 12.0%; 17-carbon cyclopropane ester, ± 9.6%; 18-carbon saturated ester, ± 7.8%; 18-carbon monoenoic ester, ± 13.6% (average error 10.26%).

Using the macrobiotic measures, the microbial parameters, and the 9 physico-chemical parameters, a regression analysis of paired variables was performed. For this analysis, both totals from all three substrates (by station and combined) and substrate-specific data sets (*Vallisneria*, leaf, plastic) were used. Also, 3-way (station x surface x exposure time) and 4-way (station x surface x exposure time x time of collection) analyses of variance of the different biological variables (microbial oxygen uptake, microbial lipid phosphate and PHB and N, S, D, H'', $E(S_N)$, D_1) were carried out with and without log transformations of biomass, numbers of individuals, and specific physical features. The paired sub-samples were combined to form a total of 60 samples which were used for a cluster analysis (Rho measure of affinity, flexible grouping strategy with beta = -0.25) to determine assemblage relationships.

RESULTS AND DISCUSSION

Comparison of the Effects of Detrital Surface and Biodegradability on the Microflora and Associated Fauna

It proved possible to test the effects of biodegradability of surfaces by comparing extruded polyethylene needles (from an artificial Christmas tree) to needles from *Pinus elliottii*. From this comparison it was possible to show that multiple measures of microbial activity demonstrate a 2- to 10-fold greater biomass on the biodegradable surface, measured in terms of extractible ATP, muramic acid, phosphodiesterase, alkaline phosphatase, or respiratory activity (Fig. 2). We have shown that there is a remarkable degree of parallelism among the estimates of microbial biomass measured as ATP, respiratory activity, and six esterase activities on detritus incubated for 14 weeks in the Apalachicola Bay estuary (*White et al.*, 1979a).

Not only is the biomass different, but the population structure is also different. Muramic acid is a cell-wall component found uniquely in bacteria and cyanophytes (*Salton*, 1960). Extractible ATP is a universal measure of biomass. While the non-degradable surface yields a ten-fold smaller amount of total ATP, it supports a population with a much lower muramic acid to ATP ratio than the population found on the pine needles (Fig. 3). This ratio is indicative of a predominantly bacterial population, as confirmed by examination of the surface by scanning electron microscopy (Fig. 4), which shows the paucity of filamentous (fungal) or algal structures. These qualitative differences between populations can also be demonstrated in the activities of the glycoesterases (Fig. 5). The glycoesterase enzyme activities were not detected in the population growing on the artificial surface.

Thus it was possible to show that a detrital microflora different in biomass and population structure developed in the same water column and on a structure which presumably offered the same physical substrate shape. Did the associated macrofauna differ significantly?

Fig. 3.   (A) Weekly weight of pine and PVC needles expressed as the mean percentage of the initial dry weight remaining after drying at 80°C *in vacuo*; and the levels of (B) extractable ATP, (C) respiration, (D) alkaline phosphatase activity, (E) phospho-diesterase activity, and (F) muramic acid of pine needles (solid symbols) and PVC needles (open symbols) exposed in Apalachicola Bay, Florida. PNP = p-nitrophenol. (*Bobbie et al.*, 1978). ⟶

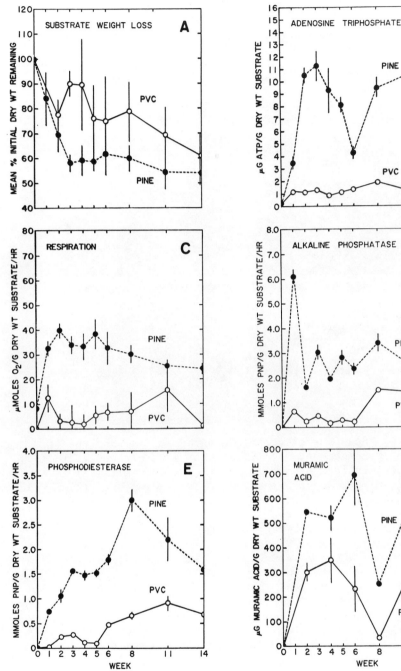
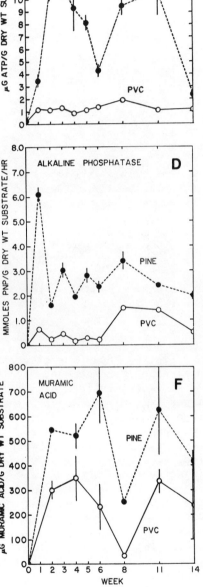

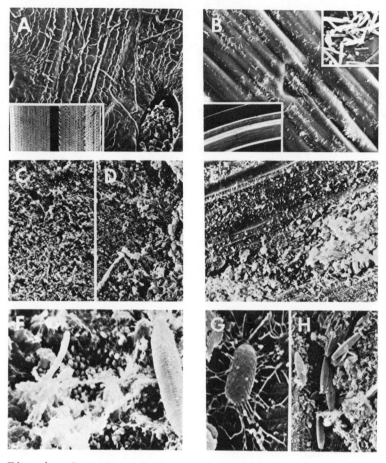

Fig. 4. Scanning electron micrographs. (A) Surface of
uncolonized pine needle, magnification X 490, picture width (p.w.)
133 μm, accelerating voltage 10 kV; inset: smooth (left) and
ridged (right) surfaces, X 10.5, p.w. 3.33 mm, 5 kV. (B) Surface
of uncolonized PVC needle, X 620, p.w. 104 μm, 5 kV; upper inset:
closeup of abiological structures on surface, X 7,240, p.w. 2.9
μm, 20 kV; lower inset: PVC needle surface showing ridges and
grooves, X 26.4, p.w. 1.09 mm, 5 kV. (C) Colonized pine needle,
ridged surface, after 3 weeks of incubation in estuary, X 500,
p.w. 64 μm, 10 kV. (D) Colonized pine needle, smooth surface,
after 6 weeks in estuary, X 185, p.w. μm, 20 kV. (E) Colonized
PVC needle surface after 14 weeks in estuary, X 235, p.w. 275 μm,
30 kV. (F) Representative surface microbiota of colonized pine
needle: (b) bacteria, (c) cyanobacterium, (d) pennate diatom; X
2,360, p.w. 27.7 μm, 10 kV. (G) Bacterial colonizer with attachment fibrils on PVC needle, X 15,880, p.w. 2 μm; 20 kV. (H)
Diatoms *Navicula* sp. and *Amphora* sp. on surface of colonized PVC
needle, X 395, p.w. 81 μm, 20 kV. (*Bobbie et al.*, 1978). (Reduced
40%.)

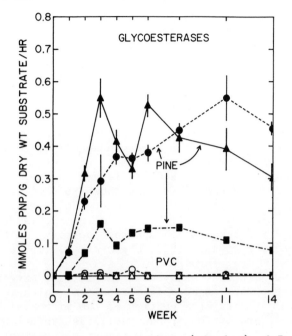

Fig. 5. Weekly β-D-galactosidase (circles), β-D-glucosidase (triangles), and α-D-mannosidase (squares) activities of the microbial population associated with pine (solid symbols) and polyvinyl chloride (open symbols) substrates exposed in Apalachicola Bay, Florida. (*Bobbie et al.*, 1978).

Regression analyses comparing the number of individuals, number of species, relative dominance, Margalef richness, Shannon diversity index, and Hurlbert formulation between the artificial pine substrate and the pine needles showed no significant correlations (p < 0.05). Despite marked differences in several measures of microbial biomass and activity, there was no detectable difference in mass or population structure between the macrofauna attracted to the baskets by the microbes on the biodegradable and non-biodegradable surfaces.

Comparison of the Effects of Water Column Chemistry, the
Detrital Surface and the Time of Exposure on the
Detrital Microflora and its Associated Macrofauna

An experiment was conducted in which samples of three detrital surfaces (hardwood leaves recovered from the flood plain, living uprooted *Vallisneria*, and polyethylene cut to the size and shape of hardwood leaves) were incubated at two stations in Apalachicola

Bay for 2 and 4 week periods during a 10-week interval in
August, September, and October at stations 3 and 5A.

For an assessment of the detrital microflora three measures
were chosen: extractable lipid phosphate, respiratory activity,
and poly-β-hydroxybutyrate.

Microbial biomass was estimated from the extractable lipid
phosphate. Bacteria contain a high proportion of phospholipid
in the membrane (*Kates*, 1964); their phospholipid content is a
direct measure of dry weight in a number of bacterial monocultures (*Morman and White*, 1970; *White and Tucker*, 1969). The
extractable lipid phosphate has been shown to parallel the
extractible ATP, the rate of ^{3}H-thymidine incorporation into DNA
in sediments (*White et al.*, 1979b; *White et al.*, 1979c), and the
extractible ATP and alkaline phosphatase activity in the detrital
microflora (*Nickels et al.*, 1979; *White et al.*, 1979a).

Respiratory activity assayed as the rate of oxygen utilization measures the rate of aerobic metabolism, which is certainly
related to the activity of a portion of the microbial biomass.
That the highest proportion of phospholipids typical of anaerobic
bacteria are found in the aerated horizons of sediments (*White
et al.*, 1979c) indicates that microcolonies of mixed metabolic
types are enormously important. Thus respiratory activity may be
a key to the metabolism of the anaerobic-aerobic consortia that
make up the mixed microcolony.

Poly-β-hydroxybutyrate (PHB) is an endogenous storage
polymer found in some species of bacteria. It is formed in monocultures during conditions of unbalanced growth, when growth is
limited by some key nutrient in the presence of adequate carbon
sources and terminal electron acceptors (*Dawes and Senior*, 1973).
PHB has been detected in the detrital microflora and a quantitative
assay has been developed (*Herron et al.*, 1978). In studies of
the detrital microflora, conditions of unbalanced growth have
been shown to increase the rate of PHB synthesis and to depress
its rate of utilization (*Nickels et al.*, 1979). From these
studies PHB appears to reflect the recent nutritional history of
these microbes. The effective natural inducers of PHB production
by the detrital microflora were associated with the brown
tannin-rich runoff from upland pine plantations above station 5A
in East Bay (*Nickels et al.*, 1979).

                                Field Results

The results of the field experimentation using the three
parameters which measure microbial mass, aerobic activity, and
nutritional status are given in Table 1. Three three-way

analyses of variance are illustrated in Table 2 which examine the
effects of the three surfaces, the two locations and the two
lengths of incubation on the microflora data from Table 1.
Essentially all the variance in the oxygen, PHB, and lipid
phosphate was accounted for by the nature of the surface, except
in the measurement of PHB, which showed a significant relationship with location (Table 2). When the data were treated in a
series of two way (surface exposure) analyses of variance, the
difference in surface again accounted for the major portion of
the variance. When the dates of the exposures were added to the
three variables used in the analyses of Table 2 and four-way
analyses of variance were performed, the nature of the surface
still accounted for nearly all of the differences. The over-
whelming significance of the surface in accounting for the
variance was not affected by logarithmic transformation of the
data or by exclusion of outliers (data showing marked differences
from the mean).

A series of three-way station date and time of exposure
analyses of variance on each surface with the three microbial
measures showed no significance above the 0.005 level. Only the PHB
level showed a trend toward an effect of station (Table 2).
Station 5A, which receives local upland runoff water, showed
higher values of PHB on occasion. The increase in PHB can be
induced readily by the addition of upland runoff to laboratory
microcosms of detrital microbes (*Nickels et al.*, 1979).

Pairwise linear regression analyses among the three measures
of microbial activity showed significant relationships, but no
pair showed a correlation greater than 40% (Table 3). Factoring
out the surface effects (which showed such a marked influence on
the analysis of variance) did not substantially increase the
level of significance of R for any pair. These three measures
clearly reflect different components of regulation of activities
in the complex association of the detrital microflora.

When linear regressions were conducted of the three microbial parameters on the measures of physicochemical functions
defined in the Methods and Materials, only a low negative correlation of turbidity with PHB and river flow with the lipid
phosphate was apparent (Table 3). Comparisons between the microbial parameters and the measures of the macrofauna of the detrital
food web by linear regression analysis showed only poor correlation
of the PHB level with the total number of animals, the Margalef
richness and the Hurlbert formulation (Table 3).

An analysis of the detritus-associated macrofauna was also
performed. A list of species taken in the litter baskets is given
in Table 4. In terms of individuals, amphipods (e.g.
Grandidierella bonnieroides, *Corophium louisianum*, and various

TABLE 1: The poly-β-hydroxybutyrate, extractible phospholipid, and respiratory activity of the detrital microflora, and the number of individuals, biomass, and number of species of the macrofauna detected on hardwood leaves (0), *Vallisneria* (V), and polyethylene leaves (A) incubated in Apalachicola Bay for 2 or 4 weeks.

A. Local runoff-dominated station

Date	Surface	Exposure (weeks)	Microflora			Macrofauna		
			PHB µg/g dry wt	Lipid P µmoles/g	Respiration µmoles O$_2$/ hr/g dry wt	Number of Individuals	Biomass g	Number of Species
8/17	0	2	5.7	2.8	33.8	349	2.3	7
	V	2	58.5	14.6	84.4	268	2.7	12
	A	2	0.7	0.3	0.9	217	2.7	12
8/31	0	2	5.5	1.9	23.6	77	2.5	12
	V	2	28.0	13.1	27.8	103	0.6	9
	A	2	0.5	1.1	0.73	203	0.4	11
	0	4	1.4	2.1	28.0	171	2.8	10
	V	4	19.0	18.9	21.1	89	5.2	8
	A	4	0.6	1.0	1.3	203	2.3	11
9/14	0	2	9.0	0.7	25.4	248	5.0	11
	V	2	31.0	8.2	70.1	193	2.6	10
	A	2	3.4	0.2	0.8	208	1.7	14
	0	4	52.0	1.5	19.9	238	4.3	9
	V	4	75.0	7.9	64.2	206	2.6	10
	A	4	0.75	0.19	1.6	188	2.3	8

TABLE 1 (continued)

Date	Surface	Exposure (weeks)	Microflora			Macrofauna		
			PHB μg/g dry wt	Lipid P μmoles/g	Respiration μmoles O₂/ hr/g dry wt	Number of Individuals	Biomass g	Number of Species
9/28	0	2	22.0	1.4	18.7	107	9.0	10
	V	2	61.0	9.2	48.6	66	7.3	10
	A	2	2.9	0.1	2.4	118	1.6	10
	0	4	20.0	2.1	21.1	104	2.5	10
	V	4	52.0	10.1	41.4	55	6.6	11
	A	4	5.5	0.1	3.4	32	2.4	6
10/12	0	2	23.0	2.1	18.4	214	1.8	13
	V	2	44.0	-	64.2	76	8.3	10
	A	2	-	-	-	239	6.5	12
	0	4	10.0	1.7	14.7	204	5.4	14
	V	4	50.0	12.7	50.7	208	1.0	7
	A	4	1.7	0.1	2.4	113	3.3	9
10/26	0	4	14.9	1.4	18.7	312	4.8	11
	V	4	103.0	6.9	64.6	201	2.8	13
	A	4	1.0	0.2	3.3	341	3.2	14

TABLE 1 (continued)

B. River-dominated station

Date	Surface	Exposure (weeks)	Microflora PHB μg/g dry wt	Lipid P μmoles/g	Respiration μmoles O₂/ hr/g dry wt	Macrofauna Number of Individuals	Biomass g	Number of Species
8/17	0	2	19.5	3.7	39.2	196	8.4	12
	V	2	54.6	33.7	89.7	414	13.2	13
	A	2	1.9	0.27	1.1	142	9.7	13
8/31	0	2	7.7	1.8	22.4	62	0.9	9
	V	2	29.0	11.4	31.0	120	11.1	10
	A	2	0.58	2.9	1.3	287	11.4	17
	0	4	3.1	3.2	38.1	207	9.2	8
	V	4	41.0	15.8	22.6	150	14.0	16
	A	4	1.2	1.0	1.5	288	12.7	15
9/14	0	2	14.0	1.3	21.3	96	2.7	8
	V	2	1.7	3.9	128.6	111	3.6	13
	A	2	1.3	0.28	1.1	204	6.0	9
	0	4	3.6	1.1	24.9	119	3.1	10
	V	4	2.3	45.0	77.8	80	1.3	11
	A	4	1.5	0.22	1.1	355	5.4	12

TABLE 1 (continued)

Date	Surface	Exposure (weeks)	Microflora			Macrofauna		
			PHB µg/g dry wt	Lipid P µmoles/g	Respiration µmoles O_2/ hr/g dry wt	Number of Individuals	Biomass g	Number of Species
9/28	0	2	20.0	1.5	18.1	132	8.8	11
	V	2	41.0	10.3	45.6	76	7.9	13
	A	2	3.9	0.1	3.5	75	5.1	11
	0	4	18.0	2.1	24.4	85	4.8	8
	V	4	–	–	–	40	2.7	10
	A	4	4.3	0.2	3.6	86	7.2	10
10/12	0	2	9.4	1.5	23.0	114	2.2	14
	V	2	44.0	–	68.2	415	9.6	11
	A	2	0.3	0.2	2.1	210	10.3	13
	0	4	5.5	1.8	27.7	222	11.0	8
	V	4	37.0	12.7	50.7	164	5.6	13
	A	4	0.3	0.1	2.3	156	15.7	10
10/26	0	4	5.7	1.4	16.9	175	5.3	11
	V	4	39.0	6.0	47.2	155	6.9	11
	A	4	0.9	0.3	2.5	479	3.2	11

TABLE 2: Three way analyses of variance of microbial parameters comparing 3 surfaces, 2 stations and 2 lengths of exposure.

Variable	Source of Variation	dF	MS	F	Sig. of F
Oxygen	Station	1	241	0.97	0.32
	Surface	2	15,200	61.3	0.001
	Exposure	1	0.339	1.36	0.248
PHB	Station	1	1,208	5.58	0.022
	Surface	2	8,154	37.7	0.001
	Exposure	1	33.6	0.15	0.695
Lipid P	Station	1	0.64	0.004	0.949
	Surface	2	623.2	39.76	0.001
	Exposure	1	18.2	1.66	0.285

TABLE 3: Interactions between the microbial parameters themselves, the microbial parameters and water quality measures, and the microbial parameters and the detrital macrofauna measures.

Parameter A	Parameter B	r	r^2	Sig. of F
Oxygen	PHB	0.6	0.36	0.00001
Oxygen	Lipid P	0.59	0.35	0.00001
PHB	Lipid P	0.62	0.38	0.00001
PHB	Turbidity	-0.20	0.04	0.069
Lipid P	Flow	-0.21	0.04	0.063
PHB	Numbers	-0.21	0.045	0.055
PHB	Margalef richness	-0.21	0.04	0.05
PHB	Hurlbert formulation	0.23	0.05	0.04

TABLE 4:   Macrofaunal species associated with litter baskets
            incubated in Apalachicola Bay.

Gastropoda
   Odostomia laevigata
   Littoridina sphinctostoma
   Neritina reclivata
   Nudibranch sp.

Polychaeta
   Hirudinia sp.
  Sedentaria
   Hypaniola florida
   Polydora ligni
   Streblospio benedicti
   Mediomastus californiensis
   Polydora websteri
  Errantia
   Laeonereis culveri
   Neanthes succinea
   Nereid sp.

Oligochaete sp.

Branchiura
   Argulus sp.

Mysidacea
   Mysidopsis bigelowi

Cumacean sp.

Tanaidacea
   Leptochelia rapax

Isopoda
   Erichsonella filiformis
   Munna reynoldsi
   Cassidinidea ovalis
   Probopyrus sp.

Amphipoda
   Gammarus macromucronatus
   Gammarus mucronatus
   Melita appendiculata
   Melita nitida
   Melita spp.
   Corophium louisianum

Amphipoda (con't)
   Grandidierella bonnieroides
   Gitanopsis n. sp.

Decapoda
   Palaemonetes pugio
   Palaemonetes intermedius
   Palaemonetes vulgaris
   Penaeus setiferus
   Callinectes sapidus
   Rhithropanopeus harrisii
   Neopanope texana
   Neopanope packardii
   Hippolyte zostericola

Copepod sp.

Aschelminthes
   Nematode sp.

Insecta
   Ceratopogonid sp.
   Dicrontendipes sp.
   Zygopteran sp.
   Insect larvae (species
               unknown)

Pisces
   Bairdiella chrysura
   Gobiosoma bosci
   Gobiosoma robustum
   Syngnathus floridae
   Microgobius gulosus

forms of *Melita*), tended to be dominant at both stations throughout the study. Station 5A was also characterized by numbers of gastropods (*Neritina reclivata*). *Gammarus mucronatus* was prevalent at this station during August, while *Palaemonetes pugio* and *Munna reynoldsi* were dominant during September and October, respectively. At station 3, *Gammarus mucronatus* was dominant throughout the study period with *Corophium louisianum* and *Palaemonetes pugio* tending to increase during late September and October. Biomass was dominated in all collections by the blue crab (*C. sapidus*) with lesser dominants including *N. reclivata*, *P. pugio*, and *Rhithropanopeus harissii*. During late August, the pink shrimp (*Penaeus setiferus*) was found at both stations while *Palaemonetes vulgaris* composed a significant proportion of the biomass taken at station 3 during late October.

Results of the litter basket experiments are given in Table 1. Numbers of individuals were generally higher at station 5A, at four-week exposures, and on artificial substrates, although these relationships were not statistically significant in the 3-way ANOVA. When time of collection was considered (4-way ANOVA), there was a significant difference by substrate (p < 0.009) and time of collection (p < 0.020). The numbers of individuals tended to peak in collections made on 8/17 and 10/26. Although the numbers of species were generally higher at station 5A and on artificial substrates, such relationships were not statistically significant. The S values (numbers of species) tended to be highest during collections taken on 8/17, 8/31, and 10/26. Biomass was significantly higher at station 5A (p < 0.001) and in collections made during 8/17 and 10/12 (p < 0.010). Biomass tended to be lower on the mixed deciduous leaves, although this trend was not statistically significant. While relative dominance was generally higher on artificial substrates and at station 3, this tendency only had statistical significance for time of collection (p < 0.040), being high on 8/17 and 10/26. The diversity indices (Shannon, Hurlbert), as mirror images of the relative dominance, tended to be lowest during these periods (p < 0.024).

The regression analysis of the principal biological functions indicated significantly (p < 0.05) strong inverse relationships of the diversity indices with numbers of individuals and relative dominance. The numbers of individuals were directly associated with relative dominance (p < 0.00003). Species richness (S) was significantly associated with numbers of individuals (p < 0.004), biomass (p < 0.007) Margalef richness (p < 0.00001), and Shannon diversity (p < 0.009), while the Margalef richness was associated with relative dominance (p < 0.036), and biomass (p < 0.018). There was a direct relationship between the numbers of individuals and relative dominance (p < 0.023). This analysis thus confirmed that major increases in numbers of individuals and

biomass strongly influenced by one or two dominant species
which, in turn, effectively reduced the diversity indices thus
rendering such functions useful only as an indication of relative
dominance. Given the above biological relationships, the data
tend to indicate that station and time of collection were the
primary determinants of the associations of litter organisms,
and that substrate type was relatively unimportant, although numbers
of individuals were significantly higher on the artificial
(plastic) substrates.

   The physico-chemical variation in the area of study is given
in Fig. 6. Local rainfall was relatively low, with scattered
daily peaks in August and September. River flow showed a progressive decrease throughout the sampling period. These two
functions appeared to account for local patterns of salinity,
turbidity, and pH. Salinity at station 5A was generally higher
than at station 3 with general decreases occurring during the
late August-September sampling period. Dissolved oxygen was low
following the September rainfall. There was a decrease in water

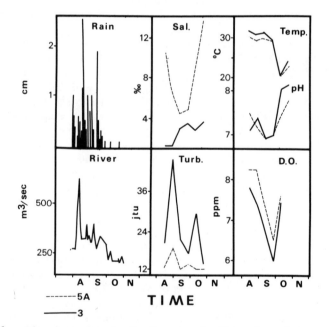

   Fig. 6. Physico-chemical changes in East Bay (stations 3,
5A) from August to November, 1977. Data include daily local rainfall totals (cm), daily Apalachicola River flow rates (m³/sec),
bottom salinity (parts per thousand), turbidity (J.T.U.), temperature (°C), pH, and dissolved oxygen (ppm).

temperature from September to October. The regression analysis
indicated that (log) river flow was associated with reduced
salinity ($p < 0.003$), increased turbidity ($p < 0.033$), and
reduced Secchi readings ($p < 0.002$). There were also positive
correlations of dissolved oxygen with salinity ($p < 0.020$),
pH ($p < 0.014$), and Secchi readings ($p < 0.023$). Increased
salinity was associated with increases in biomass ($p < 0.00003$),
Shannon diversity ($p < 0.023$), numbers of individuals ($p < 0.04$),
and numbers of species ($p < 0.050$). Dissolved oxygen and pH
were directly associated with numbers of individuals ($p < 0.00003$,
$p < 0.010$) and numbers of species (D.O., $p < 0.011$), and
were inversely correlated with relative dominance ($p < 0.022$;
$p < 0.031$). These associations are consistent with observed
habitat-specific variation and the close relationship of key
biological functions with runoff patterns and resultant water
quality conditions (salinity, pH, dissolved oxygen). Such
relationships did not appear to be seriously modified by
differences in substrate type.

The cluster analysis (Fig. 7) indicates that the primary
separation of the litter-associated assemblages was at the
station level. This difference occurred regardless of time of
collection, substrate type, or duration of exposure. The few
cases of overlap tended to occur during late September. The first
few collections at station 3 indicated that the deciduous and
plastic (artificial) substrates tended to attract similar groupings,
with *Vallisneria* tending to be separate. However, during late
September and October, no such distinction among the various
substrate surfaces could be made. At station 5A, there was no
discernible correlation of community form with substrate type,
although there were indications that time of sampling was
important. The primary groupings fell out in late August–
September and October, but this relationship was not altogether
clear. No particular difference in assemblage composition was
found with regard to exposure time (two weeks vs. four weeks).
Thus, the primary determining factor regarding the species
composition and relative abundance of the litter-associated
organisms was station location, modified to a degree by time
of sampling.

Microbial–Macrobial Relationships

A much more sensitive measure of the microbial population
structure can be found in the relative proportions of the lipid-
derived fatty acids. The fatty-acid component of the detrital
microflora was isolated from the microbial community on the
artificial surfaces, where the fatty acids of the plant from
which the detritus was derived would not present complications.
The fatty acids were isolated, esterified, then separated and

identified by gas liquid chromatography. The ratios of the
principal fatty acid methyl esters to methyl palmitate are given
in Table 5. Two-way analyses of variance of the ratios of fatty
acids (station x length of exposure) are suggestive of the
differences in the population structure between the stations.
The relevant ratios of 15 Br (15-carbon branched ester), 17 cyclo
(17-carbon cyclopropane ester), and 18:1 (18-carbon monoenoic
ester) are shown in Table 6.

The most interesting data on the detrital microflora are the
correlations between the proportion of fatty acid methyl esters
and measures of water quality, microbial biomass and the mass
and population structure of the macrofauna that was attracted to
the detritus (Table 7). A statistical analysis of the data indicates a correlation between pH and 18:0 (18-carbon saturated
ester), weak but suggestive correlations of salinity with 17 cyclo
and 18:0, and negative correlations of temperature with 17 cyclo
and 18:0 esters. The microbial biomass and activity measured
as oxygen uptake and lipid phosphate show a strong negative
correlation and a weak positive correlation, respectively, with
the proportion of 15 Br. The PHB level correlated with the
16:1 (16-carbon monoenoic ester) and weakly with 18:1 content
of the population.

Most interesting are the excellent correlations between
macrofaunal biomass and 17 cyclo, 16:1, and 15 Br. The Br
content is correlated with the number of species, the Margalef
richness, and biomass. Branched 15-carbon fatty acids are
characteristic of certain gram-positive bacteria (*Lechevalier*,
1977; *Parker et al.*, 1967) with few exceptions (*Tyrrell*, 1968).
Protozoa and fungi form significant amounts of these branched
fatty acids when supplied with branch precursors in their growth
milieux (*Erwin*, 1973; *Korn et al.*, 1965; *Mayer and Holz*, 1966), a
situation extremely unlikely in nature. The presence of branched
15-carbon fatty acids has been correlated with bacterial
colonization of dead *Spartina* (*Schultz and Quinn*, 1973) and their
absence with low bacterial content of pelagic surface particulate
matter (*Schultz and Quinn*, 1973). The high positive correlations
of 15 Br fatty acid with the number of species, the Margalef
richness, and the biomass of the detrital macrofauna clearly shows

Fig. 7. Cluster analysis (Rho measure of affinity, flexible
grouping strategy with beta = -0.25) of litter-basket organisms
found at 2- and 4-week intervals from 17 August through 26
October, 1977, at stations 3 and 5A and associated with mixed leaf
litter (0), blades of *Vallisneria americana* (V), and artificial
(plastic) leaves (A). Each coded entry represents station, substrate type, date of collection, and duration of exposure. ⟶

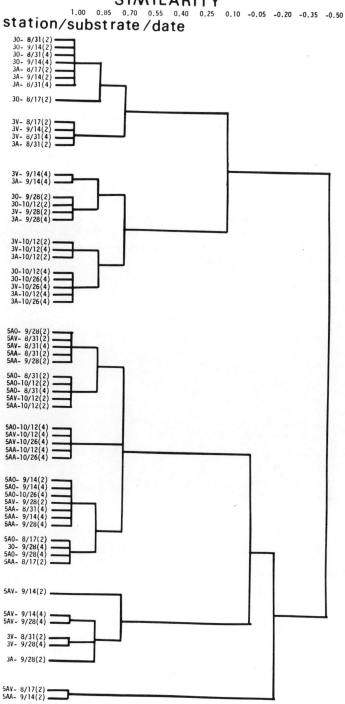

TABLE 5: Ratios of fatty acid methyl esters to methyl palmitate of the microbial lipids from the artificial surface.

Date	Exposure	15 Br.	16:0	16:1	17 Cyclo	18:0	18:1
A. Local-runoff-dominated station							
8/17	2	0.275	1.0	–	–	0.350	–
8/31	2	0.114	1.0	0.154	0.036	0.542	0.147
8/31	4	0.160	1.0	0.325	0.086	0.366	0.763
9/14	2	–	–	–	–	–	–
9/14	4	0.023	1.0	–	0.30	0.133	–
9/28	2	0.024	1.0	0.466	–	0.350	0.381
9/28	4	0.010	1.0	0.261	0.016	0.174	0.109
10/12	2	0.033	1.0	0.231	0.050	0.248	0.185
10/12	4	0.020	1.0	0.246	0.068	0.238	0.123
10/26	4	0.04	1.0	–	0.031	0.744	0.128
B. River-dominated station							
8/17	2	0.161	1.0	–	0.058	0.333	–
8/31	2	0.384	1.0	0.381	0.107	0.274	0.523
8/31	4	0.404	1.0	0.369	0.031	0.206	0.260
9/14	2	0.135	1.0	0.117	0.050	0.050	0.229
9/14	4	0.108	1.0	0.210	0.057	0.355	0.281
9/28	2	0.028	1.0	0.540	0.045	0.383	0.644
9/28	4	0.113	1.0	0.519	0.170	0.536	0.731
10/12	2	0.090	1.0	0.441	0.212	0.571	0.663
10/12	4	0.097	1.0	0.379	0.149	0.347	0.273
10/26	4	0.091	1.0	–	0.02	0.688	0.211

TABLE 6: Two-way analyses of variance of the proportions of
          microbial fatty acids comparing two stations and two
          lengths of exposure.

Variable	Source of variation	dF	MS	F	Sig. of F
15 Br	Station	1	0.042	3.14	0.096
	Exposure	1	0.002	0.12	0.73
17 Cyclo	Station	1	0.017	6.06	0.026
	Exposure	1	0.000	0.18	0.56
18:1	Station	1	0.195	3.17	0.094
	Exposure	1	0.001	0.009	0.925

a selectivity by these animals for areas rich in bacteria. Their
strong negative correlation with the rate of oxygen utilization
suggests that the 15 Br fatty acids were largely in anaerobic or
facultatively anaerobic bacteria growing at limiting oxygen
concentrations.

    Cyclopropane fatty acids are restricted to certain lipids
since they are formed from monoenoic fatty acids only when they
are esterified with those lipids (*Thomas and Law*, 1966). They
are restricted largely to gram-negative bacteria, plus the gram-
positive *Clostridia* and *Lactobacter* (*Lechevalier*, 1977; *Shaw*,
1974), the kinetosomes of some protozoa (*Meyer and Holz*, 1966) and
terrestrial plants. Cyclopropane fatty acids accumulate in
stationary growth phase (*Kates*, 1964; *Law et al.*, 1963; *Marr and
Ingraham*, 1962) or with adverse growth conditions (*Knivett and
Cullen*, 1965). The negative correlation between temperature and
proportion of 17 cyclo ester may indicate a faster turnover at
higher temperature with less accumulation of cyclopropane fatty
acids.

    Again there are excellent positive correlations between high
proportion of 17 cyclo in the microbial population and the biomass and Shannon diversity of the detritivore macrofauna.

    Monounsaturated fatty acids are found in gram negative
bacteria (and the *Lactobacter*) and in microeukaryotes (*Erwin*,
1973). The reliable negative correlation of the proportion of

TABLE 7: Interaction between proportion of fatty acid methyl esters in the detrital microflora and microbial parameters, water quality, and measures of the detrital food web.

Parameter A	Parameter B	r	r^2	Sig. of r
15 Br	Number of species	0.647	0.419	0.001
15 Br	Margalef richness	0.587	0.345	0.003
15 Br	Biomass	0.463	0.215	0.019
15 Br	Oxygen uptake	-0.475	0.226	0.017
15 Br	Lipid phosphate	0.365	0.134	0.062
16:1	Number	-0.519	0.270	0.009
16:1	Biomass	0.419	0.176	0.033
16:1	PHB	0.436	0.190	0.031
17 Cyclo	Shannon diversity	0.372	0.138	0.054
17 Cyclo	Biomass	0.643	0.414	0.001
17 Cyclo	Temperature	-0.447	0.200	0.024
17 Cyclo	Salinity	0.345	0.120	0.067
18:0	Temperature	-0.348	0.121	0.066
18:0	Salinity	0.346	0.120	0.067
18:0	pH	0.488	0.238	0.014
18:1	PHB	0.363	0.132	0.063

16:1 with the number of detritus-associated macrofaunal animals, coupled with its positive correlation with detrital animal biomass, may reflect the presence of the fatty acid in both the prokaryotes and microeukaryotes. The positive correlation of PHB content with the proportion of 16- and 18-carbon monoenoic fatty acids may reflect the high association of PHB formulation with gram negativity in the detrital bacteria.

The 18-carbon saturated fatty acid ester is found universally in microeukaryotes and prokaryotes and shows correlations not with the detrital microflora but with the temperature, salinity, and pH. This fact may reflect a correlation of the total microbial biomass with the physical parameters similar to that shown by the detrital macrofauna.

CONCLUSIONS

It seems clear that the gross measures of mass, population structure, and activity of the detrital microbial community showed little immediate direct effect on the structure, composition, and biomass of the detrital food web. The significantly larger microbial biomass on both native pine needles (Experiment 1) and hardwood or *Vallisneria* leaves (Experiment II) did not seem to be correlated with the associated macrofaunal mass, which was greater on the artificial surface.

However, when a sensitive measure of the microbial population structure was examined, it was found to be highly significantly correlated with the most sensitive measures of the detritus-associated macrofauna (Table 4). The detrital animals attracted to the cages appeared to seek out the microbial population rich in anaerobic or microaerophilic bacteria (15 Br fatty acids) and gram-negative bacteria (17 cyclo and 16:1 fatty acids). They apparently were able to discriminate between microbial populations, because total biomass, activity, and structural morphology (as measured by scanning electron microscopy) and previous nutritional history (as measured by PHB) were not correlated with the detrital macrofauna. There was suggestive evidence that, within the macrofaunal population, different animals choose different microbes, as there was a negative correlation between numbers of animals and microbial 16:1 and a positive correlation between biomass and microbial 16:1--larger animals apparently choose populations high in 16:1 fatty acids.

Clearly these field data suggest that carefully controlled experiments in microcosms where sensitive measures of the detrital microfloral population structures such as the fatty acid composition can be manipulated and the structure of the detritus-attracted macrofaunal food web can be examined can yield deeper insight into the trophodynamics of estuarine detritus. Possibly, questions of how the microbial community is shaped by the macro-invertebrates could be examined. What was detected in these experiments was the microbial community that survived predation. It is possible that the microbial community that attracted the macroinvertebrates and was consumed could be examined in controlled microcosms.

ACKNOWLEDGEMENTS

The authors wish to thank Glenn Woodsum for performing the computer calculatons. This study was supported by grant 76-19761 from the Biological Oceanography Program of the National Science Foundation and the Florida Sea Grant Administration, U. S. Department of Commerce, Contract # 04-8-M01-76.

REFERENCES

Bader, R. G., 1954. The role of organic matter in determining the distribution of bivalves in sediments, *J. Mar. Res. 13*: 31-47.

Bobbie, R. J., S. J. Morrison, and D. C. White, 1978. Effects of substrate biodegradability on the mass and activity of the associated estuarine microbiota, *Appl. Environ. Microbiol. 35*: 179-184.

Darnell, R. M., 1967. The organic detritus problem, In: *Estuaries,* edited by G. H. Lauff, 374-375, Publ. # 83, A.A.A.S., Washington, D. C.

Dawes, E. A. and P. J. Senior, 1973. The role and regulation of energy reserve polymers in microorganisms, *Adv. Microb. Physiol. 10*: 135-266.

Ekman, S., 1947. Uber die Festigheit der marinen Sedimente als faktor der Tierverbreitung, ein Beitrag zur Associations-Analyse, *Zool. Biol. Univ. Uppsala, Sweden 25*: 1-20.

Erwin, J. A., 1973. Fatty acids in eukaryotic microorganisms, In: *Lipids and Biomembranes of Eukaryotic Microorganisms,* edited by J. A. Erwin, 41-143, Academic Press, New York.

Feeny, P., 1976. Plant apparency and chemical defense, In: *Biochemical Interaction Between Plants and Insects, Vol. 10,* edited by J. W. Wallace and R. L. Mansell, 1-40, Plenum Press, New York.

Fenchel, T. M. and B. B. Jorgensen, 1977. Detritus food chains of aquatic ecosystems: The role of bacteria, *Adv. Microbial Ecol. 1*: 1-58.

Herron, J. S., J. D. King, and D. C. White, 1978. Recovery of poly-β-hydroxybutyrate from estuarine microflora, *Appl. Environ. Microbiol. 35*: 251-257.

Hurlbert, S. H., 1971. The nonconcept of species diversity: a critique and alternative parameters, *Ecology 52*: 577-586.

Jeffries, H. P., 1972. Fatty acid ecology of a tidal marsh, *Limnol. Oceanogr. 17*: 433-440.

Jeffries, H. P., 1975. Diets of juvenile Atlantic menhaden (*Brevoortia tyrannus*) as determined from fatty acid composition of gut contents, *J. Fish. Res. Bd Can. 32*: 587-592.

Jensen, P. B., 1919. Valuation of the Limfjord. I. Studies on the fish food in the Limfjord 1909-1917, its quantity, variation, and animal production, *Rep. Danish Biol. Stat. Univ. Copenhagen, Denmark 26*: 1-44.

Kates, M., 1964. Bacterial lipids, In: *Advances in Lipid Research 2*, edited by R. Paoletti and D. Kritchevsky, 17-90, Academic Press, New York.

King, J. D. and D. C. White, 1977. Muramic acid as a measure of microbial biomass in estuarine and marine samples, *Appl. Environ. Microbiol. 33*: 777-783.

King, J. D., D. C. White, and C. W. Taylor, 1977. Use of lipid composition and metabolism to examine structure and activity of estuarine detrital microflora, *Appl. Environ. Microbiol. 33*: 1177-1183.

Knivett, V. A. and J. Cullen, 1965. Some factors affecting cyclopropane acid formation in *Escherichia coli*, *Biochem. J. 96*: 771-776.

Korn, E. D., C. L. Greenblatt, and A. M. Lees, 1965. Synthesis of unsaturated fatty acids in the slime mold *Physarum polycephalum* and the zooflagellates *Leishmania tarentolae*, *Trypanosoma lewisi*, and *Crithidia* sp.: a comparative study, *J. Lipid Res. 6*: 43-50.

Law, J. H., H. Zalkin, and T. Kaneshiro, 1963. Transmethylation reactions in bacterial lipids, *Biochim. Biophys. Acta 70*: 143-151.

Lechevalier, M. P., 1977. Lipids in bacterial taxonomy - a taxonomist's view, CRC *Critical Reviews in Microbiol.* 7: 109-210.

Livingston, R. J., 1978. *Short- and Long-term Effects of Forestry Operations on Water Quality and the Biota of the Apalachicola Estuary (North Florida, U.S.A.),* Final report, Florida Sea Grant College.

Livingston, R. J., R. L. Iverson, and D. C. White, 1976. *Energy Relationships and the Productivity of Apalachicola Bay.* Final report, Florida Sea Grant College.

Margalef, R., 1958. Information theory in ecology, *Gen. Systematics 3:* 36-71.

Marr, A. G. and J. L. Ingraham, 1962. Effect of temperature on the composition of the fatty acids in *Escherichia coli, J. Bacteriol. 84:* 1260-1267.

Meyer, H. and G. C. Holz, 1966. Biosynthesis of lipids by kinetoplastid flagellates, *J. Biol. Chem. 241:* 5000-5007.

Morman, M. R. and D. C. White, 1970. Phospholipid metabolism during penicillinase production in *Bacillus licheniformis, J. Bacteriol. 104:* 247-253.

Morrison, S. J., J. D. King, R. J. Bobbie, R. E. Bechtold, and D. C. White, 1977. Evidence for microfloral succession on allochthonous plant litter in Apalachicola Bay, Florida, USA, *Marine Biol. 41:* 229-240.

Nickels, J. S., J. D. King, and D. C. White, 1979. Poly-β-hydroxybutyrate accumulation as a measure of unbalanced growth of the estuarine detrital microflora, *Appl. Environ. Microbiol. 37:* 459-465.

Odum, W. E., 1970. *Pathways of Energy Flow in a South Florida Estuary,* Ph.D. Dissertation, University of Miami.

Parker, P. L., C. van Baalen, and L. Maurer, 1967. Fatty acids in eleven species of blue-green algae: geochemical significance, *Science 155:* 707-708.

Petersen, C. G., 1913. Valuation of the sea. II. The animal communities of the sea bottom and their importance for marine zoogeography, *Rep. Danish Biol. Stat. Univ. Copenhagen, Denmark 21:* 1-44.

Petersen, C. G., 1918. The sea bottom and its production of fish
     food. A survey of the work done in connection with the
     valuation of the Danish waters from 1883–1917, *Rep. Danish
     Biol. Stat. Univ. Copenhagen, Denmark 25*: 1–62.

Rowe, G. T. and R. J. Menzies, 1969. Zonation of large benthic
     invertebrates in the deep sea off the Carolinas, *Deep Sea
     Res. 16*: 531–537.

Salton, M., 1960. *Microbial Cell Walls,* 94 pp., John Wiley
     and Sons, New York.

Schelske, C. L., and E. P. Odum, 1962. Mechanisms maintaining
     high productivity in Georgia estuaries, *Proc. Gulf Carib.
     Fish. Inst. 14th Ann. Sess.:* 75–80.

Schultz, D. M. and J. G. Quinn, 1973. Fatty acid composition of
     organic detritus from *Spartina alterniflora, Estuarine and
     Coastal Marine Science 1*: 177–190.

Shannon, E. C. and W. Weaver, 1963. *The Mathematical Theory of
     Communication,* 125 pp., Univ. Illinois Press, Urbana.

Shaw, N., 1974. Lipid composition as a guide to the classification
     of bacteria, *Adv. Appl. Microbiol. 17*: 63–108.

Thomas, P. J. and J. H. Law, 1966. Biosynthesis of cyclopropane
     compounds. IX. Structural and stereochemical requirements
     for the cyclopropane synthetase substrate, *J. Biol. Chem.
     241*: 5013–5018.

Tyrrell, D., 1968. The branched–chain fatty acids in *Conidiobius
     denaesporus* Drechsl, *Lipids 3*: 368–372.

White, D. C., R. J. Bobbie, J. S. Herron, J. D. King, and
     S. J. Morrison, 1979a. Biochemical measurements of microbial
     mass and activity from environmental samples, In: *Native
     Aquatic Bacteria, Enumeration, Activity and Ecology ASTM
     STP 695,* edited by J. W. Casterton and R. R. Colwell, in
     press.

White, D. C., R. J. Bobbie, J. D. King, J. Nickels, and P. Amoe,
     1979b. Lipid analysis of sediments for microbial biomass
     and community structure, In: *Methodology for Biomass
     Determinations and Microbial Activities in Sediments,* edited
     by C. D. Litchfield and P. L. Seyfried, American Society for
     Testing and Materials, ASTM STP 673, in press.

White, D. C., R. J. Bobbie, S. J. Morrison, D. K. Oosterhof, C. W. Taylor, and D. A. Meeter, 1977. Determination of microbial activity of estuarine detritus by relative rates of lipid biosynthesis, *Limnol. Oceanogr.* *22*: 1089-1099.

White, D. C. and R. H. Cox, 1967. Identification and localization of the fatty acids in *Haemophilus parainfluenzae*, *J. Bacteriol.* *93*: 1079-1088.

White, D. C., W. M. Davis, J. S. Nickels, J. D. King, and R. J. Bobbie, 1979c. Determination of the sedimentary microbial biomass by extractible lipid phosphate, *Oecologia* *40*: 51-62.

White, D. D. and A. N. Tucker, 1969. Phospholipid metabolism during changes in the proportions of membrane-bound respiratory pigments in *Haemophilus parainfluenzae*, *J. Bacteriol.* *97*: 199-209.

Wieser, W., 1960. Benthic studies in Buzzards Bay. II. The meiofauna, *Limnol. Oceanogr.* *5*: 121-137.

Wigley, R. L. and A. D. McIntyre, 1964. Some quantitative comparison of offshore meiobenthos and macrobenthos, *Limnol. Oceanogr.* *9*: 485-493.

Williams, R. B. and L. K. Thomas, 1967. The standing crop of benthic animals in a North Carolina estuarine area, *J. Elisha Mitchell Sci. Soc.* *93*: 135-139.

DEPOSIT-FEEDERS, THEIR RESOURCES, AND THE STUDY OF RESOURCE
LIMITATION

Jeffrey S. Levinton

State University of New York at Stony Brook

ABSTRACT

Deposit-feeding invertebrates consume particles, digest and assimilate a fraction of the microbial community living on those particles, and often defecate the particles as compact fecal pellets. The niche of deposit-feeders may therefore be defined as: (1) particle size spectrum ingested, (2) depth of feeding and living position below the sediment-water interface, (3) range of sediments over which the deposit-feeder occurs, (4) possible differences in utilization of microorganisms. Because variation in occurrence of deposit-feeders may be simply related to at least the first 3 niche parameters by simple morphological features such as inhalent siphonal opening and size of buccal apparatus, one can imagine character evolution related to ecologically significant resource parameters.

Laboratory microcosm studies show that the following parameters affect resource availability for the mobile deposit-feeding genus Hydrobia: *(1) Fecal pellet breakdown. Snails avoid ingestion of intact fecal pellets. (2) Renewal of microbial resources such as diatoms and bacteria. (3) Space.* Hydrobia *individuals feed more slowly under crowded conditions and increase emigration. (4) Particle size. Feeding rate decreases with increasing particle diameter, and switching to scraping occurs on large particles. Experiments show that resource limitation by renewable resources affects Hydrobiids within the range of maximal field densities. A theoretical model of resource renewal considering microbial recovery and fecal pellet breakdown permits a prediction of carrying capacity in field populations. Thus multifactorial microcosm studies*

combined with theoretical models permit the assessment of the
importance of competition as a potential evolutionary force.

INTRODUCTION

Deposit-feeders are a dominant component of soft-bottom
nearshore faunas and are responsible for extensive sediment
reworking and microbial grazing. Dense populations of deposit-
feeders alter (1) texture and resuspension of the sediment
(*Rhoads*, 1967, 1973), (2) vertical distribution of chemical pore
water properties (see *Jørgensen*, 1977), (3) bacterial and micro-
algal standing stock (*Fenchel and Kofoed*, 1976; *Lopez and
Levinton*, 1978), (4) microbial population dynamics (*Lopez et al.*,
1977), and (5) patterns of larval recruitment (*Rhoads and Young*,
1970). The very density of deposit-feeders suggests an important
role in acceleration of detrital breakdown and regulation of
deposit-feeding populations by resource availability. Control
by resources would provide a feedback mechanism limiting population growth. This limitation, if present, would further
regulate detrital grazing and the detrital pathway, sediment
reworking, and effects of sediment disturbance on larval
recruitment. I intend to discuss the evidence for resource
limitation of deposit-feeders and the limiting resources that
affect deposit-feeder populations. I then hope to demonstrate
an effective strategy of studying experimental populations of
deposit-feeders and their limiting resources through the use
of simple models and measurements of resource availability and
renewal.

EVIDENCE FOR RESOURCE LIMITATION OF DEPOSIT-FEEDERS

Three lines of evidence suggest that resources are limiting
to deposit-feeding benthic populations: (1) correlations of
population density with food-related parameters, such as organic
fraction of the sediment, percent clay, and microbial abundance,
(2) patterns of niche structure of coexisting deposit-feeding
species which show little niche overlap, and (3) experimental
demonstrations of intraspecific and interspecific interactions
and resource depression by deposit-feeders.

Food-related parameters

Soft-bottom communities of benthic invertebrates living
in silt-clay-rich sediments are usually dominated by deposit-
feeding bivalves and polychaetes (*Sanders*, 1956, 1958).
Deposit-feeders ingest sedimentary particles and generally
digest and partially assimilate the microbial community living

on the detrital particles (see below). Several studies relate
deposit-feeding population abundance to sedimentary properties
related to the availability of microbial food. Deposit-feeders
are found more abundantly in fine-grained sediments, where
organic detritus and its concomitant microbial community are
more abundant (*Zobell and Feltham*, 1938; *Sanders*, 1958; *Newell*,
1965; *Longbottom*, 1970; *Rhoads and Young*, 1970; *Dale*, 1974;
Levinton, 1977; *Tunnicliffe and Risk*, 1977). Because surface
area per unit volume of sediment increases with decreasing
particle diameter, more surface-bound bacteria or microalgae
can be gained by ingesting a given volume of finer particles.
If ingested volume of sediment is rate-limiting (see *Levinton
and Lopez*, 1977) then we must consider the number of diatoms
or bacteria per unit surface area of a particle, which is, in
the case of a sphere, $4\pi r^2 K$. *Kofoed and Lopez* (unpublished)
show with direct bacteria counts that K is a constant over a
wide range of particle diameters. The volume of a sphere is
$4/3\pi r^3$, and the number of particles in volume V is therefore
$(3V/4)\pi r^3$. The number of diatoms or bacteria per unit volume
ingested is then $3VK/r$. Thus the food per unit volume is
inversely proportional to particle diameter by the factor K.

This inference is substantiated in studies relating abundance of the mud snail, *Hydrobia ulvae*, and sediment particle
diameter (*Newell*, 1965). *Sanders* (1958) found the percent clay
to be a particularly good positive correlate of deposit-feeder
abundance. Direct bacterial counts employing epifluorescence
microscopy show a strong positive correlation between bacterial
abundance and population density of the deposit-feeding bivalve
Macoma balthica (*Tunnicliffe and Risk*, 1977). However, grazing
on bacteria can be a positive stimulatory factor, making the clam
populations control bacterial abundance and not the reverse (see
Rahn, 1975; *Lopez et al.*, 1977; *Barsdate et al.*, 1974).

                Niche Structure and Partitioning Mechanisms

Coexistence of species requiring similar limiting resources
implies a mechanism of niche partitioning. The great number of
studies relating deposit-feeding population abundance to food-
related parameters has motivated an investigation of the
differences in resource exploitation among coexisting species
(*e.g.*, *Fenchell*, 1975a, b; *Levinton*, 1977; *Whitlatch*, 1976).
A convenient framework for such studies is the organization of
species occurrence into niche axes with each species occupying
an ecological hypervolume (*Hutchinson*, 1957). Unfortunately
theoretical efforts at determining the amount of tolerable
niche overlap do not lend themselves to strong conclusions of
direct use to field biologists (see *Christiansen and Fenchel*,
1977). Furthermore, such an effort often leads to a static

description of several coexisting species' respective ecological
occurrences, with little gained insight on the qualitative nature
or dynamics of species interactions. But the exercise of establishing niche dimensions does organize our thoughts about
potential limiting resources and their possible effects on
deposit-feeders. We can envisage possible niche separation
along the following resource axes.

(1) *Sediment Type*. Deposit-feeders might be best adapted
to sediments ranging from low silt-clay content to very fine
muds. In the Buzzards Bay region, the deposit-feeding bivalve
Nucula proxima occurs principally in muddy sands of 50% silt-
clay or less. However, its smaller congener, *N. annulata*,
occurs in fine oozes of usually greater than 90% silt-clay
(*Levinton*, 1977). By contrast, intertidal flat populations
of *Hydrobia totteni* and *Ilyanassa obsoletus* in Long Island
Sound, New York, both occur over a wide range of silt-clay
contents.

(2) *Living or Feeding Depth Below the Sediment-Water
Interface*. Species feeding or living at different depths
within the sediment might be able to coexist (*Levinton and
Bambach*, 1975; *Levinton*, 1977). *Levinton* (1977) examined
niche structure of two deposit-feeding communities in lateral
contact in Quisset Harbor, on Cape Cod, Massachusetts. Both
communities showed co-occurrence of species but a stratification
of living depths below the sediment-water interface in each
community. Dominant deposit-feeding species did not coexist
at the same depth within a community type. The parallelism
of stratification in these two communities with exchange of,
for example, a deep-feeding species for another when one
crosses the intercommunity boundary circumstantially suggests
the importance of interspecific competition in structuring the
community.

Laboratory observations show direct interference interactions between dominants not typically co-occurring in the
field. Time lapse X-radiography of thin-walled aquaria
demonstrate that the relatively deep-burrowing *Yoldia limatula*
frequently disrupted the Y-shaped burrow of the bivalve *Solemya
velum*. A similar pattern of stratification characterizes
Silurian communities of Nuculoid bivalves (*Levinton and
Bambach*, 1975). A stratified niche structure has been observed
by *Nicolaisen and Kanneworf* (1969), who found a narrowing of
an amphipod species' vertical occurrence when it coexisted
with a second amphipod species. In Barnstable Harbor,
Massachusetts, deposit-feeding polychaetes tend to live at
different depths below the sediment-water interface (*Whitlatch*,
1976). Although the effect was demonstrated with suspension-
feeders, *Peterson* (1977) found that a field experimental removal

of one bivalve species living at a given depth resulted in its replacement by another bivalve species.

(3) *Particle Size*. Two species might live in the same sediment at the same depth below the sediment-water interface, but consume particles of differing diameters. This difference might stem from the limitations of a given feeding organ, or from a change in particle size preference through a concomitant evolutionary change of the feeding organ. The latter can be simply accomplished, for example, in many groups with a change in body size. The tellinacean bivalve *Abra tenuis* shows an increase in ingested particle diameter, >30μ, with increasing inhalent siphon diameter (*Hughes*, 1973). The ctenidium accomplishes no further size sorting. Siphon diameter increases with increasing body size. There is positive correlation between body length and ingested particle diameter for the polychaete *Pectinaria gouldii* (*Whitlatch*, 1974), and for three species of the mud snail *Hydrobia* (*Fenchel*, 1975a, b). In localities of the Limfjord where *H. ulvae* and *H. ventrosa* co-occur, the former species is always larger than the latter. However, when *H. ventrosa* occurs alone, it is similar in length to its conspecific in single species localities. This pattern has been explained as character displacement in response to competition for limiting resources. In creatures such as bivalve mollusks, body size usually correlates with depth of burial below the sediment water interface. Therefore an evolutionary change of body size causes simultaneous ecological change along two niche axes. However, it must be remembered that juveniles pass through the ecological space occupied by adults of smaller species. Therefore patterns of niche overlap probably change with age (and size).

(4) *Food Type*. Coexisting deposit-feeders might conceivably feed on different types of food. There is general agreement that most deposit-feeders efficiently digest and assimilate microorganisms but poorly digest non-living organic plant detritus (*Fenchel*, 1970; *Hargrave*, 1970a, b; *Yingst*, 1976; *Lopez et al.*, 1977). However, among-species differences in digestive ability of different food has been inadequately investigated (Table 1). *Kofoed* (1975a, b) demonstrated efficient assimilation by *Hydrobia* of bacteria and diatoms, with poorer assimilation of two blue-green algal species. Sea-cucumbers have similarly broad abilities to assimilate material derived from microalgae and bacteria (*Yingst*, 1976). *Hylleberg's* (1976) investigation of Hydrobiid carbohydrases shows differences in activity but no qualitative among-species differences in types of carbohydrase. Thus there is little evidence of microbial specialization.

TABLE 1. Digestion or assimilation of different components of the microbial community by various deposit-feeders. Digestion refers to the percent of the living food killed in the gut, whereas assimilation includes the percent of the food not passed as feces, but incorporated.

Organism	ATP	Bacteria	Flagellates, Diatoms	Blue-green Algae	Plant Detritus	Animal Detritus	Type of Uptake	Reference
Hydrobia ventrosa	--	75	60–71	8, 50	--	--	Assimilation	Kofoed, 1975a
Hydrobia ventrosa	--	42.5	52	--	--	--	Digestion	Lopez and Levinton, 1978
Orchestia grillus	60.5	--	--	--	< 1	--	Digestion	Lopez et al., 1977
Parastichopus parvimensis	--	43.2	52.5	--	0	86.8	Assimilation	Yingst, 1976
Ancylus fluviatilis	--	--	55.6–74.7	--	--	--	Digestion	Calow, 1975
Planorbis contortus	--	74–94	--	--	--	--	Digestion	Calow, 1975
Hyalella azteca	--	60–82.5	45–75	5.5–15	8.5 (22.1*)	--	Assimilation	Hargrave, 1970a

*protein

The question of assimilation of dead organic detritus deserves some special attention, as it often comprises the bulk of organic matter in some nearshore sediments. *Hargrave* (1970a) measured an assimilation of 8.5% of the organic matter from elm leaves in the benthic freshwater amphipod *Hyalella azteca*. Thus a poorer assimilation efficiency for plant detritus might sometimes be compensated for by its abundance relative to microorganism biomass. But *Lopez et al.* (1977) calculated that digestion of *Spartina* plant detritus by the amphipod *Orchestia grillus* was less than one percent.

The differential trophic value of different kinds of detritus and different stages of degradation of detritus has been emphasized by several authors (*e.g.*, *Bader*, 1954; *Lopez et al.*, 1977; *Tenore*, 1977a, b; *Bobbie et al.*, 1978). With assumed C:ATP and N:ATP ratios, *Lopez et al.* (1977) demonstrated a transfer of nitrogen from *Spartina* detritus to the attached microbial fraction as the latter colonized the detritus. Further, import of nitrogen from the dissolved nitrogen pool may permit a microbial community to develop on aging detritus and increase the nutritive value of the detrital complex (*Newell*, 1965).

The nutritive value of different forms of detritus complicates the subject of food limitation of deposit-feeding populations. Although algal detritus was poorly assimilated, the sea-cucumber, *Parastichopus parvimensis*, efficiently assimilated animal detritus (crustacean; *Yingst*, 1976). When the deposit-feeding polychaete *Capitella capitata* was fed detritus from different sources it was found that the best correlate of standing crop was nitrogen (*Tenore*, 1977a). As no counts of the microbial fraction were made, it is not clear whether the increased N of say, *Gracilaria* detritus, is directly incorporated or must be converted into microbial tissue before efficient assimilation can occur. *Bobbie et al.* (1978) measured microbial colonization on refractory polyvinyl chloride needles and needles from slash pine (*Pinus elliottii*) and found 2-10 fold higher levels of microbial indicators such as ATP, alkaline phosphatase, oxygen utilization, muramic acid, and phosphodiesterase on the latter substrate. *Tenore's* results may therefore reflect a differential development and mineralization of detritus by the microbial community.

Coprophagy has been suggested by many as a primary nutritive source for deposit-feeders (*e.g.*, *Newell*, 1965; *Frankenberg and Smith*, 1967; *Johannes and Satomi*, 1967). We can distinguish between interspecific coprophagy and intra-specific coprophagy. *Levinton and Lopez* (1977) and *Levinton et al.* (1977) examined deposit-feeding in two species of

Hydrobia and found that coprophagy was rare in sediments of particle diameter less than 62 microns. Intact pellets were usually not ingested, but pellets mechanically disrupted into constituent particles were fed upon immediately (*Levinton and Lopez*, 1977). When dismembered pellets were immediately refed to snails, the percent digestion of bacteria and diatoms were both greatly reduced (*Lopez and Levinton*, 1978). Thus there is adaptive sense in the avoidance of coprophagy, as the level of microbial food in fecal pellets is low. As fecal pellets are broken down the colonizing microbial community makes the constituent sedimentary grains more nutritionally useful. This study, however, neglects the possibility that some forms of pellets may still have large amounts of digestible material. Further, one species may dismember and ingest the pellets of another species with a substantial nutritive return (*e.g.*, *Lopez and Levinton*, 1978). Because fecal pellet formation is a major means of importing organic matter by suspension-feeders from the overlying water (*Verwey*, 1952; *Haven and Morales-Alamo*, 1966; *Frankenberg et al.*, 1967) this pathway cannot be ignored. But a distinction must be made between immediate reingestion of fresh fecal pellets and recycling of fecal material through detrital browsers. The latter process is nutritionally valuable when microbial recolonization has been sufficient to permit assimilation of enough food for maintenance and growth.

LIMITING RESOURCES FOR DEPOSIT FEEDERS

The major limitation of studies of competition and resource limitation is a paucity of data from experiments investigating limiting resources and determining the extent to which measured resource availability affects deposit-feeder populations. Most studies provide, instead, circumstantial evidence of niche partitioning (e.g., *Nicolaisen and Kanneworf*, 1969; *Levinton and Bambach*, 1975; *Whitlatch*, 1976). Manipulations of predator populations are not commonplace in investigations of deposit-feeding communities. However, recent evidence suggests that predation does play a role in controlling population size in soft sediments (*Schneider*, 1978; *Young et al.*, 1976; *Virnstein*, 1977). Therefore, it is more than a moot point to claim that competitive effects can hardly be inferred without experimental evidence demonstrating resource depression in dense deposit-feeding populations and limitation of population size by available resources. The following are possible limiting resources for nearshore deposit-feeders.

(1) *Detrital influx*. In areas of high sedimentation of plant detritus, a concomitant increase in microbial biomass might increase the food supply for deposit-feeders (*e.g.*, see

Wieser, 1963; *Tietjen*, 1969). This possibility, however, must
only have broad-scale regional implications as the number of
bacteria per unit of sedimentary grain surface area has been
found to be similar over a variety of grain sizes in several
shallow marine localities in Denmark (*Kofoed and Lopez*,
unpublished). The bulk of organic matter in a sand flat in
the Bay of Fundy was found to be in the form of bacteria
(*Tunnicliffe and Risk*, 1977). But qualitative differences in
detrital nitrogen content in different regimes (e.g., mussel
fecal pellets vs. *Spartina* detritus) must affect microbial
standing crop.

   (2) *Microbial Biomass and Renewal*. Bacteria and microalgae are renewable resources whose grazing is followed by
microbial recolonization of sedimentary grain surfaces. The
grazing rate of deposit-feeders could be sufficient to outstrip
the recolonization rate. *Fenchel and Kofoed* (1976) investigated
diatom population productivity and biomass at different densities
of *Hydrobia*. Photosynthetic productivity and biomass of large
diatoms were both depressed at high grazer densities (comparable
to dense field populations). The average individual growth of
H. ventrosa was linearly related to biomass of large diatoms
(>20 μm).

   (3) *Space*. Crowding may affect movement, feeding, and
emigration of deposit-feeders. *Woodin* (1974) demonstrated
that mobile polychaetes may emigrate from sediment dense with
tube-builders. Some deposit-feeding Tellinacean bivalves are
territorial, presumably to maintain a foraging space for the
inhalent siphon (*Holme*, 1950). *Levinton* (1979) showed that
feeding is inhibited with increased population density in
Hydrobia ventrosa.

   Direct interference between deposit-feeding species may
influence living position and behavior. The deposit-feeding
bivalve *Yoldia limatula* disrupts the burrows of *Solemya velum*
(*Levinton*, 1977). E. *Hatfield* (pers. comm.) has experimentally
demonstrated changes in an amphipod's living depth below the
sediment-water interface when another amphipod species is
introduced.

                KINETIC MODELS OF RESOURCE AVAILABILITY

   As described above, renewable resources may be limiting to
deposit feeders. The inference of food limitation is therefore
a problem of resource supply and renewal. As deposit-feeders
consume sedimentary grains and strip off a fraction of the
microflora, it is the microbial recovery rate that is of most
concern. We here consider a model of resource availability

for the deposit-feeding snail *Hydrobia ventrosa* (see *Fenchel and Kofoed,* 1976; *Levinton and Lopez,* 1977; *Levinton et al.,* 1977). In this model we therefore consider the further complication of avoidance of coprophagy. As sedimentary grains are tied up in fecal pellets for a period, reingestion of sedimentary grains is not possible until the pellets have broken down. This period has a beneficial effect, however, in providing an incubation period for microbial recovery. It is unlikely, in my opinion, that pellet formation evolved for the purpose of making microbial gardens. The mode of fecal compaction seems most likely related to constraints of gut morphology and waste disposal. However, avoidance of coprophagy is adaptive as immediate reingestion provides little additional nutrition (*Lopez and Levinton,* 1978). *Ilyanassa obsoletus* does not make compact fecal pellets but avoids grazing in recently browsed areas (R. *Wetzel,* pers. comm.).

We can relate P, the fraction of the sediment pelletized $(0 \leq P \leq 1)$, to the daily population pelletization rate p $(0 < p < 1)$ and the daily survival rate of pellets, a $(0 \leq a < 1)$ (*Levinton and Lopez,* 1977).

A series can be generalized as:

$$P = p \Sigma a^i$$

where i is the number of days of pelletization. This species has the following convergent solution when a is less than unity:

$$P = p / (1-a) \qquad\qquad\qquad\qquad (1)$$

Thus at population densities corresponding to the inequality $p \geq (1-a)$ the sediment will be completely pelletized and no ingestible particles will be available. Both p and a have been measured, and predicted maximum densities fall within the range of dense *Hydrobia* field populations (*Levinton and Lopez,* 1977). In order to relate pellet breakdown to microbial recovery we use the following parameters:

σ: pelletization rate per snail $(0 < \sigma < 1$ per day).

N: population size per square meter.

μ: pellet breakdown rate $(0 < \mu \leq 1$ per day).

η: microbial recovery rate $(0 < \eta \leq 1$ per day).

$\hat{P}$:   equilibrium fraction, pelletized.

M: fraction of maximum microbial recovery $(0 \leq M \leq 1)$.

Analogous to the series discussed above:

$$\frac{dP}{dt} = \sigma N - \mu P$$

$$P = \frac{\sigma N}{\mu} \qquad\qquad (2)$$

$$\frac{1}{P} = \frac{\mu}{\sigma} \cdot \frac{1}{N}$$

Particle ingestion rate might decrease when most of the sediment is pelletized. Using a linear approximation:

$$\frac{dP}{dt} = \sigma N(1-P) - \mu P$$

$$P = \frac{\sigma N}{\sigma N + \mu} \qquad\qquad (3)$$

$$\frac{1}{P} = 1 + \frac{\mu}{\sigma} \cdot \frac{1}{N}$$

Both derivations predict a linear relationship between $1/N$ and $1/P$, with different intercepts and identical slope of μ/σ. Figure 1 shows data from *Levinton and Lopez* (1977) plotted in this manner. A good linear fit is obtained (ANOVA on regression: $p < .001$), but the data are obviously too limited to estimate the intercept adequately. As we know the slope, μ can be calculated by multiplying the slope by σ. This value was measured to be 4.2×10^{-6} (*Levinton and Lopez*, 1977). Therefore μ is estimated to be 0.365. From equation (1) a maximum population density of 87,000 snails m^{-2} is predicted. Other measurements of μ (see *Levinton and Lopez*, 1977) predict densities in the range of 27 to 36×10^4 snails m^{-2}. The present estimate is considerably higher but still within the commonly observed range of 1 to 10×10^4 (see *Anderson*, 1977). It is therefore possible that particle availability can limit food availability for *Hydrobia*. As *Hydrobia* often moves laterally when the sediment is pelletized, a compensatory mechanism regulating density is possible.

As Hydrobiid snails rely upon microorganisms, food availability must be computed as a balance between consumption and microbial recolonization. The pelletization of the sediment prevents an average ingestable sedimentary grain from being reingested, perhaps increasing the microbial standing stock on edible grains. As in pellet breakdown we assume a linear microbial recovery rate, η. This assumption would be

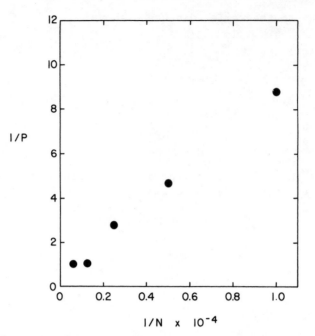

Fig. 1.   Fit of linear pelletization–pellet breakdown model for *Hydrobia ventrosa* feeding to data taken from *Levinton and Lopez* (1977).

justifiable if space or a correlate were a limiting factor for the microbial community.   Thus the present recovery, M, would be ($0 \leq M \leq 1$):

$$M = 1 - e^{-\eta t} \tag{4}$$

Calculating the fraction of the microbial community regenerated on edible particles requires the integration of the joint probabilities of pellet breakdown and microbial recovery.   I thank Robert A. Armstrong for the following derivation.   Microbial recovery, M, should be:

$$M = \int_0^\infty f(b) \; db \int_0^b \mu e^{-\mu a} \sigma N e^{-\sigma N(b-a)} \, da,$$

where $\mu e^{-\mu a}$ = probability {pellet breaks down at time a after ingestion}

$\sigma N e^{-\sigma N(b-a)}$ = probability {particle is eaten in time $a < t < b$}

$$f(b) = 1-e^{-\eta b}$$

The solution to this integral is (details available on request):

$$M = 1 - \frac{\mu}{\mu+\eta} \cdot \frac{\sigma N}{\sigma N+\eta} \tag{5}$$

With this relationship it is possible to calculate the relative increase in percent microbial recovery when fecal pellets are not reingested until they break down. Figure 2 shows percent microbial recovery when μ is 0.18 per day (estimated in *Levinton and Lopez*, 1977) and when pellet breakdown is immediate ($\mu = 1$). The difference between the curves shows a modest benefit from the strategy of not dismembering and reingesting intact fecal pellets. However, it seems more likely that avoidance of coprophagy is selected mainly to reduce feeding upon indigestable remains of material recently passed through another snail's gut (*Lopez and Levinton*, 1978).

Data of *Fenchel and Kofoed* (their Fig. 10, 1976) on standing stock of larger (>20 μm) diatoms in equilibrium with varying grazing pressures of *Hydrobia ventrosa* can be used to test the model. Maximum diatom standing stock is taken to represent $M = 1$, while those at higher snail densities are fractions of unity. Although there are unfortunately only four usable points, Fig. 3 shows that the linear recovery model is probably adequate. Using a consumption rate of sediment estimated in *Levinton and Lopez* (1977) and assuming approximately 50% digestion (see *Lopez and Levinton*, 1978), we estimate a diatom consumption rate, analogous to σ, to be 2.1 x 10^{-6}. Therefore η is 0.006. We now can test the combined pellet breakdown–microbial recovery model (equation 5) with the following estimated values: $\sigma = 4.2$ x 10^{-6}, $\eta = 0.006$, $\mu = 0.3$ (approximated average of estimates from Fig. 1 and that of *Levinton and Lopez*, 1978). As can be seen in Fig. 4, an adequate fit is obtained.

Some important details are left out of this simple model. At very low grazing pressure, diatoms are out-competed by blue-green algae, which are in turn inadequate food for *Hydrobia ventrosa* (*Fenchel and Kofoed*, 1976; *Kofoed*, 1975a). The disturbance of the sediment during browsing has therefore a stimulating effect on diatom growth. Similarly, high grazing by the amphipod *Orchestia grillus* accelerates the colonization rate of the microbial community on *Spartina* detritus (*Lopez et al.*, 1977).

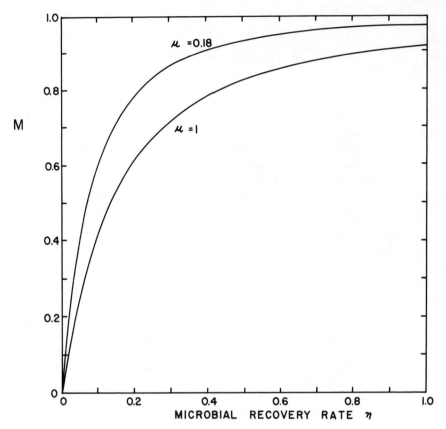

Fig. 2. Equilibrium microbial recovery balanced against consumption, when pellet breakdown is 0.18 and immediate ($\mu=1$), and *Hydrobia* density is 4 snails cm^{-2} (see *Levinton and Lopez*, 1977, for other assumptions).

A further complexity not modeled is a varying digestion efficiency with microbial standing crop (*Lopez and Levinton*, 1978). Frequent grazing might select for indigestible forms. Deposit-feeders might also compensate for poor food quality by acceleration of feeding rate and increased digestive capacity (*Calow*, 1975).

In these calculations, the good fit obtained assumes ingestion only of the silt-clay sized particles. The fit is surprising, as *Lopez and Levinton* (1978) have demonstrated that *Hydrobia ventrosa* can scrape larger sand grains if fine ingestible particles are not available. Unfortunately, data do not yet exist that directly measure a switch to scraping after fine particles are found in pellets. *Kofoed and Lopez* (in preparation)

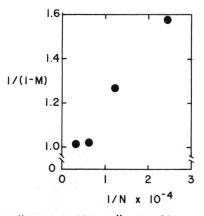

Fig. 3. Fit of "larger diatom" standing stocks under different grazing pressures of *Hydrobia ventrosa* to linear microbial recovery model.

argue that *Hydrobia* feeding is far more complex than described in *Fenchel and Kofoed* (1976).

THE ROLE OF SPACE

Kinetic models of resource renewal assume simple models of resource recovery and no spatial interactions among individuals. However, crowding may affect the pattern of resource exploitation and rates of resource utilization. *Hydrobia ventrosa* shows no territoriality but crowding affects feeding and movement. Figure 5 shows that movement is restricted in crowded populations. This effect is significant (p < .05; see *Levinton,* 1979, for details) at densities above 4 snails cm^{-2}. With increasing density, feeding rate is also reduced (Fig. 6). At low densities animals move a few millimeters and then ingest sediment, but with increased population density, more frequent encounters between snails result in continued movement without feeding. This movement, when random, will result in emigration from local areas of high density. *Hydrobia ventrosa* exhibits floating behavior similar to that described by *Newell* (1962, 1964). On encountering a vertical or inclined surface snails move upward to the intersection of the surface with the air-sea interface and launch themselves, suspended by the extended foot. Thus increased movement under high density (as opposed to feeding) increases the probability of encountering a rock or seaweed with subsequent emigration, when enough vertical or inclined surfaces are available (Fig. 7). The interaction of crowding and this behavior may serve as a mechanism to regulate population density,

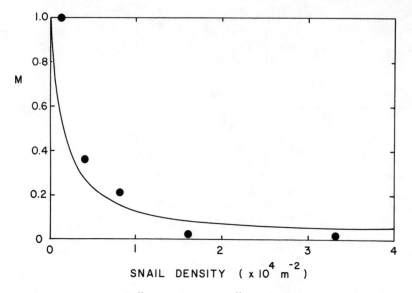

Fig. 4. Fit of "larger diatom" microbial recovery in equilibrium with varying sediment ingestion-grazing pressures of *H. ventrosa* to combined pelletization-pellet breakdown microbial recovery model.

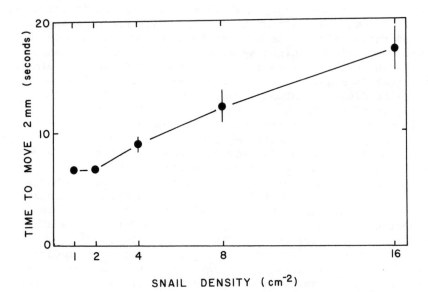

Fig. 5. Crawling rate of *Hydrobia ventrosa* individuals as a function of crowding (after *Levinton*, 1979).

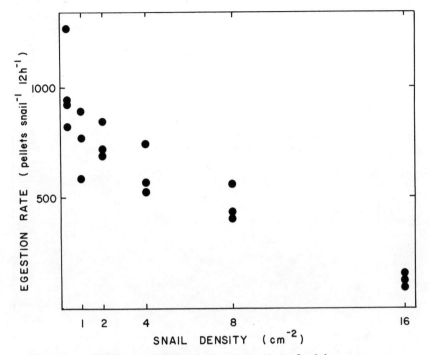

Fig. 6. Sediment ingestion rate of *Hydrobia ventrosa* estimated by pellet egestion, as a function of crowding (after *Levinton*, 1979).

or to homogenize densities on a sand flat. Table 2 shows an analysis of densities at Lendrup Strand, Denmark (see *Levinton*, 1979) with statistically homogeneous densities over a wide area.

The influence of crowding on resource renewal is complex. Crowding reduces the volume of sediment ingested per unit time per snail, but this reduction also permits a greater degree of microbial recovery between times that an average particle is ingested.

PARTICLE SIZE SELECTIVITY

The work of *Fenchel* (1975a, b) and *Whitlatch* (1974) suggests mechanisms of coexistence based upon differential utilization of particle size fractions. *Fenchel and Kofoed* (1976) and *Lopez and Levinton* (1978) both show a decline of *Hydrobia* ingestion rate with increasing particle size. Both studies use rates of feeding on separated size fractions to reconstruct how selective feeding might be in mixed sediments. This approach might be invalid, however, as the snails might

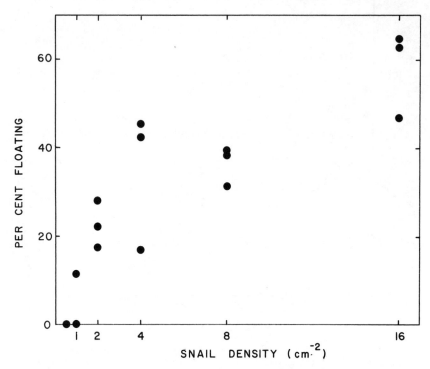

Fig. 7. Percent of the *H. ventrosa* population floating after six hours as a function of crowding in beakers with sediment (after *Levinton*, 1979).

TABLE 2. Analysis of variance of benthic densities of *Hydrobia ventrosa* at five widely separated (>30 m) localities at Lendrup Strand, Denmark; each with 10 samples (0.00049 m^2 each).

Source of Variation	df	s.s.	m.s.	F_s
Among localities	4	22.48	5.62	1.94(ns)
Within locality	45	130.10	2.89	
Total	49	152.58		

(ns) = not significant, p > .05

not be able to discriminate amongst particles of differing
diameters in mixed size fractions.

To compensate for this problem I have used mixed size
fractions of glass beads (Potter Industries) incubated in nutrient
medium with an inoculum of natural substrate. The beads are
colonized in 5-10 days with a rich coating of diatoms and
bacteria. They are then fed to the snails permitting measurement of digestion rate and size selectivity. Controls and
disaggregated pellets are measured with a Coulter Corporation
model TAII, multiple-channel particle counter. Figure 8 shows
an example of a feeding experiment with 3-4 mm long *Hydrobia
totteni*. As can be seen, snails prefer fine particles and consume such grains 2-3 times faster than their relative availability
in the sediment would predict on a random basis. However, large
particles are fed upon much less often than their proportional
availability would predict. In individual size fractions there

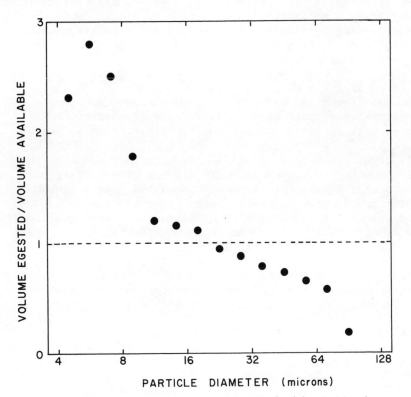

Fig. 8. Relative volume egested by *Hydrobia totteni*
divided by volume available in the sediment, for different
particle diameters of glass beads.

A

is a significant qualitative difference in fecal pellets produced when the beads are >80μ in diameter. Here pellets are 40-50% diatoms by volume, whereas in smaller fractions, pellets contain more than 95% by volume glass beads. This suggests that snails are not ingesting large beads, but scraping diatoms off bead surfaces. This confirms the switching experiments performed by *Lopez and Levinton* (1978). These results greatly complicate the simple explanation for character displacement in *Hydrobia* spp. proposed by *Fenchel* (1975b). Furthermore ingestion is apparently most rapid in the particle size range assumed by *Fenchel and Kofoed* (1976, p. 371) to be beyond the lower size limit of Hydrobiid feeding capacity (8-16μ).

CONCLUSIONS

As deposit-feeders digest and assimilate microorganisms, the kinetic approach using simple resource recovery models seems appropriate for studying resource limitation. However, even in *Hydrobia*, where no aggression is detected, spatial interactions probably play a major role in density regulation. Studies of deposit-feeding populations must therefore be multifactorial, with experimental manipulation of microbial, detrital, particle size, and crowding factors.

In combination with the value of a given microbial food type to somatic growth and reproduction, resource recovery rate is an essential component of available food. *Hargrave* (1970b) inferred that bacteria recover too rapidly ever to be greatly reduced by a browsing fresh-water amphipod. *Fenchel and Kofoed* (1976) argued that because small diatoms do not decrease in abundance under increased *Hydrobia* grazing pressure, it must be a shortage of the more grazing-responsive larger diatoms that limit body growth. Such reasoning comes from an appreciation of the kinetics of differing microbial resources. It is finally clear that these resource effects are not best studied by simple field manipulations of population density, as in rocky intertidal studies (see *Dayton*, 1979). It is just as important to understand the respective rates of renewal and nutritive value of the different microbial components. In any case, responses to manipulations are liable to be differences in somatic growth rate and reproduction rather than dramatic shifts of population density. Adult-larval interactions will also have a strong overprint on benthic distributions. Sediment disturbance by deposit-feeders has a negative effect on settling of larvae and on juvenile success of both deposit-feeding and suspension-feeding species (see *Levinton and Bambach*, 1970; *Rhoads and Young*, 1970). These factors will complicate field measures of resource limitation. Therefore microcosms are an essential tool in the study of resource renewal. Detrital mineralization can also be effectively modeled using a theoretical-microcosm combined approach.

ACKNOWLEDGMENTS

Robert A. Armstrong, Tom Fenchel, Jørgen Hylleberg and Glenn R. Lopez all provided valuable advice and criticism. I thank Mamre Wilson for typing the manuscript. Supported in part by National Science Foundation grant OCE 78-09047, this paper is contribution number 307 to the Program in Ecology and Evolution State University of New York at Stony Brook.

REFERENCES

Anderson, A., 1971. Intertidal activity, breeding and the floating habit of *Hydrobia ulvae* in the Ythan estuary, *J. Mar. Biol. Assoc. U. K. 51:* 423-437.

Bader, R. G., 1954. The role of organic matter in determining the distribution of bivalves in sediments, *J. Mar. Res. 13:* 32-47.

Barsdate, R. J., R. T. Prentki, and T. Fenchel, 1974. Phosphorous cycle of model ecosystems: significance for decomposer food chains and effect of bacterial grazers, *Oikos 25:* 239-251.

Bobbie, R. J., S. J. Morrison, and D. C. White, 1978. Effects of substrate biodegradability on the mass and activity of the associated estuarine microbiota, *Appl. Environ. Microbiol. 35:* 179-184.

Calow, P., 1975. The feeding strategies of two freshwater gastropods, *Ancylus fluviatilis* Müll. and *Planorbis contortus* Linn. (Pulmonata), in terms of ingestion rates and absorption efficiencies, *Oecologia (Berl.) 19:* 33-49.

Christiansen, F. B. and T. Fenchel, 1977. *Theories of Populations in Biological Communities,* 144 pp., Springer, Berlin.

Dale, N. G., 1974. Bacteria in intertidal sediments: factors rated to their distribution, *Limnol. Oceanogr. 19:* 509-518.

Dayton, P. K., 1979. Ecology: a science and a religion, In: *Ecological Processes in Coastal and Marine Systems,* edited by R. J. Livingston, Plenum, N. Y. (this volume).

Fenchel, T., 1970. Studies on the decomposition of organic detritus derived from the turtle grass *Thalassia testudinum, Limnol. Oceanogr. 15:* 14-20.

Fenchel, T., 1975a. Factors determining the distribution patterns of mud snails (Hydrobiidae), *Oecologia (Berl.) 20:* 1–17.

Fenchel, T., 1975b. Character displacement and coexistence in mud snails (Hydrobiidae), *Oecologia (Berl.) 20:* 19–32.

Fenchel, T. and L. H. Kofoed, 1976. Evidence for exploitative interspecific competition in mud snails (Hydrobiidae), *Oikos 27:* 367–376.

Frankenberg, D. and K. L. Smith, 1967. Coprophagy in marine animals, *Limnol. Oceanogr. 12:* 443–450.

Frankenberg, D., S. L. Coles, and R. E. Johannes, 1967. The potential trophic significance of *Callianassa major* fecal pellets, *Limnol. Oceanogr. 12:* 113–120.

Hargrave, B. T., 1970a. The effect of deposit-feeding amphipods on the metabolism of benthic microflora, *Limnol. Oceanogr. 15:* 21–30.

Hargrave, B. T., 1970b. The utilization of the benthic microflora by *Hyalella azteca, J. Anim. Ecol. 39:* 427–437.

Haven, D. and R. Morales-Alamo, 1966. Aspects of biodeposition by oysters and other invertebrate filter-feeders, *Limnol. Oceanogr. 11:* 487–498.

Holme, N. A., 1950. Population dispersion in *Tellina tenuis* da Costa, *J. Mar. Biol. Assoc. U. K. 29:* 267–280.

Hughes, T. G., 1973. Deposit feeding in *Abra tenuis* (Bivalvia: Tellinacea), *J. Zool. 171:* 499–512.

Hutchinson, G. E., 1957. Concluding remarks, *Cold Spring Harbor Symp. Quant. Biol. 22:* 415–427.

Hylleberg, J., 1976. Resource partitioning on basis of hydrolytic enzymes in deposit-feeding mud snails (Hydrobiidae). II. Studies on niche overlap, *Oecologia (Berl.) 23:* 115–125.

Johannes, R. E. and M. Satomi, 1967. Composition and nutritive value of fecal pellets of a marine crustacean, *Limnol. Oceanogr. 11:* 191–197.

Jørgensen, B. B., 1977. Bacterial sulfate reduction within reduced microniches of oxidized marine sediments, *Mar. Biol. 41:* 7–17.

Kofoed, L. H., 1975a. The feeding biology of *Hydrobia ventrosa*
    (Montagu). I. The assimilation of different components of
    the food, *J. Exp. Mar. Biol. Ecol. 19:* 233-241.

Kofoed, L. H., 1975b. The feeding biology of *Hydrobia ventrosa*
    (Montagu). II. Allocation of the components of the carbon
    budget and the significance of the secretion of dissolved
    organic material, *J. Exp. Mar. Biol. Ecol. 19:* 243-256.

Levinton, J. S., 1977. The ecology of deposit-feeding
    communities: Quisset Harbor, Massachusetts. In:
    Ecology of Marine Benthos, edited by B. C. Coull, 191-
    228, Univ. of S. Carolina Press, Columbia, S. C.

Levinton, J. S., 1979. The effect of density upon deposit-
    feeding populations: movement, feeding and floating of
    Hydrobia ventrosa Montagu (Gastropoda: Prosobranchia),
    submitted to *Oecologia.*

Levinton, J. D. and R. K. Bambach, 1970. Some ecological aspects
    of bivalve mortality patterns, *Amer. J. Sci. 268:* 97-112.

Levinton, J. S. and R. K. Bambach, 1975. A comparative study
    of Silurian and recent deposit-feeding bivalve communities,
    Paleobiology 1: 97-124.

Levinton, J. S. and G. R. Lopez, 1977. A model of renewable
    resources and limitation of deposit-feeding benthic
    populations, *Oecologia (Berl.) 31:* 177-190.

Levinton, J. S., G. R. Lopez, H. H. Lassen, and U. Rahn, 1977.
    Feedback and structure in deposit-feeding marine benthic
    communities, *11th Europ. Symp. Mar. Biol.,* edited by
    B. F. Keegan, P. O. Ceidigh, and P. J. S. Boaden, 409-
    416, Pergamon, Oxford.

Lopez, G. R. and J. S. Levinton, 1978. The availability of
    microorganisms attached to sediment particles as food for
    Hydrobia ventrosa Montagu (Gastropoda: Prosobranchia),
    Oecologia (Berl.) 32: 236-275.

Lopez, G. R., J. S. Levinton, and L. B. Slobodkin, 1977. The
    effect of grazing by the detritivore *Orchestia grillus* on
    Spartina litter and its associated microbial community.
    Oecologia (Berl.) 30: 111-127.

Longbottom, M. R., 1970. The distribution of *Arenicola marina*
    (L.) with particular reference to the effect of particle
    size and organic matter of the sediments, *J. Exp. Mar.
    Biol. Ecol. 5:* 138-157.

Newell, R., 1962. Behavioural aspects of the ecology of *Peringia* (= *Hydrobia) ulvae* (Pennant) (Gasteropoda, Prosobranchia), *Proc. Zool. Soc. Lond. 140:* 49–75.

Newell, R., 1964. Some factors controlling the upstream distribution of *Hydrobia ulvae* (Pennant) (Gastropoda, Prosobranchia), *Proc. Zool. Soc. Lond. 142:* 85–106.

Newell, R. C., 1965. The role of detritus in the nutrition of two marine deposit feeders, the prosobranch *Hydrobia ulvae* and the bivalve *Macoma balthica, Proc. Zool. Soc. Lond. 144:* 25–45.

Nicolaisen, W. and E. Kanneworf, 1969. On the burrowing and feeding habits of the amphipods *Bathyporeia pilosa* Lindstrom and *Bathyporeia sarsi* Watkin, *Ophelia 6:* 231–250.

Peterson, C. H., 1977. Competitive organization of the soft-bottom macrobenthic communities of southern California lagoons, *Mar. Biol. 43:* 343–359.

Rahn, V., 1975. *Phosphorous Turnover in Relation to Decomposition of Detritus in Aquatic Microcosms,* Dissertation, Institute of Ecology and Genetics, University of Aarhus, Denmark (in Danish).

Rhoads, D. C., 1967. Biogenic reworking of intertidal and subtidal sediments in Barnstable Harbor and Buzzards Bay, Massachusetts, *J. Geol. 75:* 461–474.

Rhoads, D. C., 1973. Organism-sediment relationships on the muddy sea floor, *Oceanogr. Mar. Biol. Ann. Rev. 12:* 263–300.

Rhoads, D. C. and D. K. Young, 1970. The influence of deposit-feeding organisms on sediment stability and community trophic structure, *J. Mar. Res. 28:* 150–178.

Sanders, H. L., 1956. Oceanography of Long Island Sound, 1952–1954. X. Biology of marine bottom communities. *Bingham Oceanogr. Coll. Bull. 15:* 345–414.

Sanders, H. L., 1958. Benthic studies in Buzzards Bay. I. Animal-sediment relationships, *Limnol. Oceanogr. 3:* 245–258.

Schneider, D. C., 1978. Equalisation of prey number by migratory shorebirds, *Nature 271:* 353–354.

Tenore, K. R., 1977a. Growth of the polychaete *Capitella capitata* cultured in different levels of detritus derived from various sources, *Limnol. Oceanogr. 22:* 936-941.

Tenore, K. R., 1977b. Utilization of aged detritus derived from different sources by the polychaete, *Capitella capitata*. *Mar. Biol. 44:* 51-55.

Tiejen, J. H., 1969. The ecology of shallow water meiofauna in two New England estuaries, *Oecologia (Berl.) 2:* 251-291.

Tunnicliffe, V. and M. J. Risk, 1977. Relationships between the bivalve *Macoma balthica* and bacteria in intertidal sediments: Minas basin Bay of Fundy, *J. Mar. Res. 35:* 499-507.

Verwey, J., 1952. On the ecology of distribution of cockle and mussel in the Dutch Waddensee, their role in sedimentation, and the source of their food supply with a short review of the feeding behavior of bivalve mollusks, *Arch. Neerl. Zool. 10:* 172-239.

Virnstein, R. W., 1977. The importance of predation by crabs and fishes on benthic infauna in Chesapeake Bay, *Ecology 58:* 1199-1217.

Whitlatch, R. B., 1974. Food resource partitioning in the deposit-feeding polychaete *Pectinaria gouldii, Biol. Bull. 147:* 227-235.

Whitlatch, R. B., 1976. Methods of resource allocation in marine deposit-feeding communities, *Amer. Zool. 16:* 195.

Wieser, W., 1960. Benthic studies in Buzzards Bay. II. The meiofauna, *Limnol. Oceanogr. 5:* 121-137.

Woodin, S. A., 1974. Polychaete abundance patterns in a marine soft-sediment environment: the importance of biological interactions, *Ecol. Monogr. 44:* 171-187.

Yingst, J. Y., 1976. The utilization of organic matter in shallow marine sediments by an epibenthic deposit-feeding Holothurian, *J. Exp. Mar. Biol. Ecol. 23:* 55-69.

Young, D. K., M. A. Buzas, and M. W. Young, 1976. Species densities of macrobenthos associated with seagrass in a field experimental study of predation, *J. Mar. Res. 34:* 577-592.

Zobell, C. E. and C. B. Feltham, 1938. Bacteria as food for certain marine invertebrates, *J. Mar. Res. 1:* 312-327.

CYCLIC TROPHIC RELATIONSHIPS OF FISHES IN AN UNPOLLUTED,

RIVER-DOMINATED ESTUARY IN NORTH FLORIDA

Peter F. Sheridan[1] and Robert J. Livingston

Florida State University

ABSTRACT

Regular patterns in seasonal occurrence of dominant fishes were observed over a six-year period in the Apalachicola estuary of north Florida. Examination of potential physico-chemical and biological community determinants has led to the hypothesis that trophic relationships and underlying physical-biological inter-actions structure this estuarine fish community. Six species (Anchoa mitchilli, Micropogonias undulatus, Leiostomus xanthurus, Cynoscion arenarius, Brevoortia patronus, *and* Bairdiella chrysura) *comprise 85% of the trawl-susceptible fishes in the Apalachicola system, and each is characterized by distinctive seasonal abundances and trophic spectra. Two benthic omnivores* (Micropogonias *and* Leiostomus) *exhibit high spatial and temporal overlap but differ in prey type and size. These two species utilize the estuary subsequent to high river discharge/detritus input and concurrent with maximum benthic standing crops. Two epibenthic carnivores* (Cynoscion *and* Bairdiella) *also use the estuary but differ in times of peak abundances and in prey types. Two planktivores* (Brevoortia *and* Anchoa) *also frequent the estuary but during different seasons (spring and fall, respec-tively), yet neither cooccurs with the maximum zooplankton standing crop (summer).* Anchoa *is prevented from doing so by the piscivorous* Cynoscion *population. Thus, regular seasonal progressions of dominant fishes are linked to available trophic resources, competition, and predation,*

---

[1]Present address: U. S. Environmental Protection Agency, Bears Bluff Field Station, S. C.

*which are in turn dependent upon such factors as river flow,
detrital input, plankton production, and offshore processes.
The data indicate that the trophic organization of the Apalachicola
estuary is highly structured and is not simply the result of a
series of physically forced events superimposed over a network
of trophic opportunism.*

*The influence of offshore shelf areas upon the observed
estuarine community presently remains an enigma. Little is known
concerning the interactions of shelf and estuary, and it is hoped
that future research will elucidate the connections.*

INTRODUCTION

Estuaries are important nursery grounds for juvenile forms
of various coastal fishes. Past studies indicate that most such
populations are relatively short-lived and euryhaline; as such they
represent one phase of an inshore-offshore migratory life history
pattern. Estuarine fish communities are characterized by high
numerical dominance by a small number of species. In the past
few years, estuarine studies have emphasized the relative
importance of key environmental factors which affect such populations. Most have concluded that although physico-chemical
factors are important, biological functions such as reproduction,
food habits, and interspecific interactions may be extremely
influential in structuring the estuarine fish community (*McErlean
et al.*, 1973; *Oviatt and Nixon*, 1973; *Gallaway and Strawn*, 1974;
Haedrich and Haedrich, 1974; *Livingston et al.*, 1976b).

Various studies have described trophic interactions of
estuarine and coastal fishes with particular emphasis on
spatial/temporal abundance patterns, inter- and intraspecific
competition, and resource partitioning. *Tyler* (1972) noted
resource partitioning among demersal fishes in Canadian coastal
waters. *Oviatt and Nixon* (1973) found little overlap in feeding
habits of dominant demersal flatfishes in a Rhode Island estuary,
and trophic resource partitioning was viewed as a mechanism to
reduce competition. *Haedrich and Haedrich* (1974) found trophic
partitioning among dominant resident fishes of a Massachusetts
estuary. *Stickney et al.* (1974) found that estuarine flatfishes
of Georgia showed temporal partitioning of trophic resources,
with morphological adaptation (relative mouth size) as a functional determinant of feeding relationships. *Stickney et al.*
(1975) investigated sciaenid trophic patterns and found a direct
relationship between prey size and predator size. *McEachran
et al.* (1976) and *Kravitz et al.* (1976) found little trophic
interaction among various continental shelf and slope fishes.

Overall, the analysis of resource partitioning in coastal fishes is incomplete at this time and, although various studies have indicated that the physico-chemical environment determines the population structure in estuarine/coastal systems, the evidence remains largely indirect and, in fact, is hampered by the growing contention that specific environmental variables cannot be directly associated with long-term population distributions (*Dahlberg and Odum*, 1970; *McErlean et al.*, 1973; *Oviatt and Nixon*, 1973; *Livingston et al.*, 1976b, 1978). This contention is based on the hypothesis that biological functions (food habits, predation, competition) are key determinants of such population functions, with potential indirect links to the microhabitat organization through trophic response and prey distribution.

This paper addresses several interrelated problems: 1) the food habits of the six dominant fishes in the Apalachicola estuary of Florida, 2) the observed seasonal periodicity of these fishes, and 3) the relationship of the trophic structure of dominant fishes in the estuary relative to variations in productivity, habitat distribution, and interspecific competition.

METHODS AND MATERIALS

Methods of data collection have been described in detail elsewhere (*Livingston*, 1976; *Livingston et al.*, 1976b; *Livingston et al.*, 1977). Physico-chemical and biological (trawl) samples were taken monthly from March, 1972, through February, 1978, at a series of 10 stations in the Apalachicola estuary (Fig. 1, stations 1, 1A, 1B, 1C, 2, 3, 4, 5, 5A, 6). Included were surface and bottom determinations of temperature, salinity, light penetrance (Secchi disk), dissolved oxygen, color, turbidity, and pH. River flow data were provided by the U. S. Army Corps of Engineers (Mobile, Alabama), while local climatological data were provided by the Environmental Data Service, NOAA.

Methods used to determine phytoplankton standing crop, productivity, and associated controlling factors have been described by *Livingston et al.* (1974) and *Iverson and Myers* (1976). These data were taken by Dr. Richard L. Iverson (Department of Oceanography, Florida State University). Detailed methods used to determine the distribution of detritus in the Apalachicola estuary have been given by *Livingston et al.* (1976b). Macro-particulate detritus was sampled at various bay stations with 5-m otter trawls (19 mm mesh wing and body, 6.5 mm mesh liner) while microdetritus (45 μ–2.0 mm) was sampled at the mouth of the Apalachicola River by pumping 250–1000 liters of river water through a series of sieves. All detritus samples were taken at monthly intervals from December, 1974, through March, 1977.

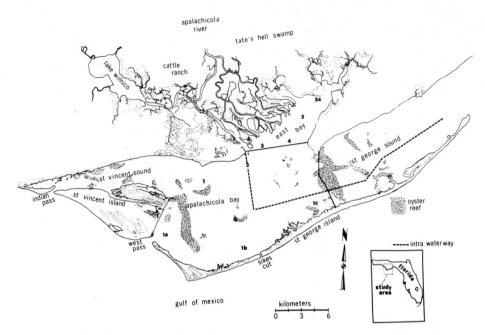

Fig. 1.   The Apalachicola estuary showing station distribution for long-term sampling program.

        Estuarine fishes were collected with a 5-m otter trawl
(19 mm mesh wing and body, 6.5 mm mesh liner).   Stomach contents
were analyzed by the gravimetric method (*Carr and Adams*, 1972,
1973).   Food habits were analyzed with the ρ index of faunal
similarity (*Matusita*, 1955; *Van Belle and Ahmad*, 1974) and
flexible group cluster analysis (β = -0.25) (*Lance and Williams*,
1967).

RESULTS AND DISCUSSION

        The Apalachicola estuary is a shallow, bar-built system
dominated physically by the Apalachicola River, which delivers
peak discharge rates during winter-spring months (Fig. 2).   During
such periods, there are maximum influxes of dissolved organics,
inorganic nutrients, and detritus (*Livingston et al.*, 1976a).
At this time there are coincident peaks in the numbers of epibenthic fishes and invertebrates and benthic infauna (*Livingston*,
1978; *Livingston and Loucks*, in press).   Phytoplankton densities
peak during early summer months and remain at relatively high
levels through the summer (*Estabrook*, 1973; *Iverson and Myers*,
1976).   The zooplankton community (numbers and biomass) increases

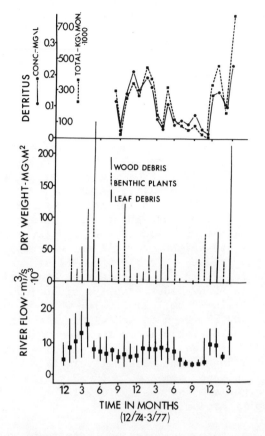

Fig. 2.   Apalachicola River flow (monthly ranges and means),
macrodetritus (wood, leaf, and benthic macrophyte debris) and
microdetritus from the river mouth (totals per month; concentration) from December, 1974, through March, 1977.

in the spring, and biomass peaks during early summer months.
Zooplankton remain relatively abundant into the fall (*Edmiston*,
1979), when there also are relatively high numbers of epibenthic
organisms.

The six dominant fish species in the Apalachicola estuary
undergo seasonal and annual fluctuations in abundance, and comprise 85% of all trawl-susceptible fishes collected over the six
years of study.   They are, in decreasing order of numerical
abundance, the bay anchovy (*Anchoa mitchilli*), the Atlantic
croaker (*Micropogonias undulatus*), the spot (*Leiostomus xanthurus*),
the sand seatrout (*Cynoscion arenarius*), the Gulf menhaden
(*Brevoortia patronus*), and the silver perch (*Bairdiella chrysura*).
While each species has regular seasonal peaks in abundance,

statistical analyses remain inadequate for direct correlations of
population distributions with key physico-chemical parameters such
as temperature or salinity (*Livingston et al.*, 1976b; *Livingston
et al.*, 1978). However, river fluctuations appear to be strongly
correlated with fish abundance (*Livingston et al.*, 1978).

Trophic Relationships

The intra- and interspecific trophic relationships of the
four most abundant fishes in the Apalachicola estuary have been
examined in detail (*Sheridan*, 1978). Food habits for the remaining two species (menhaden and silver perch) were obtained from
published studies of their habits in similar estuarine areas.
Anchovies are planktivorous throughout all life-history stages
(Fig. 3), preying mainly on calanoid copepods (averaging 69% of
stomach contents by dry weight over all collections); however,
copepod predation decreased steadily from 98% of total diet in
the smallest fish to 49% in the largest fish. All size classes
were moderately related by cluster analyses, but two broad groupings were noted: 1) 10-39 mm individuals, which consumed mainly
small food items, and 2) 40-69 mm individuals, whose diet reflected
increased consumption of large items such as mysids, insect larvae,
and small fishes. There was only minor spatial variability in
feeding patterns although some seasonal variation was found.
During the winter, copepod numbers in the bay were low and predation on other organisms such as mysids, insects, cladocerans,
and barnacle nauplii predominated.

Croakers are benthic omnivores and prey mainly on polychaetes
(32% mean dry weight of stomach contents over all collections).
Similarity analyses (Fig. 4) demonstrated two distinct size-
related feeding groups: 1) 10-69 mm individuals had a relatively
diverse diet which was dominated by polychaetes, detritus, and
insect larvae, and 2) 70-159 mm individuals tended towards
specialization on only a few prey types, mainly polychaetes,
infaunal shrimp (*Ogyrides limicola*), crabs, and/or juvenile
fishes. Further analysis demonstrated that temporal patterns in
croaker feeding habits corresponded to life history functions,
such as entry and eventual emigration from the estuary, and
spatial relationships of this species which related differences
in prey availability between shallow, oligohaline areas and deeper,
mesohaline sites (*Sheridan*, 1978).

Spot, also a benthic omnivore, primarily consumed polychaetes
and harpacticoid copepods (23% and 21% of stomach contents,
respectively, by dry weight over all collections). Three size-
related feeding categories were evident (Fig. 5): 1) 20-69 mm
individuals, whose diet was dominated by insect larvae, harpacticoid
copepods, and polychaetes, 2) 70-99 mm individuals, which fed mainly

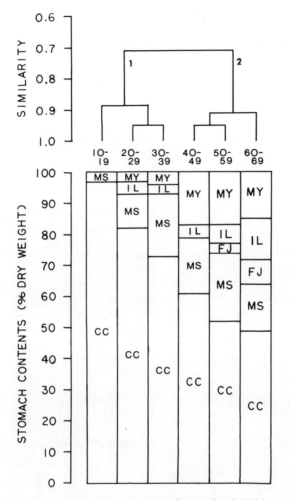

Fig. 3.   Feeding patterns of *Anchoa mitchilli*. Data
clustered by size class (mm) over all stations and dates with
food items indicated by two-letter codes (Table 1).

on detritus and harpacticoid copepods, and 3) 100-109 mm indi-
vididuals, which preyed mainly upon bivalves. Temporal variations
were not distinctive, but spatial analyses showed differences
between shallow, oligohaline stations and deeper, mesohaline
areas (*Sheridan*, 1978).

Young sand seatrout (10-39 mm) are epibenthic carnivores
and prey mainly on mysids (Fig. 6). This species became increasingly piscivorous with growth. Juvenile fishes comprised 65%
(by dry weight) overall of the sand seatrout diet, while mysids
made up an additional 26%. Of the fishes found in trout stomachs,

TABLE 1.   Alphabetical listing of two-letter codes for food items.

Code	Food Item
AM	Amphipods
BJ	Juvenile bivalves
BV	Bivalve veligers
CC	Calanoid copepods
CR	Juvenile crabs
CU	Cumaceans
CZ	Crab zoeae
DE	Detritus
FJ	Juvenile fish
FL	Fish larvae
HC	Harpacticoid copepods
IE	Invertebrate eggs
IL	Insect larvae
IS	Isopods
MY	Mysids
NE	Nematodes
PO	Polychaetes
SA	Sand grains
SH	Shrimp
SP	Shrimp postlarvae
SZ	Shrimp zoeae
MS	Miscellaneous, which includes all items individually amounting to less than 3% of the total stomach contents.

78% were identified as anchovies (*A. mitchilli*).   There was reduced fish consumption in late summer when the trout were in areas of reduced salinity.

Previous studies indicate that silver perch are epibenthic carnivores.   The young prey on copepods, amphipods and mysids while larger individuals eat shrimp and fishes (*Reid*, 1954; *Darnell*, 1958; *Springer and Woodburn*, 1960).   Menhaden are known to be filter feeders: this species consumes phytoplankton, small zooplankton and detritus (*Darnell*, 1958).

In order to determine the distinctiveness of the food habitats of each species, diet similarity among size classes and species was examined.   Of particular interest were croakers and spot, benthic omnivores with high temporal and spatial over-lap in utilization of the estuary.   Although polychaetes were the

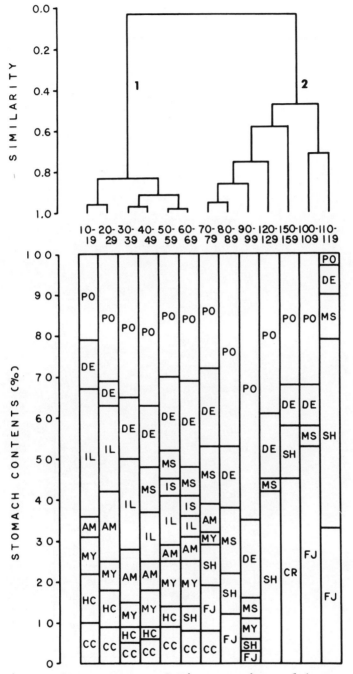

Fig. 4. Feeding patterns of *Micropogonias undulatus*. Data clustered by size class (mm) over all stations and dates with food items indicated by two-letter codes (Table. 1).

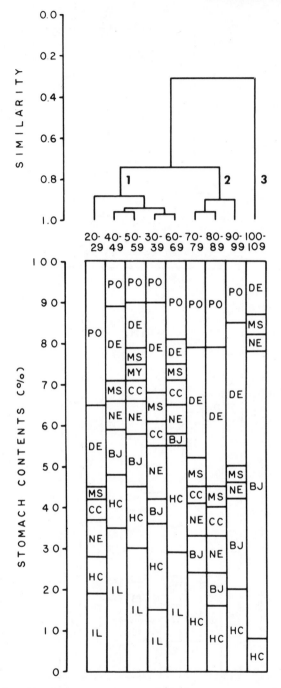

Fig. 5. Feeding patterns of *Leiostomus xanthurus*. Data clustered by size class (mm) over all stations and dates with food items indicated by two-letter codes (Table 1).

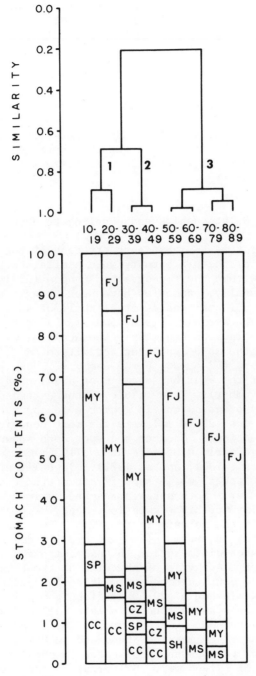

Fig. 6.   Feeding patterns of *Cynoscion arenarius*.   Data clustered by size class (mm) over all stations and dates with food items indicated by two-letter codes (Table 1).

main food for both species, quantities consumed were generally
different, averaging 32% in croakers and 23% in spot (by weight).
Secondary food items were distinct. This difference has been
associated with morphology (size) of the feeding apparatus and
behavior (*Roelofs*, 1954; *Chao and Musick*, 1977). Comparison of
the four dominant species in the Apalachicola estuary (Fig. 7)
demonstrated that although there were overlaps in diet components,
the individual feeding patterns were different, and each species
was charactized by a distinctive cluster. Thus, among species
which occur concurrently in this estuarine system, there is
considerable evidence of trophic resource partitioning. Feeding
differences are related to food particle size spectra (Fig. 8)
as well as to food type.

Life History Relationships and Coupling to Physical Processes

        The life history and abundance patterns of the dominant
fishes in the Apalachicola estuary were similar to known patterns
in other areas within their geographic ranges (*Livingston et al.*,
1977). Traditionally, anchovies have been thought to spawn from
spring through fall near passes or slightly offshore (*Gunter*,
1938; *Springer and Woodburn*, 1960; *Dunham*, 1972; *Hoese*, 1973).
However, ichthyoplankton studies in the Apalachicola estuary
have shown that the major spawning activity of A. *mitchilli* is
within the estuary (*Blanchet*, pers. comm.). Thus, anchovies are
taken throughout the year but occur most abundantly in the fall
(October and November). As a result of the extended spawning,
no clear monthly progressions in growth were observed. Instead,
mid-sized anchovies (20-40 mm) were the most conspicuous group
in monthly collections. Croaker and spot both spawn offshore
from fall through early winter (*Pearson*, 1928; *Gunter*, 1945;
Reid, 1954; *Perret*, 1971; *Hoese*, 1973). Juvenile croaker were
first noted in October, while spot first entered the Apalachicola
estuary in December. Both species reached maximum abundance in
March, and regular monthly growth increments were noted before
each species left the estuary in the fall. Sand seatrout spawn
offshore in spring and summer (*Gunter*, 1938, 1945) and exhibited
a bimodal abundance distribution in Apalachicola Bay, with peaks
in May and August. Analysis of the size of trout during these
periods indicated continued recruitment of juveniles, which
resulted in predominance of 20-50 mm fish from March through
October. Menhaden spawn offshore in late fall and winter
(*Suttkus*, 1956; *Springer and Woodburn*, 1960; *Hoese*, 1965;
Perret, 1971) and reach maximum abundance in March. The
majority (99%) of the menhaden collected in the Apalachicola
estuary were small (21-30 mm). Silver perch are thought to
spawn in or near estuaries in late spring (*Gunter*, 1938, 1945;
Reid, 1954; *Tabb and Manning*, 1961; *Moe and Martin*, 1965). This
species exhibited a bimodal abundance distribution, with peaks

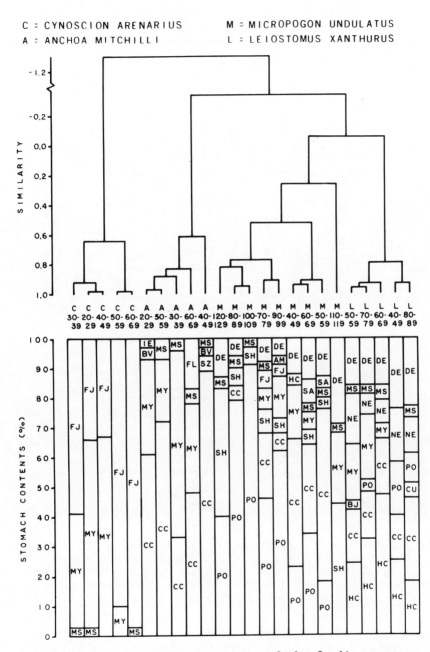

Fig. 7. Interspecific comparison of the feeding patterns of *Anchoa mitchilli* (A), *Cynoscion arenarius* (C), *Leiostomus xanthurus* (L), and *Micropogonias undulatus* (m) at stations 1A, 1C, and 1 (combined) in May, 1976. Food items are indicated by two-letter codes (Table 1).

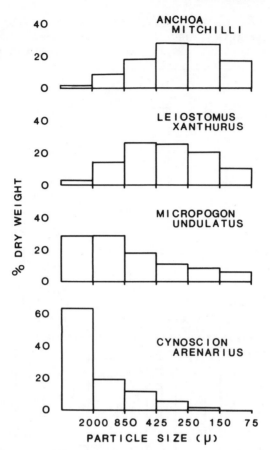

Fig. 8.   Food particle size spectra of *Anchoa mitchilli*,
Cynoscion arenarius, *Leiostomus xanthurus*, and *Micropogonias*
undulatus expressed as mean percent dry weight in each of six
particle size fractions. Data are averaged over all size
classes for each species.

in July and November. Analysis of monthly size distributions
shows that 30-50 mm fish peak in July while the November peak
is composed of fish greater than 100 mm in length.

The coupling of the trophic and distributional information
on these fishes with the chief forcing functions of the Apalachicola
estuary should be considered. There is a seasonal progression of
developmental stages of the dominant fishes which is associated with
the type and amount of food available in the estuary. The benthic
omnivores (croaker and spot) reach peak abundance at the precise
period (winter-early spring) when peak numbers of benthic fauna
occur. The planktivores are abundant in spring (menhaden) and

fall (anchovies); since neither of these maxima coincides with
the zooplankton peak of early summer it is possible that the
zooplankton peaks are in part associated with predator-prey
activities of these populations. The depression of the anchovy
population occurs in the summer during an influx of piscivorous
sand seatrout; this fact could reflect predation pressure on the
generally increasing anchovy population. When sand seatrout
leave the estuary in early fall, anchovies increase to a seasonal
maximum. The silver perch may initially exploit the zooplankton
and small epibenthos in the summer and switch to the larger
epibenthos in the fall. Such energy distributions are controlled
by winter/spring river flooding, summer phytoplankton blooms,
late summer/fall die-offs of benthic macrophytes, and, possibly,
changes associated with emergent (marsh) vegetation. Peak levels
of biological activity appear to be correlated with hydrological
events (river flooding, local rainfall patterns) and associated
physico-chemical phenomena (salinity, turbidity, dissolved
nutrients and organics). In this way temporal changes of the
trophic spectrum are determined by energy relationships within
the context of physical microhabitat gradients (temperature and
salinity) and biological interactions (predation, resource
partitioning). Such interaction of disparate factors would
help to explain the general lack of success of statistical
verification of direct association of population changes with
key physico-chemical functions and the failure of such models
to define and ultimately predict even the simplest biological
functions.

Future Research Directions

Studies of estuarine processes seem to be in the forefront
of current research, from viewpoints of both "purely" scientific
knowledge and economic/environmental impact. A great deal of
information now exists on abundance, distribution, life history,
resource utilization, and general ecological prerequisites of
estuarine organisms. At present, though, there is a paucity of
information concerning the link between estuaries and the continental shelf. It has long been known (or postulated) that many
estuarine fish populations actually breed in shelf areas. However, little is known concerning the functional aspects of such
processes. Where do the fishes spawn? Do the fishes share
common spawning areas, and at what depths does spawning occur?
What environmental and biological factors affect choice of
spawning sites, reproduction, and larval success? How do
the subadults and adults feed as they utilize the shelf? How
do these migrants interact with shelf residents? Estuarine
observations only document the aftermath of such processes.
Future research should be directed towards integrated investigations of these and other questions involving inshore/offshore

interactions, with particular attention to trophic phenomena and the relative importance of biological interactions in the determination of inshore communities.

ACKNOWLEDGMENTS

The authors acknowledge G. C. Woodsum, Jr., for his help with the computer programs used for this project. Portions of the data collection were funded by NOAA Office of Sea Grant, Department of Commerce, under Grant No. 04-3-158-43. Data analysis was supported by EPA Program Element No. 1 BA025 under Grant No. R-803339. Matching funds came from the people of Franklin County, Florida.

REFERENCES

Carr, W. E. S. and C. A. Adams, 1972. Food habits of juvenile marine fishes: evidence of the cleaning habit in the leatherjacket, *Oligoplites saurus*, and the spottail pinfish, *Diplodus holbrooki*, *Fishery Bull.* 70: 1111-1120.

Carr, W. E. S. and C. A. Adams, 1973. Food habits of juvenile marine fishes occupying seagrass beds in the estuarine zone near Crystal River, Florida, *Trans. Amer. Fish. Soc.* 102: 511-540.

Chao, L. N. and J. A. Musick, 1977. Life history, feeding habits, and functional morphology of juvenile sciaenid fishes in the York River estuary, Virginia, *Fishery Bull.* 75: 657-702.

Dahlberg, M. D. and E. P. Odum, 1970. Annual cycles of species occurrence, abundance, and diversity in Georgia estuarine fish populations, *Amer. Midl. Nat.* 83: 382-392.

Darnell, R. M., 1958. Food habits of fishes and larger invertebrates of Lake Pontchartrain, Louisiana, an estuarine community, *Publ. Inst. Mar. Sci. Univ. Texas* 5: 353-416.

Dunham, F., 1972. A study of commercially important estuarine-dependent industrial fishes, Tech. Bull. No. 4, 63 pp. La. Wildl. Fish. Comm.

Edmiston, H. L., 1979. *The Zooplankton of Apalachicola Bay*, M. S. Thesis, Florida State University.

Estabrook, R. H., 1973. *Phytoplankton Ecology and Hydrography of Apalachicola Bay*, M. S. Thesis, Florida State University.

Gallaway, B. J. and K. Strawn, 1974. Seasonal abundance and distribution of marine fishes at a hot-water discharge in Galveston Bay, Texas, *Contrib. Mar. Sci. 18:* 71-137.

Gunter, G., 1938. Seasonal variations in abundance of certain estuarine and marine fishes in Louisiana, with particular reference to life histories, *Ecol. Monogr. 8:* 313-346.

Gunter, G., 1945. Studies on marine fishes of Texas, *Publ. Inst. Mar. Sci. Univ. Texas 1:* 1-90.

Haedrich, R. L. and S. O. Haedrich, 1974. A seasonal survey of the fishes in the Mystic River, a polluted estuary in downtown Boston, Massachusetts, *Est. Coastal Mar. Sci. 2:* 59-73.

Hoese, H. D., 1965. *Spawning of Marine Fishes in the Port Aransas, Texas, Area as Determined by the Distribution of Young and Larvae,* Ph.D. Dissertation, University of Texas.

Hoese, H. D., 1973. A trawl study of the nearshore fishes and invertebrates of the Georgia coast, *Contrib. Mar. Sci. 17:* 63-98.

Iverson, R. L. and V. B. Myers, 1976. Phytoplankton ecology of Apalachicola Bay, Florida, In: *Energy Relationships and the Productivity of Apalachicola Bay,* edited by R. J. Livingston, R. L. Iverson, and D. C. White, Florida Sea Grant Technical Paper, June, 1976.

Kravitz, M. J., W. G. Pearcy, and M. P. Guin, 1976. Food of five species of co-occurring flatfishes on Oregon's continental shelf, *Fishery Bull. 74:* 984-990.

Lance, G. N. and W. T. Williams, 1967. A general theory of classificatory sorting strategies. I. Heirarchical systems, *Computer J. 9:* 373-380.

Livingston, R. J., 1976. Diurnal and seasonal fluctuations of organisms in a north Florida estuary, *Est. Coastal Mar. Sci. 4:* 373-400.

Livingston, R. J., 1978. Effects of clearcutting and upland deforestation activities on water quality and the biota of the Apalachicola estuary (North Florida, U. S. A.). Final Report to Florida Sea Grant College.

Livingston, R. J., R. L. Iverson, R. H. Estabrook, V. E. Keys, and J. Taylor, Jr., 1974. Major features of the Apalachicola Bay system: physiography, biota, and resource management. *Florida Scientist 37:* 245-271.

Livingston, R. J., R. L. Iverson, and D. C. White, 1976a. Energy
    relationships and the productivity of Apalachicola Bay.
    Florida Sea Grant Technical Paper, June, 1976.

Livingston, R. J., G. J. Kobylinski, F. G. Lewis, III, and
    P. F. Sheridan, 1976b. Long-term fluctuations of epibenthic
    fish and invertebrate populations in Apalachicola Bay, Florida,
    Fishery Bull. 74: 311-321.

Livingston, R. J. and O. L. Loucks. Productivity, trophic
    interactions, and food web relationships in wetlands and
    associated systems. Proceedings of the American Water
    Resources Association National Symposium on Wetlands, Lake
    Buena Vista, Florida, November, 1978. In press.

Livingston, R. J., P. F. Sheridan, B. G. McLane, F. G. Lewis, III,
    and G. J. Kobylinski, 1977. The biota of the Apalachicola
    Bay system: functional relationships, In: *Proceedings of
    the Conference on the Apalachicola Drainage System,* edited
    by R. J. Livingston and E. A. Joyce, Jr., 75-100, Fla. Dept.
    Nat. Res., Mar. Res. Publ. 26.

Livingston, R. J., N. P. Thompson, and D. A. Meeter, 1978.
    Long-term variation of organochlorine residues and assem-
    blages of epibenthic organisms in a shallow north Florida
    (USA) estuary. *Mar. Biol. 46:* 355-372.

McEachran, J. D., D. F. Boesch, and J. A. Musick, 1976. Food
    division within two sympatric species-pairs of skates
    (Pisces: Rajidae). Mar. Biol. 35: 301-317.

McErlean, A. J., S. G. O'Connor, J. A. Mihursky, and C. I. Gibson,
    1973. Abundance, diversity and seasonal patterns of estuarine
    fish populations, *Est. Coastal Mar. Sci. 1:* 19-36.

Matusita, K., 1955. Decision rules based on the distance for
    problems of fit, two samples and estimation, *Ann. Math.
    Statist. 26:* 631-640.

Moe, M. A., Jr., and G. T. Martin, 1965. Fishes taken in monthly
    trawl samples offshore of Pinellas County, Florida, with new
    additions to the fish fauna of the Tampa Bay area, *Tulane
    Stud. Zool. 12:* 129-151.

Oviatt, C. A. and S. W. Nixon, 1973. The demersal fish of
    Narragansett Bay: an analysis of community structure,
    distribution and abundance, *Est. Coastal Mar. Sci. 1:*
    361-378.

Pearson, J. C., 1928. Natural history and conservation of the redfish and other commercial sciaenids on the Texas coast, *Bull. U. S. Bur. Fish. 44:* 129-214.

Perret, W. S., 1971. Cooperative Gulf of Mexico estuarine inventory and study, Louisiana. Phase IV, Biology, 175 pp., La. Wildl. Fish. Comm., New Orleans.

Reid, G. K., Jr., 1954. An ecological study of the Gulf of Mexico fishes in the vicinity of Cedar Key, Florida, *Bull. Mar. Sci. Gulf Caribb. 4:* 1-94.

Roelofs, E. W., 1954. Food studies of young sciaenid fishes, *Micropogon* and *Leiostomus,* from North Carolina, *Copeia* 1954: 151-153.

Sheridan, P. F., 1978. *Trophic Relationships of Dominant Fishes in Apalachicola Bay (Florida),* Ph.D. Dissertation, Florida State University.

Springer, V. G. and K. D. Woodburn, 1960. An ecological study of the fishes of the Tampa Bay area. Mar. Res. Lab. Prof. Papers Ser. 1. 104 pp., Fla. Dept. Nat. Res.

Stickney, R. R., G. L. Taylor, and R. W. Heard, III, 1974. Food habits of Georgia estuarine fishes. I. Four species of flounders (Pleuronectiformes: Bothidae), *Fishery Bull. 72:* 515-525.

Stickney, R. R., G. L. Taylor and D. B. White, 1975. Food habits of five species of young southeastern United States estuarine Sciaenidae, *Chesapeake Sci. 16:* 104-114.

Suttkus, R. D., 1956. Early life history of the Gulf menhaden, *Brevoortia patronus,* in Louisiana, *Trans. N. Amer. Wildl. Conf. 21:* 309-407.

Tabb, D. C. and R. B. Manning, 1961. A checklist of the flora and fauna of northern Florida Bay and adjacent brackish waters of the Florida mainland collected during the period July, 1957, through September, 1960, *Bull. Mar. Sci. Gulf Caribb. 11:* 552-649.

Tyler, A. V., 1972. Food resource division among northern, marine demersal fishes, *J. Fish. Res. Bd Canada 29:* 997-1003.

Van Belle, G. and I. Ahmad, 1974. Measuring affinity of distributions, In: *Reliability and Biometry: Statistical Analysis of Life Length,* edited by F. Proschan and R. J. Serfling, 651-668, S. I. A. M., Philadelphia.

SECONDARY PRODUCTION MECHANISMS OF CONTINENTAL SHELF COMMUNITIES

Lawrence R. Pomeroy

The University of Georgia

ABSTRACT

A compartmental model of energy flux through a continental shelf ecosystem is presented which examines the potential for significant energy flow through detritus, microorganisms, and dissolved material to terminal consumers. A number of assumptions are examined which have bearing on the outcome of the modeling exercise. Evidence is reviewed that primary production may be substantially higher than the usual ^{14}C method indicates. If this were the case there would be great latitude possible for the other assumptions inherent in the model, but for the present it is assumed that current estimates of photosynthesis in the sea are approximately correct. Evidence is presented that the input of primary detritus (new plant material) is a significant one, although relatively little of it is present in samples of particulate matter. Secondary detritus, principally fecal material, takes many forms and is involved in both benthic and pelagic food webs. The observations that bacteria mediate the utilization of detritus by metazoans, but that there are very few bacteria on detritus particles, are not necessarily in contradiction. However, little is known about the rate of growth of bacteria on particles and the frequency with which particles pass through the guts of detritivores. Similar information is also lacking about the population of free-living bacteria in the water.

Since there is little agreement about the significance of energy flux through detritus, dissolved material, and microorganisms, a model was constructed so that all, one, or none of them could move a major fraction of the energy fixed by primary production. If the steps in such a model are assumed to be trophic levels, and the ecological efficiency of transfer of energy is assumed to be 10% per trophic level, such an anastomosing model will not carry enough energy to the terminal consumers. If we accept the proposition that the compartments in such a model do not fit the concept of trophic levels, and if we use existing data on gross growth efficiencies to estimate the efficiency of transfer of energy between the compartments, any of the several pathways can be a major energy conduit. However, large changes in energy flux from one pathway to another result in changes in the amount of energy available to various terminal consumer groups. The results indicate that either conventional assumptions about ecological assimilation efficiency are low by a factor of 2-3, or current measurements of photosynthesis are low by a factor of 5-10. While this modeling exercise in no way proves that alternative energy pathways are important in real continental shelf ecosystems, it shows that there is no underlying reason why they may not be. Situations in the real world are described which may result in shifts of energy flux from direct consumption of phytoplankton by grazers to consumption of microorganisms which are growing on detritus and dissolved materials.

INTRODUCTION

The transfer of energy through marine communities has been a concern of scientists for a century. It is a fundamental problem in ecosystem function, as well as a matter of concern in the management of fisheries. Models of energy transfer in marine communities were developed before the rise of computer simulation techniques. Even without simulation, models can tell us much about the limits and possibilities for growth of a community, provided we have some fundamental information about the trophic structure, energy input, and efficiency of transfer of energy through the system. Early models assumed a simple, linear food chain of the *Lindeman* (1942) type, consisting of diatoms, copepods, benthos, and fishes (*Clarke*, 1946; *Riley et al.*, 1949), and the acceptance of this type of model persists. The compartments of such models generally are equated with trophic levels, and ecological transfer efficiencies (*Slobodkin*, 1960, 1961) usually are applied to evaluate energy flux. *Ryther* (1969) attempted in this way to show how the production of fishes is limited by the number of transfers of energy. He suggested that the high production of fishes typical of upwellings was the result not only of a high rate of primary production but also of a short food chain. *Steele* (1974) developed a bifurcated food chain model of the North Sea in which fecal pellets from the copepods transfer energy to the benthos and the demersal fishes. His use of the energy in feces from zooplankton to support the benthos

and demersal fishes is a departure from the *Lindeman* (1942) model
in which decomposers serve no energetic role. The pelagic portion of his model, leading to pelagic fishes, is essentially the
same as that of Clarke. While Steele did develop a simulation
model, the results of his computer runs appear to be of less
interest than his choice of compartments and his assumptions
about the structure and the transfer of energy. In this paper
alternative compartmental structures are developed and alternate
assumptions about energy transfer are examined.

Simple, highly aggregated models have been satisfying and
widely accepted, not only because of their simiplicity but also
because the numbers attached to them usually indicate a sufficient,
albeit minimal, transfer of energy to fishes. Since there is no
excess energy at terminal consumer levels in the existing models,
it has been intuitively troublesome to see how a more complex
model with additional energy transfer could satisfy the fundamental criterion of supporting the consumers. In the real world
there are other energy pathways involving such entities as microorganisms, detritus, and dissolved organic materials (DOM). In
nearly every case, modelers have dismissed these as quantitatively
trivial or unproven. However, the major evidence offered is the
null hypothesis that longer food chains cannot work because not
enough energy reaches the end. Other lines of evidence suggest
that alternate pathways not only exist but may sometimes dominate
marine food webs (*Pomeroy*, 1974). A model of energy flux on
continental shelves is presented here which is based on the
hypothesis that there are alternate energy pathways in marine
food webs, and that fluxes may vary substantially through one
or another. The success of the model depends, however, on
reconsideration of several basic assumptions: the amount and
nature of primary photosynthate, the production of primary
detritus, the fate of fecal material, the roles of the microbial
community, and the efficiency of the transfer of energy through
the system.

THE SUPPLY OF PRIMARY PHOTOSYNTHATE

A large body of data has been accumulated on the rates of
photosynthesis in the ocean, including the continental shelves
(*Koblentz-Mischke et al.*, 1970; *Finenko*, 1978). Over the years
there have been occasional suggestions that the method may be in
error (*Odum*, 1967; *McAllister*, 1969; *Sieburth*, 1976, 1977).
Sheldon and Sutcliffe (1978) present evidence that the ^{14}C method
as it is commonly used may underestimate photosynthesis by an
order of magnitude, especially in mid to low latitudes at times
when the dominant phytoplankton populations consist of small
organisms having generation times of less than six hours. If
grazers are not excluded from bottles in which the rate of

photosynthesis is measured, much of the fixed photosynthate will
pass along the food chain, with substantial recycling of CO_2.
This conclusion is implicit in the findings of *Beers and Stewart*
(1971), although they did not point out that protozoan and small
metazoan grazers (<100 or 200 μm) are indeed present in the
bottles in which photosynthesis is measured. Sheldon and
Sutcliffe propose that the excess primary production which they
detect is part of a "closed system," which is not available to
terminal consumers. However, the grazers are most likely
ciliates and nauplii, with gross growth efficiencies of 30-50
per cent. Therefore these findings, if substantiated by further
research, may in fact have important implications for our paradigm of marine food webs.

PRODUCTION OF PRIMARY DETRITUS

In order to differentiate some pathways in the food web, it
is necessary to distinguish between primary and secondary detritus.
The term *primary detritus* is used here to designate dead plant
material which has never been eaten by a grazer. Observations of
particulate matter in the ocean always reveal a population of
dead, usually broken phytoplankton (*Gordon,* 1970a, b; *Riley,*
1970; *Wiebe and Pomeroy,* 1972). There is no reason to believe
that this population is the result of collection methods,
particularly since nets are not involved. *Marshall and Orr*
(1962) estimated that 25% of the phytoplankton was broken but
not eaten by copepods under experimental conditions. Therefore,
a portion of the primary detritus probably is the result of
inefficient or unsuccessful feeding by zooplankton. *Cushing*
(1959) says that *Calanus* destroys more than it eats, but he
does not make a clear distinction between unsuccessful feeding
and superfluous feeding. *Porter* (1977) has shown that some
species of phytoplankters, because they are noxious or difficult
to eat, are consumed only by a few specialized zooplankters.
These prove to be the organisms commonly forming phytoplankton
blooms, such as some of the blue-green algae and some dinoflagellates. While this is not the only cause of bloom formation,
it appears that some phytoplankton blooms are the result of long-
term, relatively slow growth of a phytoplankter in the absence
of a grazer that will accept it. *Roman* (1978) has shown that
marine *Oscillatoria (Trichodesmium)* is eaten by the harpactacoid
copepod *Macrostella gracilis,* but evidently by little else.
Because of their limited acceptability, blooms of dead or dying
Oscillatoria are frequently found on the continental shelf off
the southeastern U. S. *Lasker* (1973) has shown that the dinoflagellate *Gymnodinium splendens* is important in the early diet
of larval anchovies. Similar relationships may exist between
other dinoflagellates and larval fishes. However, dinoflagellate
blooms sometimes persist and remain uneaten to the point of their

death and disintegration (*Pomeroy et al.*, 1956). Despite the
occasional persistence of phytoplankton blooms and the less than
perfect feeding efficiency of zooplankton, primary detritus is
not ordinarily a major component of the standing stock of particulate organic matter in the ocean. Although a substantial
amount may be produced, it is quickly consumed, but probably
by bacteria rather than by grazers, for it is now a part of the
detritus food web.

Models of estuarine systems usually contain a primary
detritus component because those systems contain macrophytes
which are not efficiently consumed by direct grazing (*Miller
et al.*, 1971; *Wiegert et al.*, 1975). In most terrestrial and
estuarine systems, 90% of macrophyte production becomes primary
detritus. However, primary detritus has been omitted from previous condensed models of oceanic systems on the grounds that
grazing efficiency of zooplankton consuming phytoplankton
approaches 100% (*Steele*, 1974). Both field observations on the
relative standing stocks of phytoplankton and grazing zooplankton
and laboratory observations on feeding rates of zooplankton
support the postulate that most phytoplankton populations are
overgrazed and are limited at least as much by grazing as by
the availability of nutrients. But grazing cannot approach
100% efficiency. It seems reasonable to conclude that while
the production of primary detritus on continental shelves is
probably substantially less than the 90% of primary production
found in estuarine and terrestrial systems, it is probably substantially greater than zero, perhaps approaching 25% most of
the time and exceeding 50% in late stages of some phytoplankton
blooms.

THE FATE OF FECAL MATERIAL

Fecal material, which is one component of secondary detritus,
may be consumed by the plankton or the benthos, depending upon the
depth of the water and composition of the zooplankton which produces it. Planktonic crustacea usually produce discrete fecal
pellets encased in a tube of chitin. Our own observations of
fecal pellets of adult copepods collected aboard ship indicate
that the pellets remain intact for about two days. Therefore,
the pellets would likely fall to the bottom on continental shelves,
except for the fraction intercepted by detritus feeders. Feces of
other plankters lack a chitinous envelope and disintegrate more
rapidly into small particles of low settling velocity. They can
be expected to remain in the water column for a considerable time,
long enough to develop an indigenous community of bacteria and
protozoans. Particles of this kind probably are those which have
been shown to increase in per cent nitrogen over time (*Odum and
de la Cruz*, 1967) as bacteria scavenge nitrate or ammonia from

the water to supplement nitrogen-deficient dissolved and particulate growth substrates. After some time in the water column such
particles and their flora are a nutritionally good source of food
for filter feeders. Because most such particles are 5-50 μm in
diameter (*Lenz*, 1972), they are potentially accessible to most
grazing, filtering, and net casting populations.

The proportions of fecal material going to the plankton and
the benthos will vary according to the composition of the zooplankton, water depth, and turbulent transport. Where crustacean
zooplankton dominate in terms of the rate of grazing, there will
be a substantial transfer of fecal material to the bottom on
continental shelves. Once on the bottom, this material forms
the basis for an extensive food web involving bacteria and other
microbenthos, meiobenthos, and macrobenthos. We know less about
the fate of other forms of fecal material which remain in suspension in the water column. When seen with the microscope, the
particles are amorphous, chemically diverse, and indistinguishable
from organic aggregates. Amorphous particles, which may be
aggregates or feces, make up a substantial part of the standing
stock of particulate organic matter in most ocean water samples.
However they must be removed by degradation (*Wangersky*, 1977) and
are a source of energy for some part of the food web which
processes detritus. It is not certain what organisms use this
energy in the real world, but other trophic groups, such as
mucus-net feeders, merit consideration both as consumers and as
producers of fecal material and discarded houses (*Aldredge*, 1976)
which remain in the water column. They are included in the model
as potential recipients of energy from detritus.

THE MICROBIAL COMMUNITY ON PARTICLES OF DETRITUS

Detritus particles are flakes or flocculent aggregations
which may originate from fecal material, mucus, or the condensa-
tion of dissolved organic materials. They are sufficiently
amorphous at this stage of degradation to be of ambiguous origin,
with few identifiable organismal fragments. In most parts of the
ocean the typical particle has a sparse community of bacteria and
protozoans, at most one or two active bacteria and an occasional
non-pigmented flagellate. In situations of exceptional productivity, bacteria and heterotrophic flagellates sometimes
are more abundant (*Pomeroy and Johannes*, 1968; *Wiebe and Pomeroy*,
1972).

The paucity of the community on particles in the ocean may
not be indicative of its rate of production. On particles freshly
taken from the ocean and held at *in situ* temperature, after a lag
period of about 12 hours bacteria appear, grow, and reproduce
(*Wiebe and Pomeroy*, 1972). Within 24-48 hours, bacteria have

overgrown the particle and are being consumed by a rising population of flagellates and sometimes ciliates. If this is representative of what happens in the ocean, virtually all particles must have passed through the gut of a filter feeder on a daily basis, where they were stripped of microbial flora. The observations of *Sheldon and Sutcliffe* (1978), previously mentioned, suggest that a high grazing rate on nannoplankton and, by inference, on detritus and free bacteria, is possible.

The standing stock of non-living particulate organic material is several times that of living phytoplankton in most water samples, but that is not necessarily a good indication of its relative food value. The transformation into useful bacterial protoplasm takes some time and probably several gut passages. Particulate organic matter, like litter in a forest, is a part of the environment as well as the food source of microorganisms. Therefore, it cannot be totally consumed without destruction of the environment (*Wiegert and Owen*, 1971), and its consumption is dependent on passage through the guts of detritus feeders as well as on the attack of microorganisms. Therefore, the presence of a standing stock does not indicate that detritus goes unused. Production and consumption rates are not necessarily a function of the size of the standing stock and must be estimated independently. That has not been done. As *Wangersky* (1977) has pointed out, we cannot assume, as many have, that most particulate organic matter, like most dissolved organic matter, is not readily available as energy to the planktonic and benthic communities.

Protozoa are often suggested as a link between bacteria and larger consumer organisms. Using an elaborate and unusual collecting system, *Beers and Stewart* (1967, 1970, 1971) demonstrated the presence of substantial populations, especially of non-loricate ciliates, both in coastal and in oceanic waters. They suggested that ciliates have a major impact on nannoplankton populations, but it is reasonable to expect that the ciliates will also graze both free and attached bacteria. To the extent that this grazing occurs, the efficiency of transfer of energy from bacteria to larger organisms will be reduced, while the regeneration of nutrients for phytoplankton will be accelerated. The major deficiency in our knowledge of the particle community, however, remains the rate of production of bacteria. If they are indeed growing at the rate found by *Sorokin* (1978) and by *Sieburth* (1976; *Sieburth et al.*, 1977), on the order of 40 mg C m^{-3} day^{-1}, as much energy is flowing through bacteria as through zooplankton. Even if their rate of production is but a fraction of that suggested, bacteria should be included as secondary producers in models of marine ecosystems.

FREE-LIVING BACTERIA IN THE WATER COLUMN

In the ocean the standing stock of free-living bacteria in the water is larger than that attached to particles (*Wiebe and Pomeroy*, 1972; *Azam and Hodson*, 1977). Although the free bacteria are recognizable morphologically and by acridine orange fluorescence, they do not appear to be dividing as actively as bacteria associated with particles. We have no separate estimates of the growth rates of free and attached bacteria, and we can only infer that in the ocean and most outer continental shelf water the free bacteria are limited either by soluble carbon sources or by available nitrogen. If they are indeed growing slowly, the relative abundance of free bacteria can be explained by the refuge value of small size. Single bacteria free in the water can be removed effectively only by mucus-net feeders. Therefore, although they are more abundant than bacteria attached to particles, the rate of production of free bacteria may be lower, and their routing through the food web may be distinct from that of the populations on particles. Therefore, we must clearly separate these populations in food web models. Moreover, we must always keep in mind that the size of the standing stock of a population is not necessarily indicative of its metabolic activity or reproductive rate. Ultimately we need generally accepted estimates of the rates of production of both free and attached bacteria in the ocean.

THE FOOD WEB

Alternate pathways in the marine food web are quite generally acknowledged to exist, but there is not agreement on their relative potential for energy flux. Estimates of the microbial portion of oxygen consumption in the sea range from less than 10% (*Steele*, 1974) to 50% (*Riley*, 1972), and as high as 90% (*Pomeroy and Johannes*, 1966; *Joiris*, 1977). The flux of energy through DOM is variously believed to approach 50% of total photosynthate (*Andrews and Williams*, 1971; *Derenbach and Williams*, 1974) or to be trivial (*Steele*, 1974). This question may be complicated by the multiplicity of chemical species of DOM. Clearly, most of the standing stock of DOM in the ocean is highly refractory and is not a significant part of the food web (*Williams et al.*, 1969). It may be slowly accumulating, despite *Wangersky's* (1977) assertion that it must be in steady state. The labile, low-molecular-weight compounds seldom accumulate to concentrations of more than a few tens of micrograms per liter. These are significant, however, provided their rate of production is a substantial fraction of total primary production, and that remains a subject of almost polemical disputation in the literature.

Given such uncertainties about the alternate pathways in marine food webs, it is understandable that modelers have limited their attention to the best-defined pathways. A model of alternate pathways cannot tell us which are significant ones in the real world, but it can help us define the upper and lower limits to energy flux through them, and in this way it can shape future research. The model presented is a generalized one of a continental shelf food web upon which various special features can be imposed. It is a highly aggregated model, and all elements represented can be expected to be present to a greater or lesser degree in all shelf waters. The model is not of the linear, Lindeman type. Rather, it follows the structural concept of *Wiegert and Owen* (1971) in which heterotrophs are biophages, saprophages, or a combination of both. Since there may be parallel levels of biophages and saprophages, the trophic level concept is no longer tenable, and attempts to estimate energy degradation by trophic level sequence are futile.

Since the model is generalized, the rate of primary production is set arbitrarily at 1000 KCal m^{-2} yr^{-1}. That is realistic if we accept the ^{14}C measurements as approximately correct. This arbitrary value makes it possible for us to follow by inspection the percentages of primary production which remain at various points in the web. The pathways in *Steele's* (1974) model are a subset of the present model (the upper half, Fig. 1). However, not all energy now passes down that chain. Instead, variable amounts of energy may pass through primary detritus or fecal detritus to populations of bacteria. Some fraction of both phytoplankton and bacteria is utilized by mucus-net feeders. Some fraction of primary photosynthate is released in dissolved form, and will be utilized by bacteria. The bacteria are operationally divided into three communities: those attached to particles in suspension, those free in the water column, and those associated with fecal material on the bottom. If the latter two are ever significant pathways, it is important to separate them, as we shall see when a few simple variations in energy flux are performed.

There is much that a highly aggregated model cannot do. Events are time averaged and space averaged. Therefore, it cannot tell us about local concentrations of organisms or the development of cohorts or populations. It is sometimes argued that, because populations impose their properties on ecosystems, realistic models are impossible without detailed treatment of the dynamics of species populations. However, populations are a different hierarchial level of aggregation. What is fundamentally important in an ecosystem model is to separate levels of function in a realistic and pragmatic way. It seems unlikely that ecosystem function is controlled primarily by population interactions. Dramatic failures of species populations occur, as in the cases of

the El Niño phenomenon or the dynamics of the Pacific sardine
(*Lasker*, 1977), but there is no reason to believe that those
changes are necessarily reflected in net energy flux. Realistically, ecosystem-level models must be limited to the structure
and function of guilds or more generalized feeding groups. Much
of the controversy surrounding ecosystem modeling arises from a
basic misconception of what such models can be expected to include
or to show. They are useful principally as heuristic devices for
the study of system properties and should not usually be viewed
as a means of prediction or decision-making, or to show the
interaction of species populations.

Some of the uncertainties in the detritus food web become
particularly apparent when we attempt to model it. It has been
assumed that the assimilation of detritus by metazoans must be
facilitated by bacteria and protozoans in a sort of external-
rumen process in which relatively refractory materials are
assimilated by enzymatically versatile bacteria that are then
consumed by protozoans or small metazoans. There appears to be
no clear demonstration of direct utilization of naturally occurring detritus by a metazoan without bacterial mediation.
Therefore, the detritus pathway in the model is represented
as a bacterially mediated one, although some direct utilization
cannot be ruled out and efficiency would improve if it did occur.

ASSIMILATION EFFICIENCY

Assimilation efficiency imposes stringent limitations on
performance of the system. This fact has been recognized for
many years, but there is some lack of agreement both about the

---

Fig. 1. Energy flux model for continental shelves. Values
directly beneath each compartment are state variables (standing
stocks) in KCal m^{-2}. Energy flux values, in KCal m^{-2} yr^{-1}, are
beneath or beside arrows. Compartments are numbered (large
numerals). To simplify the diagram, some fluxes are designated
by an arrow followed by the compartment number. Gross growth
efficiencies are: bacteria, 50%; mucus-net feeders, 17%;
grazing zooplankton, 19%; benthic invertebrates, 33%; carnivorous zooplankton, 13%; fishes and other carnivores, 10%.
In (a) grazing zooplankton receive all but a minor part of the
primary production. In (b) most phytoplankton production is
nannoplankton, and is utilized largely by mucus-net feeders.
In (c) the late stage of a phytoplankton bloom finds remaining
species largely inedible, production of dissolved organic
material is substantial, and detritus is exploited by benthic
invertebrates and mucus-net feeders. ⟶

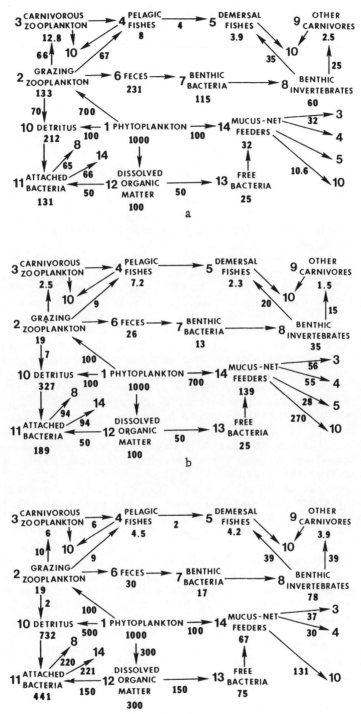

trophic efficiency of populations in nature and about how to
represent efficiency properly in a food web model. Most
previous models of marine systems are based on the assumption
that the compartments represent trophic levels, and that ecological assimilation efficiencies should be used. That problem
has been compounded by the assumption that all ecological
assimilation efficiencies (between trophic levels) are close
to 10%. This assumption apparently originated from a misinter-
pretation of a very limited data set by *Slobodkin* (1960, 1961).
Both *Gulland* (1970) and *Steele* (1965) have made it clear that
the assumption of a 10% ecological efficiency between compartments
in the simplest food webs will hardly yield known fish production.
Slobodkin more recently (1970) remarked that the extant data do
not support the use of an assumed 10% ecological efficiency in
marine food webs. In fact the efficiency is highly variable
(Table 1), and lack of detailed knowledge of it is a major
obstacle to modeling.

    In the present model, classically defined trophic levels,
if they exist at all, are obscured by omnivory and microbial
processes, and the most realistic values to assign to energy
transfers between compartments are those of gross growth
efficiency, that is, energy from one compartment converted into
new tissue in the next. There is a body of experimental and
observational data which gives us an idea of the gross growth
efficiency of most of the compartments in the model (Table 1).
As *Reeve* (1970) has already observed, a number of detailed
laboratory estimates of gross growth efficiency fall in the
vicinity of 30%. Estimates based principally on field methods
(*Petipa et al.*, 1970) tend to be lower. One might expect adults
to be less efficient than rapidly growing juveniles, and while
that is dramatically true for chaetognaths (*Reeve*, 1970), it is
not always true for copepods (*Corner and Cowey*, 1968). Efficiencies are thought to be greater at high latitudes, where the mean
water temperature is lower, but that is not the case. While an
organism which passes from warm to cold water during a day's
time, as in vertical migration, may experience a significant
change in metabolic rate (*McLaren*, 1963), local races of species
populations grown for generations at widely differing temperatures tend to have similar metabolic rates (*Rao*, 1953). Certain
groups of organisms tend to have relatively high gross growth
efficiencies. Actively growing bacteria are reported to have
efficiencies of 60% (*Payne*, 1970; *Ho and Payne*, 1979). At the
other extreme, some adult metazoans, growing little and putting
much energy into reproduction, may even show negative efficiency
values (*Reeve*, 1970). Pelagic fishes, such as sardines, have
an efficiency over their lifetime of less than 10% (*Lasker*,
1970), while sedentary plaice achieve about 15% (*Hempel*, 1970).
For a condensed food web model the most valuable datum is an
estimate of median gross growth efficiency for the entire life

TABLE 1. Gross growth efficiency of species populations, based on field and laboratory observations.

Organism	gross growth efficiency, %	Source
Marine bacteria	30-80	*Payne, 1970; Ho & Payne, 1979*
Oikopleura dioica	23-50*	*Paffenhöfer, 1975*
Oikopleura sp.	4-16	*Petipa et al., 1970*
Sagitta hispida, subadult	25-50	*Reeve, 1970*
Sagitta setosa	9-18	*Petipa et al., 1970*
Calanus hyperboreus, adult	30	*Conover, 1962*
Calanus hyperboreus, adult	13-18	*Corner and Cowey, 1968*
Calanus hyperboreus, stage IV-V	15-50	*Corner and Cowey, 1968*
Calanus finmarchicus, adult	14	*Corner and Cowey, 1968*
Calanus finmarchicus, subadult	25-50	*Corner and Cowey, 1968*
Rhinocalanus nasutus, nauplius to adult	25-55	*Mullin and Brooks, 1970*
Calanus helgolandicus, nauplius to adult	18-72	*Mullin and Brooks, 1970*
Euphausia pacifica, life cycle	30	*Lasker, 1970*
Nereis diversicolor, adult	14-43	*Ivleva, 1970*
Leander adspersus	1-10	*Ivleva, 1970*
Mactra sp., 1 & 2 yr.	55	*Hempel, 1970*
Asterias sp.	55	*Hempel, 1970*
0-stage plaice	30	*Edwards et al., 1969*
Plaice whole life history	15	*Hempel, 1970*
Pacific sardine, 6 yr. life	7	*Lasker, 1970*

*Calculated from the data given

history of a population. It would be helpful to have several of
these in each trophic group in the model, but in fact we are
usually fortunate to have one. At present there is no open sea
or continental shelf community for which we know the gross growth
efficiency of most of the dominant organisms. At the same time,
we must avoid the pitfall of assuming some standard value, in view
of the variability which is known to exist.

AN ENERGY FLUX MODEL FOR CONTINENTAL SHELVES

The generalized model of energy flux through continental
shelf communities has been assembled using data taken worldwide.
While this fact obviously introduces uncertainty, the extant data
suggest that variations within one population or one community as
a result of changing local conditions are as great as geographical
variations. The model includes both pelagic and benthic communities and incorporates both direct grazing and detritus energy
pathways (Fig. 1). The compartments are neither guilds nor
trophic levels, but trophic groups. This model may be compared
with earlier ones which lacked the detritus pathway by assigning
that part of the model zero values and comparing the result with
values greater than zero. Of course there must be some detritus,
and the question is one of relative significance. Therefore, flux
through the detritus pathway can be varied and different assumptions about its fate at higher levels tested. Another feature of
this model is the division of zooplankton into grazers and mucus-
net feeders. The mucus-net feeders are usually of some importance
and occasionally dominant (*Paffenhöfer*, 1975). While they may
compete with grazers for some types of food, they are probably
the major consumers of nannoplankton and bacteria, both attached
and free in the water. On the other hand, mucus-net feeders
cannot as a rule ingest large organisms (*Paffenhöfer*, 1975). If
large organisms are captured, they will be rejected and may clog
the feeding mechanism. Therefore, there is interplay between
grazers and mucus-net feeders, depending on the rate of production
of nannoplankton versus larger phytoplankton and also depending
on the production of bacteria.

Initially the model was tested with efficiencies similar to
those used in *Steele's* (1974) model of energy flux in the North
Sea to ascertain the effect of the addition of detrital and DOM
pathways. Nominal amounts of energy flux were assigned to DOM
production, primary detritus production, and to consumption of
nannoplankton by mucus-net feeders. Then, each of these in turn
was assigned a major portion of primary production. DOM production might be as high as 30% if conditions were adverse for the
phytoplankton (*Thomas*, 1971). Production of primary detritus
might be high at the end of a bloom of relatively inedible
organisms. Consumption of nannoplankton is high when nannoplankton

production is high, a common occurrence in continental shelf
waters (*Pomeroy*, 1974). Higher efficiencies were then assumed
for grazers, carnivorous zooplankton, and fishes, based on the
evidence presented in Table 1. Again, the flux of energy from
primary producers was divided in the various ways just described.
The gross growth efficiency of bacteria was assumed to be 50%
throughout. We know very little about the efficiency of bacteria
in nature, but *Ho and Payne* (1979) have shown in culture that
gross growth efficiency of bacteria increases as substrate
concentration decreases toward that which would be expected
in the natural environment. Cultures of oceanic bacteria show
maximal efficiency at much lower substrate concentrations than
do *E. coli*, for example. One possible deficiency of the model
is the lack of consideration of protozoa and small metazoans.
These may constitute a large fraction of the standing stock
(*Banse*, 1962). They, together with bacteria, probably mediate
much energy and are a link in the secondary production system.

A mathematical model has been written in collaboration with
a colleague, and the results of simulations will be reported
elsewhere. For the present purpose it suffices simply to examine
the effect of varied assumptions about food web structure and
efficiency on the flux of energy to terminal consumers. All
the essential information has been condensed into Figures 1 and
2, which contain three sets of numbers. The large numbers are
the identifying compartment numbers. An arrow pointing to such
a large number indicates a flux to that compartment, which merely
serves to simplify the diagrams. The smaller numbers beneath
the compartments are the state variables, or standing stocks,
expected in steady state. The small numbers beneath the arrows
are the fluxes between compartments. The latter are assigned
arbitrarily, and the resulting effect on the succeeding compartments is then calculated.

From many variations tried with the energy flux model, three
are used to illustrate the range of results when efficiencies
comparable to those used by *Steele* (1974) in his North Sea model
are assumed. In the first case (Fig. 1, a) nominal portions of
primary production are assigned to mucus-net feeders and the
detritus and DOM pathways. No matter how the flux of energy from
the detritus-DOM pathways is directed, little influence on production at the carnivore level is evident. In the second case
(Fig. 1, b) most phytoplankton production is assumed to be
nannoplankton, relatively unavailable to crustacean grazers but
available to mucus-net feeders. Although there is probably some
predation on appendicularians by larval demersal fishes
(*Paffenhöfer*, 1975), it is assumed here that most production
from mucus-net feeders goes to pelagic carnivores. This
assumption results in reduced production of benthos. In the
real world, it might result in a shift in demersal and benthic

populations to those which are filter feeders, mucus-net feeders, and predators on that trophic group. In the third case (Fig. 1, c) conditions which might be found in the late stages of a phytoplankton bloom are represented. Most of the phytoplankton are dying but uneaten by both grazers and mucus-net feeders. Detritus production is high, and so is release of DOM. This configuration appears to reduce energy flux to planktonic carnivores, while the benthos and demersal carnivores experience favorable conditions. None of these configurations leads to the demise of any trophic group or even reduces it below a level which might pass as a bad year class. Obviously, the real causes of bad year classes of species populations (as opposed to trophic groupings) are not seen in a condensed model such as this one. The model shows that it is possible for sufficient energy to flow through either the grazers or the alternate pathways to support all major trophic groups at a reasonable level, even assuming rather low gross growth efficiencies. However, if all ecological efficiencies between compartments in the model were assumed to be 10%, energy flux would be insufficient. The efficiencies assumed by Steele approached the minimum that will support the system at the assumed level of primary production.

While many investigators still believe Steel's efficiency assumptions are too high, this writer, for reasons stated above, believes they are too low. To examine the effect of gross growth efficiency on secondary production of terminal consumers, the model was tested using 30% gross growth efficiencies for grazers, mucus-net feeders, benthic invertebrates, and carnivorous zooplankton. Bacterial efficiency remained at 50%, and efficiency of fishes and other carnivores remained at 10%. As expected, this led to increased production of terminal consumers (Fig. 2, a), with carnivorous zooplankton and pelagic fishes showing the greatest gains. Dividing the primary production more evenly

---

Fig. 2. Energy flux model for continental shelves. Values directly beneath each compartment are state variables (standing stocks) in KCal m^{-2}. Energy flux values, in KCal m^{-2} yr^{-1}, are beneath or beside arrows. Compartments are numbered (large numerals). To simplify the diagram, some fluxes are designated by an arrow followed by the compartment number. Gross growth efficiencies are: bacteria, 50%; mucus-net feeders, 30%; grazing zooplankton, 30%; benthic invertebrates, 30%; carnivorous zooplankton, 30%; fishes and other carnivores, 10%. In (a) grazing zooplankton dominate energy flux. In (b) 90% of detritus goes to mucus-net feeders and 10% goes to benthos. Grazing zooplankton get 75% of available phytoplankton. In (c) a late stage of a phytoplankton bloom finds most phytoplankton becoming detritus. 90% of total detritus goes to mucus-net feeders and 10% goes to benthos. $\longrightarrow$

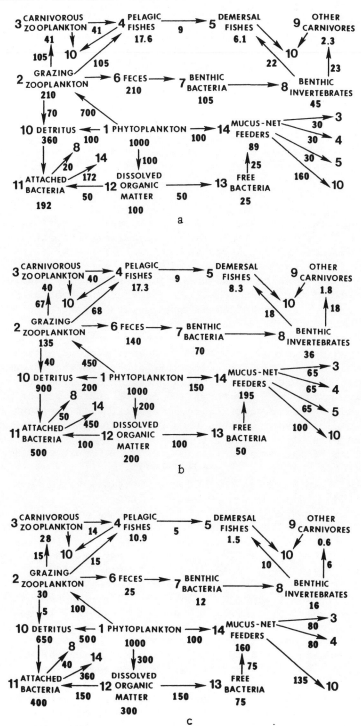

among energy pathways results in a more equal production of
pelagic and demersal fishes (Fig. 2, b). The fate of attached
bacteria begins to assume importance as the role of detritus is
increased. This phenomenon is modeled by assuming late bloom
conditions, and further assuming that 90% of bacteria attached
to detritus is consumed by mucus-net feeders (Fig. 2, c). The
result is a decline in benthic invertebrates, demersal fishes,
and other carnivores. If the benthic invertebrate community
were able to shift to populations which could utilize suspended
detritus, the balance of the system could be restored. However,
this does demonstrate that there are limits, even assuming high
efficiencies, beyond which not only species populations but
trophic groups may suffer. The model cannot tell us whether
these conditions occur in the real world; that must be decided
by observation. However, the model does help us define conditions under which we would expect catastrophic changes in
marine populations for purely energetic reasons.

The model also reveals some major gaps in our knowledge of
energy flux in marine systems. Our understanding of assimilation
efficiency clearly is deficient and is further obscured by
adherence to generalities about ecological efficiency which
have not stood the test of time. Moreover, there are links
in the food web which have not been quantified. Protozoa and
small metazoans have not received the attention which seems
appropriate for such metabolically active and reproductively
responsive organisms. Bacteria and nannoplankton are only now
being studied actively, after long neglect, with substantial
populations of organisms less than 0.5 μm in diameter now confirmed to exist in the ocean (*e.g.* the "coccoids" of *Wiebe and
Pomeroy*, 1972; the "minibacteria" of *Watson et al.*, 1977; the
"picoplankton" of *Sieburth et al.*, 1977).

As these deficiencies in our knowledge are filled, the
schism grows between the view of the marine food chain as linear,
with well known rates of primary production, and the view of the
marine food chain as anastomosing, with major unresolved dis-
crepancies between accepted rates of phytosynthesis and apparent
rates of respiration (*Joiris*, 1977), bacterial production
(*Sorokin*, 1978; *Sieburth*, 1977; *Sieburth et al.*, 1977), and
bacterial biomass (*Ferguson and Rublee*, 1976; *King and White*,
1977; *Sieburth et al.*, 1977). Modeling the system in such a
way as to include alternate pathways and alternate rates of
production and consumption cannot confirm or deny what happens
in the ocean, but it can help us make formal statements which
better define the deficiencies in our knowledge. The model
presented here shows that either conventional assumptions about
efficiency must be low by a factor of 2 to 3, or measurements
of phytosynthesis must be low by a factor of 5 to 10. The
addition of a step involving protozoans would make this

discrepancy even more apparent. By carrying the exercise on to the next step of simulation, we shall be able to incorporate additional details such as growth rates, lags in response, refuging and switching in predator-prey relations, and the interaction of nutrient supply with ecosystem-level function. At that level of detail, even condensed ecosystem models can be expected to raise testable hypotheses about the structure of the food web of continental shelves.

ACKNOWLEDGMENTS

This work was supported by U. S. Department of Energy Contract EY-76-S-09-0639. Although they do not necessarily endorse all of the speculation, G.-A. Paffenhöfer, Janet Pomeroy, W. J. Wiebe and R. G. Wiegert provided valuable suggestions during the preparation of the manuscript.

REFERENCES

Aldredge, A. L., 1976. Discarded appendicularian houses as sources of food, surface habitats, and particulate organic matter in planktonic environments, *Limnol. Oceanogr. 21:* 14-23.

Andrews, P. and P. J. LeB. Williams, 1971. Heterotrophic utilization of dissolved organic compounds in the sea. III. Measurement of the oxidation rates and concentrations of glucose and amino acids in sea water, *J. Mar. Biol. Ass. U. K. 51:* 111-125.

Azam, F. and R. E. Hodson, 1977. Size distribution and activity of marine microheterotrophs, *Limnol. Oceanogr. 22:* 492-501.

Banse, K., 1962. Net zooplankton and total zooplankton, *Rapp. Proc.-Verb. Réunions, Cons. Perm. Int. Explor. Mer 153:* 211-214.

Beers, J. R. and G. L. Stewart, 1967. Micro-zooplankton in the euphotic zone at five localities across the California current, *J. Fish. Res. Bd Can. 24:* 2053-2068.

Beers, J. R. and G. L. Stewart, 1970. Numerical abundance and estimated biomass of microzooplankton, *Bull. Scripps Inst. Oceanogr. 17:* 67-87.

Beers, J. R. and G. L. Stewart, 1971. Micro-zooplankters in the
    plankton communities of the upper waters of the eastern
    tropical Pacific, *Deep-Sea Res. 18:* 861-883.

Clarke, G. L., 1946. Dynamics of production in a marine area,
    Ecol. Mongr. 16: 323-335.

Conover, R. J., 1962. Metabolism and growth in *Calanus
    hyperboreus* in relation to its life cycle, *Rapp.
    Proc.-Verb. Réunions, Cons. Perm. Int. Explor. Mer.
    153:* 190-196.

Corner, E. D. S. and C. B. Cowey, 1968. Biochemical studies on
    the production of marine zooplankton, *Biol. Rev. 43:*
    393-426.

Cushing, D. H., 1959. On the nature of production in the sea,
    Fishery Invest. II, 22: 1-41, H. M. Stationery Office,
    London.

Derenbach, J. B. and P. J. LeB. Williams, 1974. Autotrophic
    and bacterial production: fractionation of plankton
    populations by differential filtration of samples from
    the English Channel, *Mar. Biol. 25:* 263-269.

Edwards, R. R. C., D. M. Finlayson, and J. H. Steele, 1969.
    The ecology of O-group plaice and common dabs in Loch Ewe.
    II. Experimental studies, *J. Exp. Mar. Biol. Ecol. 3:* 1-17.

Ferguson, R. L. and R. Rublee, 1976. Contribution of bacteria to
    standing crop of coastal plankton, *Limnol. Oceanogr. 22:* 141-
    145.

Finenko, Z. Z., 1978. Production in plant populations, In:
    Marine Ecology 4, edited by O. Kinne, 13-87, John Wiley &
    Sons, New York.

Gordon, D. C., 1970a. A microscopic study of non-living organic
    particles in the North Atlantic Ocean, *Deep-Sea Res. 17:*
    175-185.

Gordon, D. C., 1970b. Some studies on the distribution and
    composition of particulate organic carbon in the North
    Atlantic Ocean, *Deep-Sea Res. 17:* 233-243.

Gulland, J. A., 1970. Food chain studies and some problems in
    world fisheries, In: *Marine Food Chains,* edited by J. H.
    Steele, 296-315, University of California Press, Berkeley.

Hempel, G., 1970. Introduction, In: *Marine Food Chains*, edited by J. H. Steele, 255-260, University of California Press, Berkeley.

Ho, K. P. and W. J. Payne, 1970. Assimilation efficiency and energy contents of prototrophic bacteria, *Biotechnol. Bioengineering*, in press.

Ivleva, I. V., 1970. The influence of temperature on the transformation of matter in marine invertebrates, In: *Marine Food Chains*, edited by J. H. Steele, 96-112, University of California Press, Berkeley.

Joiris, C., 1977. On the role of heterotrophic bacteria in marine ecosystems: Some problems, *Helgo. wiss. Meeresunters. 30:* 611-621.

King, J. D. and D. C. White, 1977. Muramic acid as a measure of microbial biomass in estuarine and marine samples, *Appl. Env. Microbiol. 33:* 777-783.

Koblentz-Mishke, O. J., V. V. Volkovinsky, and J. G. Kabanova, 1970. Planktonic primary production of the World Ocean, In: *Scientific Exploration of the South Pacific*, 183-193, Nat. Acad. Sci. U. S., Washington.

Lasker, R., 1970. Utilization of zooplankton energy by a Pacific sardine population in the California current, In: *Marine Food Chains*, edited by J. H. Steele, 265-284, University of California Press, Berkeley.

Lasker, R., 1973. Field criteria for survival of anchovy larvae: the relation between inshore chlorophyll maximum layers and successful first feeding, *Fishery Bull. 73:* 453-462.

Lenz, J., 1972. The size distribution of particles in marine detritus, *Mem. Ist. Ital. Idrobiol. 29 Suppl.:* 17-35.

Lindeman, R. L., 1942. The trophic-dynamic aspect of ecology, *Ecology 23:* 399-418.

McAllister, C. D., 1969. Aspects of estimating zooplankton production from phytoplankton production, *J. Fish. Res. Bd Can. 26:* 199-220.

McLaren, I. A., 1963. Effects of temperature on growth of zooplankton, and the adaptive value of vertical migration, *J. Fish. Res. Bd Can. 20:* 685-727.

Marshall, S. M. and A. P. Orr, 1962. Food and feeding of copepods, *Rapp. Proc.-Verb. Réunions, Cons. Perm. Int. Explor. Mer 153:* 92–98.

Miller, R. J., K. H. Mann, and D. J. Scarratt, 1971. Production potential of a seaweed-lobster community in Eastern Canada, *J. Fish. Res. Bd Can. 28:* 1733–1738.

Mullin, M. M. and E. R. Brooks, 1970. Growth and metabolism of two planktonic, marine copepods as influenced by temperature and type of food, In: *Marine Food Chains,* edited by J. H. Steele, 74–95, University of California Press, Berkeley.

Odum, E. P. and A. de la Cruz, 1967. Particulate organic detritus in a Georgia salt marsh-estuarine ecosystem. In: *Estuaries,* edited by G. H. Lauff, 383–388, AAAS Publ. 83. Washington, D. C.

Odum, H. T., 1967. IBP Symposium: environmental photosynthesis, *Science 157:* 415–416.

Paffenhöfer, G.-A., 1975. On the biology of Appendicularia of the southeastern North Sea, *Proc. 10th European Symposium on Marine Biology 2:* 437–455.

Payne, W. J., 1970. Energy yields and growth of heterotrophs, *Ann. Rev. Microbiology 24:* 17–52.

Petipa, T. A., E. V. Pavlova, and G. N. Mironov, 1790. The food web structure, utilization and transport of energy by trophic levels in the planktonic communities, In: *Marine Food Chains,* edited by J. H. Steele, 142–167, University of California Press, Berkeley.

Pomeroy, L. R., 1974. The ocean's food web, a changing paradigm, *BioScience 24:* 499–504.

Pomeroy, L. R., H. H. Haskin, and R. A. Ragotzkie, 1956. Observations on dinoflagellate blooms, *Limnol. Oceanogr. 1:* 54–60.

Pomeroy, L. R. and R. E. Johannes, 1966. Total plankton respiration, *Deep-Sea Res. 13:* 971–973.

Pomeroy, L. R. and R. E. Johannes, 1968. Occurrence and respiration of ultraplankton in the upper 500 meters of the ocean, *Deep-Sea. Res. 15:* 381–391.

Porter, K. G., 1977. The plant-animal interface in freshwater ecosystems, *Amer. Sci. 65:* 159–170.

Rao, K. P., 1953. Rate of water propulsion in *Mytilus californianus* as a function of latitude, *Biol. Bull. 104:* 171-181.

Reeve, M. R., 1970. The biology of Chaetognatha. I. Quantitative aspects of growth and egg production in *Sagitta hispida*, In: *Marine Food Chains*, edited by J. H. Steele, 168-189, University of California Press, Berkeley.

Riley, G. A., 1970. Particulate organic matter in sea water, *Adv. Mar. Biol. 8:* 1-118.

Riley, G. A., 1972. Patterns of production in marine ecosystems, In: *Ecosystem Structure and Function. Oregon State University Annual Biology Colloquium 31*, edited by J. A. Wiens, 91-100, Oregon State University Press, Corvallis.

Riley, G. A., H. Stommel, and D. F. Bumpus, 1949. Quantitative ecology of the plankton of the Western North Atlantic, *Bull. Bingham Oceanogr. Coll. 12:* 1-169.

Roman, M. R., 1978. Ingestion of the blue-green alga *Trichodesmium* by the harpactacoid copepod, *Macrostella gracilis*, *Limnol. Oceanogr. 23:* 1245-1248.

Ryther, J. H., 1969. Photosynthesis and fish production in the sea, *Science 166:* 72-76.

Sheldon, R. W. and W. H. Sutcliffe, Jr., 1978. Generation times of 3 h for Sargasso Sea microplankton determined by ATP analysis, *Limnol. Oceanogr. 23:* 1051-1055.

Sieburth, J. McN., 1976. Bacterial substrates and productivity in marine ecosystems, *Ann. Rev. Ecol. System. 7:* 259-285.

Sieburth, J. McN., 1977. Convener's report on the informal session on biomass and productivity of microorganisms in planktonic ecosystems, *Helgo. wiss. Meeresunters. 30:* 697-704.

Sieburth, J. McN., K. M. Johnson, C. M. Burney, and D. M. Lavoie, 1977. Estimation of *in situ* rates of heterotrophy using diurnal changes in dissolved organic matter and growth rates of picoplankton in diffusion cultures, *Helgo. wiss. Meeresunters. 30:* 565-574.

Slobodkin, L. B., 1960. Ecological energy relationships at the population level, *Amer. Nat. 94:* 213-236.

Slobodkin, L. B., 1961. *Growth and Regulation of Animal Populations*, 184 pp., Holt, Rinehart, and Winston, New York.

Slobodkin, L. B., 1970. *Summary*. In: *Marine Food Chains*, edited by J. H. Steele, 537–540, University of California Press, Berkeley.

Sorokin, Yu., 1978. Decomposition of organic matter and nutrient regression. In: *Marine Ecology 4*, edited by O. Kinne, 501–616, John Wiley & Sons, New York.

Steele, J. H., 1965. Some problems in the study of marine resources, *Spec. Pubs. Int. Comm. N. W. Atlantic Fisheries No. 6*, 463–476.

Steele, J. H., 1974. *The Structure of Marine Ecosystems*, 128 pp., Harvard University Press, Cambridge, Mass.

Thomas, J. P., 1971. Release of dissolved organic matter from natural populations of marine phytoplankton, *Mar. Biol. 11*: 311–323.

Wangersky, P. J., 1977. The role of particulate matter in the productivity of surface waters, *Helgo. wiss. Meeresunters. 30*: 546–564.

Watson, S. W., T. J. Novitsky, Helen L. Quinby, and F. W. Valois, 1977. Determination of bacterial number and biomass in the marine environment, *Appl. Env. Microbiol. 33*: 940–946.

Wiebe, W. J. and L. R. Pomeroy, 1972. Microorganisms and their association with aggregates and detritus in the sea: a microscopic study, *Mem. Ist. Ital. Idrobiol. 29 Suppl.*: 325–352.

Wiegert, R. G., R. R. Christian, J. L. Gallagher, J. R. Hall, R. D. H. Jones, and R. L. Wetzel, 1975. A preliminary ecosystem model of a coastal Georgia *Spartina* marsh, In: *Estuarine Research Vol. 1*, edited by L. E. Cronin, 583–601, Academic Press, New York.

Wiegert, R. G. and D. F. Owen, 1971. Trophic structure, available resources, and population density in terrestrial vs. aquatic ecosystems, *J. Theor. Biol. 30*: 69–81.

Williams, P. M., H. Oeschger, and P. Kinney, 1969. Natural radiocarbon activity of the dissolved organic carbon in the north-east Pacific Ocean, *Nature 224*: 256–258.

IV. Benthic Community Organization

PERSPECTIVES OF MARINE MEIOFAUNAL ECOLOGY[1]

Bruce C. Coull and Susan S. Bell

University of South Carolina

ABSTRACT

Meiofauna research has seen an increased ecological emphasis in the last ten years. The period prior to 1970 was primarily descriptive but current research is directed more toward testing hypotheses using the meiofauna. There is a fairly large descriptive base and generalities from a worldwide data set indicate that there are unique opportunities for additional studies. In this paper we present an overview of the historical development of meiofaunal research and a synopsis of the present knowledge, and provide our perceptions as to where meiofauna research needs to be directed. Hypothesis testing in the field, the role of meiofauna in total benthic energetics, and biological control mechanisms appear to be most fruitful avenues for future research.

INTRODUCTION

In this essay we shall attempt to review the historical development of meiofaunal research and the state of the art regarding meiofaunal ecology. Further, we will present some ideas and speculation on where we perceive research in meiofaunal ecology to be headed and suggest avenues in need of additional research. It is not the purpose of our overview to be exhaustive,

---

[1]Contribution No. 276 from the Belle W. Baruch Institute for Marine Biology and Coastal Research. Research supported, in part, by Oceanography Section, National Science Foundation, NSF Grant OCE-76-17584.

but rather to illustrate some major trends in the ecology of these omnipresent organisms.

Although research on meiofauna was ongoing in the early 1900's (*Remane*, 1933) the term "meiobenthos" (meiofauna) was first coined by *Mare* (1942) to describe those benthic metazoans of intermediate size. These are animals smaller than those traditionally called "macrobenthos," but larger than the "microbenthos," (i.e., algae, bacteria, and protozoa). There is a wide diversity in the habitats of meiofauna. Some meiofauna are interstitial (moving between the sediment particles), others are burrowing (moving through or displacing sediment particles), while still others have an epibenthic or phytal existence. The meiofauna may further be broken down into: 1) temporary meio-benthos, those which spend only their larval stages as part of the meiobenthos (usually larvae of the macrofauna), and 2) permanent meiobenthos, including Rotifera, Gastrotricha, Nematoda, Archiannelida, Tardigrada, Copepoda, Ostracoda, Mystacocarida, Turbellaria, Acarina, Gnathostomulida, and some specialized membranes of the Hydrozoa, Nemertina, Bryozoa, Gastropoda, Aplacophora, Holothuroidea, Tunicata, Priapulida, Polychaeta, Oligochaeta and Sipunculida. Despite the fluidness of the definition, it is generally accepted that the term "meiofauna" refers to those animals which pass through a 0.5 mm sieve and are retained on a sieve with mesh widths smaller than 0.1 mm.

HISTORICAL DEVELOPMENT OF MEIOFAUNAL ECOLOGY

Early meiofaunal research has been primarily systematic and done mostly in Europe. Although taxonomic studies of marine meiobenthic organisms were ongoing in the 1800's (e.g., *Schultze*, 1853, on Gastrotricha; *Bastian*, 1865, on Nematoda; *Claus*, 1866, on Copepoda), the first quantitative paper to enumerate all taxa did not appear until 1931 (*Moore*, 1931). Table 1 summarizes those meiobenthic papers that enumerate "all" organisms in the sediments published through 1969. A preponderant number of all meiofaunal investigations to 1969 took place in Europe (Table 2) or were done by Europeans in other areas (e.g. *Wieser*, 1960; *Renaud-Debyser*, 1963; *Wigley and McIntyre*, 1964, in North America; *Renaud-Mornant and Serene*, 1967, and *McIntyre*, 1968, elsewhere). That most of the studies listed in Table 1 are primarily descriptive is not too surprising since in any new field of endeavor much descriptive work is requisite to experimental research. Additionally, much headway correlating meiofauna taxa and abundance with physical factors (such as grain size, tidal exposure) was being made during this time. We learned, for example, that many taxa are restricted to particular sediment types (*Wieser*, 1959) and that changes in

TABLE 1. Quantitative investigations of marine meiofauna (studies that enumerate all taxa) through 1969.

Year	Europe	North America	Elsewhere
1931	Moore (UK)		
1933	Remane (Germany)		
1936	Krogh & Spärck (Denmark)		
1940	Rees (UK)		
1942	Mare (UK)		
1946	Bougis (France)		
1947	Purasjoki (Finland)		
1951	Smidt (Denmark)		
1952	Remane (Germany)		
1959	Renaud-Debyser (France		
1960	Bregnballe (Denmark)	Wieser (Mass.)	
1961	McIntyre (UK)		
1962			Ganapati & Rao (India)
1963	Renaud-Debyser & Salvat (France)	Renaud-Debyser (Bahamas)	
1964	McIntyre (UK)	Wigley & McIntyre (Mass.) Tietjen (R.I.)	
1966	Riemann (Germany) Thiel (Germany-Deep Sea) Fenchel & Jansson (Sweden)	Bush (Florida)	Kikuchi (Japan)
1967	Muus (Denmark) Fenchel *et al.* (Denmark)		
1968	Guille & Soyer (France) Jansson (Sweden)	Coull (Bermuda)	Renaud-Mornant & Serene (Malaysia)
1969	Fenchel (Denmark)		McIntyre (India)

tidal exposure often are the primary factors limiting sandy beach
interstitial fauna. We view this era as an exploratory period--a
period of rather routine but necessary work which subsequently
led to the ability to ask testable questions.

In the late 60's and early 70's there was a surge of experimental ecological interest in the meiofauna and in the question
"what do they do?" Pioneers in much of this research were *Fenchel*
(his 1969 paper summarizes earlier work), *Gray* (1966a, b, c,
1967a, b, 1968; *Gray and Johnson* 1970) and *Jansson* (1967, 1968),
who used quantitative hypothesis testing to ascertain the
functional role of marine benthic ciliates (*Fenchel*) and the
behavioral responses of meiofauna to external factors (*Gray,
Jansson*). There subsequently developed a great interest in the
energetic role of meiofauna in benthic systems (see *Gerlach*, 1971,
for review) as well as in the physiological capacities, tolerances,
and preferences of these organisms (e.g. *Coull and Vernberg*, 1970;
Jansson, 1967; *Tietjen and Lee*, 1972; *Wieser et al.*, 1974). Much
of this research was meiofauna oriented but it was really more
than the study of meiofauna for meiofauna's (or the meiofaunologists') sake. Rather it was using the meiofauna to test basic
ecological hypotheses; hypotheses in need of elaboration and
amenable to being tested using meiofaunal organisms.

Various avenues of research followed the early 1970's
questioning period and, in addition to the energetic/functional
role, interest was generated in using meiofauna to evaluate
community structure mechanisms in marine soft bottom communities.
Species diversity theory, a very "in" concept in the early 1970's
ecology, was a fruitful avenue of meiofaunal research (see, for
example, *Coull*, 1970, 1972; *Ott*, 1972; *Heip and DeCraemer*, 1974).
Small scale spatial distribution also saw a flood of interest
(e.g. *Vitiello*, 1968; *Gray and Rieger*, 1971; *Heip and Engels*,
1977). More recently a significant amount of interest has been
generated in the biological control of meiofaunal populations and,
as we shall discuss in the *Prospectus* section of this paper,
meiofaunal ecology appears to be at the stage where such an
effort is not only warranted but should be of interest to all
ecologists.

Overall, the 1970's have proven an exciting period and
meiofauna research has taken on a much more ecological thrust
(Table 2). The data presented in Table 2 are admittedly biased
since the counts were made on those references in the first
author's reference catalogue. He does not have all the
meiofauna papers published, but does have a rather complete
collection. In any case, the trend is clear--more ecologically-
related meiofauna papers were published in 7 years post-1970
than in the prior 29. Further, it is evident that although all
regions of the world have seen increased ecological emphasis,

TABLE 2. Meiofauna papers published[1] from 1930–1969 and 1970–1977.

	Site of Study	TOPICS			
		Morphology/Systematics/Taxonomy	Ecology	Total	% Ecology
1930–1969	Europe	225	107	332	32
	North America[2]	76	39	115	34
	Elsewhere[3]	88	10	98	10
	TOTAL	389	156	545	29
1970–1977	Europe	281	165	446	37
	North America[2]	84	118	202	58
	Elsewhere[3]	58	18	76	24
	TOTAL	423	301	724	42

[1]Based on reference collection of first author (see text).

[2]Includes United States, Canada and Caribbean Island waters.

[3]Includes all areas not in Europe and North America, but primarily South America, Africa, Japan, India.

North America appears to be primarily ecologically oriented in
its pursuit of the elusive meiofauna.

PRESENT ECOLOGICAL KNOWLEDGE

Swedmark (1964), *McIntyre* (1969), *Gerlach* (1971, 1978) and
Coull (1973) review much of the meiofauna accomplishments (and/or
failures). We shall not present an exhaustive referencing to
particular studies, but attempt to synthesize the known generalities of meiofaunal ecology to date.

On the average one can expect to find 10^6 m^{-2} meiofaunal
organisms and a standing crop dry weight biomass of between
1 and 2 g m^{-2}. Obviously, these numbers will vary according
to season, latitude, water depth, etc., but they are reasonable
approximations of worldwide meiofauna density and biomass
(*Gerlach*, 1978). Numbers and biomass decrease with increasing
depth into the oceans and the highest values are known from
intertidal mudflats--2.6 x 10^7 m^{-2} (R. M. *Warwick*, pers. comm.).
Values tend to be highest in detritally-derived sediments and
lowest in clean sands. Nematodes and harpacticoid copepods are
usually the two most abundant taxa in all sediments, although
there are occasions when another taxon may take over first or
second place (e.g. *Hogue*, 1978).

Certain taxa are restricted to particular sediment types.
Sediments where the median particle diameter is below 125 μm
tend to be dominated by burrowing meiofauna. Interstitial
groups, e.g. Gastrotricha (with one genus expected, *Musellifer*)
and Tardigrada, are excluded from muddy substrates where the
interstitial lacunae are closed. Obviously, the converse is
true in that a burrowing taxon, e.g. Kinorhyncha (one exception,
Cateria), is excluded from the interstitial habitat. In those
taxa that have both interstitial and burrowing representatives,
e.g. Nematoda, Copepoda, Turbellaria, there is a difference in
the morphology of mud and sand dwellers. The sand fauna tends
to be slender as it must maneuver through the narrow interstitial openings, whereas the mud fauna is not restricted to a
particular morphology but is generally larger.

Vertically, the meiofauna are concentrated in the upper
levels of the sediments, decreasing to near zero in the redox
(RPD) layer, although some taxa may remain below the RPD in
sandy sediments as true anaerobes (*Fenchel and Riedl*, 1970;
Maguire and Boaden, 1975). In muddy detrital sediments of
South Carolina estuaries we have found that 94% of all the fauna
is located in the upper 1 cm of sediment. On sandy beaches the
marine fauna is known to be distributed to a depth of up to
90 cm, or to the depth of the groundwater (*McIntyre*, 1969;

McLachlan, 1977). Subtidally, most of the studies indicate that
>95% of the fauna is in the upper 7 cm and 60-70% in the upper 2
cm (e.g. *Moore*, 1931; *Wieser*, 1960; *McIntyre*, 1964, *Muus*, 1967;
Tietjen, 1969; *Coull*, 1970; *de Bovée and Soyer*, 1974). However,
McLachlan et al. (1977) report significant numbers of meiofauna
as deep as 30 cm in cores from 27 m water depth, and correlate
this distribution with oxygen availability in the sediment.
They never reached the RPD and therefore continued to encounter
fauna.

Horizontally, meiofauna are known to exhibit a patchy
distribution (*Gray and Rieger*, 1971; *Vitiello*, 1968; *Heip and
Engels*, 1977; *Hogue*, 1978) in sediments that appear to be
homogeneous. Small scale physical variations may be present
and in fact what appears as a homogeneous habitat to the
investigator may not be to the meiofauna. *Harrison* (1977)
distinguished two distinctly separable faunas within 0.09 m^2,
one in the trough and one on the crest of sand ripples. However,
even when environmental homogeneity is assumed (or known) replicate samples taken side by side may differ significantly in
numbers and species. The true causes of such patchiness are
not known but several ideas have been put forth. *Bush* (1966) and
Gerlach (1977) have suggested that the random placement of a
macrofaunal carcass (dead fish, dead crab, etc.) and its subsequent leaking of organic matter may attract meiofauna and that
the patches may thus be the result of attraction to an organic
point source. Others have suggested that the food of meiofauna
(bacteria, microflora) are patchily distributed (*Gray and Johnson*,
1970; *Giere*, 1975; *Hummon et al.*, 1976; *Lee et al.*, 1977) and that
certain "attractive" food sources, being so localized, can then
determine the horizontal distribution of the meiofauna. One
might predict that selective predation on meiofauna could also
cause patches. We can envision a predator feeding in very
localized regions of the sediment and removing particular taxa
(or size classes). This localized removal could then result in
a discordant distribution (*Levin and Paine*, 1974). *Thistle* (1979)
and *Lee et al.* (1977) have suggested that polychaete tubes may
serve to increase the structural heterogeneity of the environment,
provide refugia for the organisms, and localize the patches. *Teal
and Wieser* (1966) suggested that *Spartina* roots created micro-
oxygenated zones and that meiofaunal nematodes concentrated
around the roots. However, *Bell et al.* (1978) have shown that
Spartina roots have either a negative or no correlation with
nematode distribution and thus do not significantly affect
patch distribution.

The mode of most meiofauna reproduction, lacking a larval
dispersal mechanism, might result in aggregated patterns of
juveniles. Likewise, for adult forms, gregariousness for reproductive success was noted by *Heip and Engels* (1977), who observed

that male copepods and females without eggs occurred together,
while females with eggs were independently distributed. They
suggested a pheromonal attraction of males and females as the
cause of some of their observed patchy distributions. For
whatever reason, it is well know that the meiofauna are not
homogeneously distributed. Any investigator needs to be cognizant
of this distribution and plan sampling programs appropriately.
Obviously, one small sample cannot adequately estimate the sample
mean.

We know little of the long-term temporal variability of
meiofauna. *McIntyre and Murison* (1973), based on monthly sampling
for one year and irregular sampling for 9 years, suggested that
annual variations in collective meiofauna abundance (and subsequent biomass) were quite small and that the total assemblage was
in a relatively constant state over long periods of time (years).
Other "long-term" meiofauna studies (*Stripp*, 1969; *Tietjen*, 1969;
Coull, 1970; *de Bovée and Soyer*, 1974), did point to seasonal
variations, but the fact that each is only about a year in
duration certainly limits the predictive value of the data and
their ability to confirm or refute the *McIntyre and Murison* (1973)
hypothesis.

As part of a continuing study on the meiofauna of North Inlet,
South Carolina, we have been monitoring the long-term variability
of the meiofauna at two sites (one a subtidal sand, the other a
subtidal mud) to determine natural fluctuations of marine meiofauna. Figure 1 illustrates the monthly fluctuations in meiofaunal
density from September 1972 through November 1977 (63 months) at
the muddy site. It is not our purpose to discuss potential
causes for the observed fluctuations, but rather to illustrate
that a short interval, long-term data set on meiofaunal abundance does *not* suggest constancy in numbers (and concommitant
biomass) over time. Model I ANOVA (with significant treatment
effect of comparing abundances on a yearly basis) shows significant differences between the yearly total fauna (1973, 1974, 1975,
1976, 1977) based on each of 3 or 4 monthly replicates ($p <$
0.0001). Short-term sampling is, of course, better than no
sampling and can be very valuable, especially when one wants
to use the meiofauna to attack specific questions. However, we
would argue that short-term (a season, a year?) sampling does
not adequately characterize the meiofaunal dynamics of a region,
habitat, sediment type, etc., and thus wholeheartedly agree with
Wiens (1977) that "unless populations or communities are in
resource-defined equilibrium, it is unlikely that short-term,
single season studies . . . will produce results that are open
to clear and unambiguous interpretation. What is needed are
long-term studies of defined, closely monitored populations, with
special attention to variation . . . in relation to differing
environmental contexts."

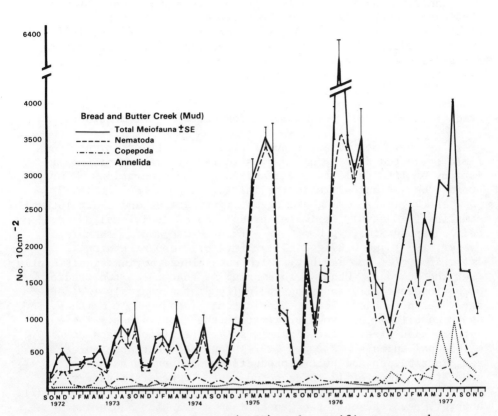

Fig. 1. Total meiofauna (± SE) and specific taxa numbers over a 63 month period at a muddy subtidal site in North Inlet, South Carolina (33°20'N, 79°10'W). Note the large amount of year to year fluctuation.

An important aspect of ecological studies is the energetic role played by the organisms under investigation. Questions such as: How much do they produce? What percentage of the total energy do they use? and What is their ultimate fate? are of particular interest. *Gerlach* (1971) suggests that, for an equivalent biomass, the meiofauna are responsible for about 5 times the total benthic metabolism of the macrofauna. Thus, a macrofauna/meiofauna standing crop biomass ratio of > 5:1 is requisite before the total energy requirements of the macrofauna exceed that of the meiofauna. Macrofauna biomass in most systems is usually greater than necessary to yield the 5:1 ratio, but in extremely shallow water (mudflats, estuaries) and the deep sea, the macro/meiofauna standing crop biomass ratio approximates 1 (*Gerlach*, 1971; *Thiel*, 1975); thus the meiofauna play a more significant role in benthic energetics in these systems.

Annual turnover rate (life cycle turnover x number of generations), a very difficult measure to quantify, has been estimated to approximate 10 for the meiofauna (*McIntyre*, 1964, *Gerlach*, 1971, 1978). Coupling turnover rate with standing crop biomass indicates that, in systems dominated by macrofaunal biomass, meiofaunal and foraminiferan production is about equal to that of the deposit-feeding macrofauna (*Gerlach*, 1978), but in systems like the deep sea or shallow water, total annual animal production will most certainly be dominated by the meiofauna.

The above mentioned production rates are based primarily on field collected density/biomass estimates and laboratory observations of life cycle turnover and number of generations. *In situ* production estimates are in the initial stages. *Feller* (1977) and *Fleeger* (unpubl.), using age class distributions and short interval sampling (2-6 days), estimated production of individual harpacticoid copepods at 1.0 g C m^{-2} yr^{-1} ($\bar{x}$ of *Huntemannia jadensis*; *Feller*, 1977) and 0.87 g C m^{-2} yr^{-1} ($\bar{x}$ of *Microarthridion littorale*; *Fleeger*, unpubl.), with the values varying significantly over the seasons. What is so amazing is that these two studies on phylogenetically and geographically distant animals (*Feller*--sandy beaches, Puget Sound, Washington; *Fleeger*--salt marshes, South Carolina) came to approximately the same values. These authors also present the first *in situ* life history data on meiofaunal animals. *Fleeger* (1979) reported that at stable age distribution and peak reproduction *Microarthidion* had generation times of 12-15 days and up to 12 generations per year: 2-4 times those predicted by *Gerlach* (1971). *Feller* (1977), however, reported only two generations a year for *Huntemannia* from Washington. Obviously, *in situ* production/life history studies are an important avenue of research to pursue if we are ever to evaluate the energetic role of the meiobenthos.

We have discussed how much the meiofauna produce and metabolize, but a primary question is that of the ultimate fate of this production. There has been much controversy in recent years as to the interaction between meiofauna and higher trophic levels. *McIntyre and Murison* (1973) and *Heip and Smol* (1975) suggested that meiobenthic prey species are consumed primarily by meiobenthic predators and thus were not available to higher trophic levels. *Lasserre et al.* (1975) speculated that there may be competition for food between species of detrital-feeding meiofauna and the European grey mullet, *Chelon labrosus*. *McIntyre* (1964) and *Marshall* (1970) concluded that there is competition for food between macrofauna and meiofauna and that meiofauna serve primarily as rapid metazoan nutrient regenerators. However, *Feller and Kaczynski* (1975) and *Sibert et al.* (1977) showed quite conclusively that juvenile salmon feed almost exclusively on meiobenthic copepods, and *Odum and Heald* (1972) reported that meiobenthic copepods comprise 45% of North American

grey mullet gut contents. Also, *Sikora* (1977) has reported that
nematodes provide a significant portion of the *in situ* food of
the grazing grass shrimp *Palaemonetes pugio*.

It is most likely, however, that meiofauna serve as food
for higher trophic levels more in muds than in sands. Table 3
lists those papers that discuss the functional role of meiofauna
in situ and in almost all cases where meiofauna are known to be
food for higher trophic levels the study has been conducted in
a muddy or detrital substrate. Conversely those papers which
suggest that meiofauna are not food for higher trophic levels
are based primarily on work done in sandy environments. In
muddy/detrital substrates, most of the meiofauna are restricted
to the top-most sediment layers and an indiscriminate browser/
ingester would inevitably collect the resident meiofauna. However, with all the available interstitial space in sands,
meiofauna generally go deeper into the sediment and would not
be as susceptible to browsing predation. Figure 2 schematically
illustrates mud and sand cores and associated redox layers with
the mean numbers of animals and biomass from 5 years of monthly
subtidal data in North Inlet, South Carolina. The mud core has
approximately twice the meiofaunal biomass as the sand core and
in the mud all the fauna is concentrated in the upper 1 cm,

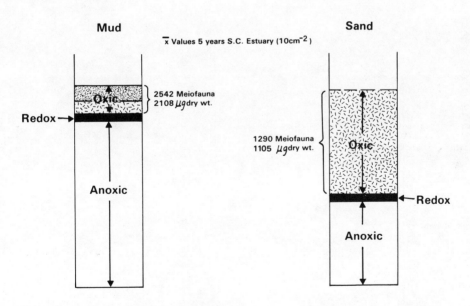

Fig. 2. A schematic diagram of the sediment at a muddy and
sandy site in North Inlet, South Carolina, with numbers and biomass at the two sites. The oxidized layer in the mud is limited
to 1 cm; in the sand it varies between 10 and 15 cm.

TABLE 3. Summary of papers which mention functional role of meiofauna *in situ*.

FOOD FOR HIGHER TROPHIC LEVELS		NUTRIENT REMINERALIZATION/ COMPETITION WITH MACROFAUNA	
Author	Substrate (Predator)	Author	Substrate (Predator)
Smidt (1951)	mud/sand (plaice)	McIntyre (1964)*	sand/mud
Bregnballe (1961)	mud/sand (plaice)	Tietjen (1967)	muddy sand
Teal (1962)	mud (*Uca*)	McIntyre (1969)	sand
Muus (1967)	mud (plaice)	Marshall (1970)	sand
Mauchline (1971a,b)	detritus (mysids)	McIntyre (1971)*	sand
Odum & Heald (1972)	detritus (several fish)	McIntyre & Murison (1973)	sand
Braber & DeGroot (1973)	? (larval plaice)	Lasserre *et al.* (1975)*	detritus-mud
Walter (1973)	mud (sipunculan)	Giere (1975)	sand
Feller & Kaczynski (1975)	epibenthic (larval salmon)	McLachlan (1977)	sand
Sibert *et al.* (1977)	detritus (larval salmon)	Tenore *et al.* (1977)	lab – sand
Sikora (1977)	mud (grass shrimp)		
Bell & Coull (1978)			

*Suggest competition with macrofauna.

whereas in the sand the fauna is distributed to a depth of 10-15
cm. Obviously, if higher trophic levels feed on meiofauna it
seems reasonable to expect predation pressures to be most pronounced in muddy/detrital substrates and not of particular
significance in sands. This, of course, does not preclude
higher level predation of meiofauna in sands, or nutrient
regeneration/mineralization/bioturbation, etc., in muds, but
does emphasize the primary role in each biotope. As a hypothesis
then, we suggest that the mud/detrital meiofauna does serve as a
significant source of food for higher trophic levels (primarily
natant browsers), whereas in sandy substrates the members of the
meiofauna serve primarily as relatively rapid metazoan nutrient
regenerators.

There is also evidence that the meiofauna plays a role in
making detritus available to macroconsumers (*Tenore et al.*, 1977).
Net incorporation rates of five-month-aged eelgrass detritus by
the polychaete *Nephthys*, cultured with and without meiobenthos,
were nearly doubled in cultures containing meiofaunal organisms.
The authors suggested that the observed increase in net incorporation of the aged detritus could be due to a combination of
factors such as "meiofaunal enhancement of microbial activity
and subsequent polychaete utilization and/or ingestion of the
meiofauna themselves."

If the mud meiofauna does serve as a significant food source
for higher trophic levels, what effects do these predators have
on the resident meiofauna? We (*Bell and Coull*, 1978) have shown
that predation/disturbance by the grass shrimp *Palaemonetes pugio*
significantly reduced meiofauna density in a salt marsh environment. The reduction in numbers was particularly acute in the
vermiform meiofauna (nematodes, polychaetes, oligochaetes) but
not in the randomly darting copepods. We suspect that a sediment
tactile "picker" like *Palaemonetes* cannot catch the copepods as
effectively and that copepods are most likely to be captured by
sight feeders, primarily fish and fish larvae.

Most research has suggested that control of meiofaunal
populations is by the physical properties of the habitat (*Coull*,
1969; *McIntyre*, 1971). While this is true in the most general
sense--interstitial forms live in sand; burrowers in mud; density
tends to fluctuate seasonally in relation to temperature; certain
species are limited by salinity, etc., etc.,--it is becoming
increasingly apparent that biological interactions, whether
they be quality and quantity of food, roots and worm tubes
increasing structural heterogeneity, or predatory controls,
play a much more significant role in structuring meiofaunal
assemblages than previously thought.

PROSPECTUS FOR MEIOFAUNAL ECOLOGY

What follows will, without doubt, indicate the biases of the authors on the direction we feel ecological research on meiofauna should take. Others may not agree with us and will obviously follow different investigative pathways, but we list our perceptions of current needs and welcome additional suggestions.

Energetically, just what role does the meiofauna play? If it does serve as food for higher trophic levels (at least in muds), how much of the requirements of its predators does it meet? One extremely promising way to attack such a question *in situ* is by immunochemistry. Many "suspected" predators grind their food and the non-cuticularized nature of most meiofauna precludes recognition of meiofaunal remains in the guts of predators. If one collects live meiofaunal organisms in bulk and homogenizes them (whether a phylum like nematodes or a specific species), a cell-free protein extract can be obtained. This extract (primary antigen) can then be injected into laboratory rabbits to obtain an antibody specific to a given taxon. The resultant antiserum is then exposed by a modified Ouchterlony double diffusion precipitin test (*Mansi*, 1958) to the gut contents of suspected meiofauna predators which were frozen in the field immediately upon collection to arrest enzymatic breakdown of the gut contents. A match between the gut contents and the extract used to produce the antiserum will produce distinct precipitin bands, indicating the presence of the same protein in the extract and in the guts of the potential predators. Such a positive result, indicating that the potential predator has preyed on the meiofaunal taxon tested, can be quantitified; the amount of meiofaunal protein in the gut can be calculated from the diffusion rate and the time of precipitin band formation. Realize, of course, that this process will not determine whether the predator assimilated the meiofauna, but it will provide a first-level look at *in situ*, and not laboratory, predation. Studies of this kind would certainly indicate whether the meiofauna is food for higher trophic levels.

Coming from the other end of the trophic spectrum, we still do not know exactly what, or how much, the meiofauna consume. *Tietjen and Lee* (1977; and other studies by their group), using radioactively tagged foods, have elegantly demonstrated specificity in the diets of several laboratory-reared marine nematodes. That meiofauna serve as "packagers" of microbial biomass has been suggested by *Sikora et al.* (1977) and *Rieper* (1978), and *Tenore et al.* (1977), as mentioned previously, have shown that detrital incorporation by *Nephthys* is in large part a function of meiofaunal interactions. *Gerlach's* (1978) suggestion that the meiofauna stimulate bacterial productivity is certainly intriguing. Further, the role of meiofaunal bioturbation in the regeneration of bottom nutrients is unknown. Meiofauna do bioturbate (*Cullen*, 1973; *Rhodes et al.*, 1977), but the extent to which they are responsible for nutrient regeneration from the bottom certainly needs to be ascertained. Quantification of these processes will not be easy, but all of these authors point to the need for increased research emphasis on meiofaunal/microbial interactions.

Coupled with the food/energetic role of meiofauna must be a concerted effort to approach *in situ* production. The studies of *Feller* (1977) and *Fleeger* (1978) have certainly started us on the right path. Short-interval cohort analyses from field data will most assuredly provide more accurate assessment of meiofauna production than use of assumed numbers for turnover rate based on laboratory generation times.

Life history studies of meiofauna can help to answer some of the pressing questions in reproductive ecology. How do small size, lack of a pelagic dispersal stage, and sediment dependence affect the reproductive success of these organisms? Are there constraints imposed because of these characteristics, and what adaptations are there that allow the meiofauna to maintain population levels of 10^6 m^{-2}? The meiofauna presents a challenge to reproductive ecologists and certainly deserve increased attention from these revered colleagues.

Because we still do not have a field-generated life table for any meiofaunal animal, population biology questions abound. For example, what effect does competition or predation have on the fecundity, natality, and mortality of an individual species? Meiofaunologists have regularly studied adult distributions, but still unknown is the age specific mortality of a developing population. Any observed adult distribution is obviously the end result, which then requires a guess as to the dynamics within the population. Meiofaunal species, with their large numbers, benthic larvae, and relatively short generation times, provide ideal opportunities for studying problems relevant to population biology. There have been several laboratory life history studies-- it is now time to ask questions as to the particular forcing function(s) and determine whether, indeed, these functions are valid for field populations. We urge this emphasis rather than the descriptive, laboratory route of the past. Obviously it is not an easy task, but short interval sampling (daily?) and very tedious counting and measuring of juveniles and adults can provide us with invaluable information.

Spacing of animal populations has been of concern to ecologists since *Elton* (1927). Meiofaunal organisms are ideal (excluding taxonomic problems) for study of spatial problems because of their high densities, and as *Heip* (1976) has pointed

out, "the meiobenthos is probably more suited for this kind of
analysis than any other group in the marine environment." As
an extreme example, if one wished to study the spatial distribution of elephants and get an accurate distributional pattern,
one might have to sample Tanzania's entire Serengeti National
Park, whereas for meiofauna one square meter will provide more
than enough animals to determine spatial patterns accurately.
One way to approach small scale spatial distribution is to use
multiple mini-cores. We have successfully employed "drinking
straws" as cores for meiofauna to elucidate small scale patch
size; this method is a useful mechanism for determination of
potential spatial niche relationships of congeners. *Ivester
and Coull* (1977), using morphological (scanning electron
microscopy), temporal, and spatial analysis, could not adequately explain the niche fractionating mechanisms of two
sympatric meiobenthic copepods. However, their spatial analysis
was based on replicate cores 3.5 cm in diameter and possibly
overlapped several species patches. Indeed, in "drinking straw"
(0.6 cm diameter) multiple samples, the two sympatric species
did *not* co-occur, indicating that niche fractionation was most
likely on the spatial axis and not on the temporal or trophic
axes of the three dimensional niche (*sensu Pianka,* 1974).
Another practical approach to understanding the trophic niche
component is through functional morphology. *Wieser* (1953) proposed a classification of nematode feeding groups based on buccal
morphology, and *Marcotte* (1977), using slow motion video taping
and SEM, defined four basic feeding groups in harpacticoid
copepods. With the availability of more sophisticated
technology such faunal morphological studies can be refined,
greatly advance our knowledge of the way animals feed, and
perhaps help in understanding how they fit into the benthic
biocenose.

Another pressing question pursuable in the framework of
meiofauna spatial distributions is the role of predation in
influencing spatial patterns. Do predators randomly crop an
area and thus create patches? Additionally, what role do
biogenic structures play in influencing the distribution of
animals? *Pielou* (1969) points out that it is futile to attempt
conclusions about mechanisms underlying a pattern by only
examining the observed frequency distribution of the number
of individuals per unit. In the past we have simply measured
the effect--it most certainly is time to look, at least correla-
tively, at the cause of the distribution; we would argue that
in the marine environment the meiofauna are most suitable for
answering these ecological questions.

Again assuming that taxonomy is not a problem, the meiofauna
is an excellent group with which to study community structure
mechanisms. Correlation analyses (see *Thistle,* 1979) are our

primary tools at the moment, and indeed these are providing us
with significant insight into community organization. Manipulative ecology coupled with laboratory observations on species
interactions is an obvious route to pursue but manipulation with
meiofauna is not an easy task. For example, the small size of
the organisms makes it very difficult to change *in situ* densities
without disturbing the sediment structure with which the organisms
are intimately tied. In any case, there is more and more evidence
that the meiofauna is not a static, physically controlled assemblage; we must experiment with the assemblages rather than simply
describe them.

     In situ visual observations on meiofauna are virtually
impossible, we feel that several potentialities exist for
manipulative meiofaunal ecology. Experiments that include/exclude
potential meiofaunal predators/disturbers (*sensu Bell and Coull,*
1978) are extremely valuable in determining the potential role of
these activities in the meiofaunal community and can provide
quantification of a biological control of meiofauna. Changing
the biogenic (e.g., tube density, root density) structure of
the habitat would also be a good meiofaunal experiment to test
structural effects. If this technique were coupled with predator exclusion, it might be possible to test the idea that
biogenic structures serve as a refuge from predators. Caging
experiments have been extremely useful in many macrofaunal
investigations, and we have been pleased with the results of
our cages in a high *Spartina* salt marsh. However, we must
caution that meiofauna caging has its limits! We do not think
that meiofaunal caging would be *apropos* in a subtidal habitat
or in a hydrodynamically active intertidal system. In the high
marsh environment, currents and sedimentation are negligible
(*Richard,* 1978). Further, the *Spartina* roots provide integrity
to the substrate, and the culms serve as a natural baffle, thus
minimizing anomalies. In other habitats the absence of these
properties may impose artificial "cage effects" upon the system.
In any case, appropriate cage and habitat controls must be run
and the data viewed with caution (see *Dayton,* 1979).

     We also feel that the meiofauna represent an assemblage
amenable to the testing of recolonization/disturbance theory.
Ecological generalists might define meiofaunal organisms as
opportunistic since they have relatively rapid generation
times and may produce a large number of young, but the lack
of a pelagic dispersal stage for rapid spreading of the species
most likely retards rapid colonization. By artificially disturbing a sediment (killing the fauna) of known meiofaunal
density and composition, and following subsequent colonization
by short-interval sampling (daily?), one should be able to
determine colonization rates of the organisms. With such an
experimental design we can envision being able to answer several

questions of ecological and meiobenthological importance.
1) How rapid is recolonization by meiofauna of a disturbed area?
2) Is the recolonization directional? Is it unidirectional
because of the organisms' being "carried" by currents or is it
multidirectional, coming from all sides of the disturbed area?
3) Are there density and species gradients into the disturbed
area? *MacArthur and Wilson* (1967) have suggested that, in general,
areas further away from the species source will have a lower number
of species, and that one should therefore see a species gradient.
4) Are there "opportunistic" meiobenthic species that will
colonize first? If so, who are they and what specific characteristic morphologies and dispersal mechanisms do they have? For
example, are they epibenthic forms (we can distinguish the mode
of existence of copepods by their morphology--see *Coull*, 1977)
or are they burrowers? 5) Are there, as would be predicted
from Island Biogeography Theory, the typical immigration and
extinction curves that eventually reach equilibrium densities?
6) How long does it take for a disturbed meiobenthic assemblage
to return to the natural state?

    Another aspect of community organization worthy of future
investigations'involves the interaction between meiofauna and
recently settled macrofauna (temporary meiofauna). *Woodin* (1976)
has shown that many adult macrofaunal organisms interact significantly with their own larvae, or with those of other macrofauna.
Even though these interactions are of great importance to the well
being of a species, if competition plays an important role in
structuring communities, competition is most likely between
similarly sized organisms. We would argue, then, that the probability of competition is greater between a larval polychaete
and a nematode, for example, than between an adult and a larval
polychaete of the same species. In fact, we see some evidence
of an inverse correlation between juvenile polychaetes and nematodes in our South Carolina data. The problem needs to be
attacked by manipulation of meiofaunal and adult macrofaunal
densities where larval macrofauna are settling. A formidable
task, but certainly worthwhile!

    We must keep in mind that the meiofauna is a dynamic
element of the marine environment. With our fairly good
general knowledge from the descriptive base it is now time
to use the meiofauna to test interactions and hypotheses
relevant to ecology. We are on the verge of a new era in
meiofaunal ecology. To ignore the potentialities of these
ubiquitous organisms would be inane and a waste of many years
of good background research. No longer should the meiofauna
simply be described and abandoned--if we ask the appropriate
questions in our research objectives, the meiofauna can effectively be used to answer many of the questions of importance to
all ecologists.

REFERENCES

Bastian, H. C., 1865. Monograph on the Anguilluidae, or free nematoides, marine, land, and freshwater; with descriptions of 100 new species, *Trans. Linn. Soc. London, 25:* 73-184.

Bell, S. S. and B. C. Coull, 1978. Field evidence that shrimp predation regulates meiofauna, *Oecologia (Berl.) 35:* 141-148.

Bell, S. S., M. C. Watzin, and B. C. Coull, 1978. Biogenic structure and its effect on the spatial heterogeneity of meiofauna in a salt marsh, *J. Exp. Mar. Biol. Ecol. 35:* 99-107.

Bougis, P., 1946. Analyse quantitative de la microfauna d'une vase marine à Banyuls, *C. R. Acad. Sci. 222:* 1122-1124.

de Bovée, F. and J. Soyer, 1974. Cycle annuel quantitative du Méiobenthos des vases terrigènes côtières. Distribution verticale, *Vie Milieu 24:* 141-157.

Braber, L. and S. J. De Groot, 1973. The food of five flatfish species (*Pleuronectiformes*) in the southern North Sea, *Nethl. J. Sea. Res. 6:* 163-172.

Bregnballe, F., 1961. Plaice and flounder as consumers of the microscopic bottom fauna, *Meddr Danm. Fisk.-og Havunders., N.S. 3:* 133-182.

Bush, L. F., 1966. Distribution of sand fauna in beaches at Miami, Florida, *Bull. Mar. Sci. 16:* 58-75.

Claus, C., 1866. Die Copepoden-Fauna von Nizza. Ein Beitrag zur Charakteristik der Formen und deren Abänderungen "im Sinn Darwins," *Schr. ges. Naturw. Marsburg, suppl. 1:* 1-34.

Coull, B. C., 1968. *Shallow Water Meiobenthos of the Bermuda Platform,* 189 pp., Ph.D. Thesis, Lehigh University.

Coull, B. C., 1969. Hydrographic control of meiobenthos in Bermuda, *Limnol. Oceanog. 14:* 953-957.

Coull, B. C., 1970. Shallow water meiobenthos of the Bermuda platform, *Oecologia (Berl.) 4:* 325-357.

Coull, B. C., 1972. Species diversity and faunal affinities of meiobenthic Copepoda in the deep sea, *Mar. Biol. 14:* 48-51.

Coull, B. C., 1973. Estuarine meiofauna: review, trophic
    relationships and microbial interactions, In: *Estuarine
    Microbial Ecology*, edited by L. H. Stevenson and R. R.
    Colwells, 499-511, Univ. S. Carolina Press, Columbia, S. C.

Coull, B. C., 1977. Marine Flora and Fauna of the Northeastern
    United States. Copepoda: Harpacticoida, *NOAA Tech. Rpts.*,
    NMFS Circ. 399, 49 pp.

Coull, B. C. and W. B. Vernberg, 1970. Harpacticoid copepod
    respiration: *Enhydrosoma propinquum* and *Longipedia
    helgolandica*, *Mar. Biol. 5:* 341-344.

Cullen, D. J., 1973. Bioturbation of superficial marine sedi-
    ments by interstitial meiobenthos, *Nature 242:* 323-324.

Dayton, P. K., 1979. Ecology: a science and a religion, In:
    Ecological Processes in Coastal and Marine Systems, edited
    by R. J. Livingston, Plenum, N. Y. (this volume).

Elton, C., 1927. *Animal Ecology*, 209 pp., Sidgwick & Jackson,
    London.

Feller, R. J., 1977. *Life History and Production of Meiobenthic
    Harpacticoid Copepods in Puget Sound*, 249 pp., Ph.D. Thesis,
    Univ. Washington.

Feller, R. J. and V. W. Kaczynski, 1975. Size selective
    predation by juvenile chum salmon *(Oncorhynchus keta)* on
    epibenthic prey in Puget Sound, *J. Fish. Res. Bd Can.
    32:* 1419-1429.

Fenchel, T., 1969. The ecology of marine microbenthos IV.
    Structure and function of the benthic ecosystem, its
    physical factors and the microfauna communities with
    special reference to the ciliated Protozoa, *Ophelia 6:*
    1-182.

Fenchel, T. and B.-O. Jansson, 1966. On the vertical distri-
    bution of the microfauna in the sediments of a brackish
    water beach, *Ophelia 3:* 161-177.

Fenchel, T. M. and R. J. Riedl, 1970. The sulfide system: a
    new biotic community underneath the oxidized layer of
    marine sand bottoms, *Mar. Biol. 7:* 255-268.

Fenchel, T., B.-O. Jansson, and W. v. Thun, 1967. Vertical and
    horizontal distribution of the metazoan microfauna and of
    some physical factors in a sandy beach in the northern part
    of the Øresund, *Ophelia 4:* 227-244.

Fleeger, J. W. Secondary production of some meiobenthic copepods in a South Carolina salt marsh (unpublished).

Fleeger, J. W., 1979. Population dynamics of three estuarine meiobenthic harpacticoids (Copepoda) in South Carolina, *Mar. Biol.* 52: 147-156.

Ganapati, P. N. and G. C. Rao, 1962. Ecology of the interstitial fauna inhabiting the sandy beaches of the Waltair coast, *J. Mar. Biol. Ass. India* 4: 44-57.

Gerlach, S. A., 1971. On the importance of marine meiofauna for benthos communities, *Oecologia (Berl.)* 6: 176-190.

Gerlach, S. A., 1977. Attraction to decaying organisms as a possible cause for patchy distribution of nematodes in a Bermuda beach, *Ophelia* 16: 151-165.

Gerlach, S. A., 1978. Food chain relationships in subtidal silty sand, marine sediments and the role of meiofauna in stimulating bacterial productivity, *Oecologia (Berl.)* 33: 55-69.

Giere, O., 1975. Population structure, food relations and ecological role of marine oligochaetes, with special reference to meiobenthic species, *Mar. Biol.* 31: 139-156.

Gray, J. S., 1966a. The attractive factor of intertidal sands to *Protodrilus symbioticus*, *J. Mar. Biol. Ass. U. K.* 46: 627-645.

Gray, J. S., 1966b. The response of *Protodrilus symbioticus* (Giard) (Archannelida) to light, *J. Anim. Ecol.* 35: 55-64.

Gray, J. S., 1966c. Selection of sands by *Protodrilus symbioticus* (Giard), *Veröff. Inst. Meeresforsch. Bremerh.* 11: 105-116.

Gray, J. S., 1967a. Substrate selection by the archiannelid *Protodrilus rubropharyngeus*, *Helgolander wiss. Meeresunters.* 15: 253-269.

Gray, J. S., 1967b. Substrate selection by the archiannelid *Protodrilus hypoleucas* Armenante, *J. Exp. Mar. Biol. Ecol.* 1: 47-54.

Gray, J. S., 1968. An experimental approach to the ecology of the harpacticoid *Leptastacus constrictus* Lang, *J. Exp. Mar. Biol. Ecol.* 2:278-292.

Gray, J. S. and R. M. Johnson, 1970. The bacteria of a sandy
    beach as an ecological factor affecting the interstitial
    gastrotrich *Turbanella hyalina* Schultze, *J. Exp. Mar. Biol.
    Ecol. 4:* 119-133.

Gray, J. S. and R. M. Rieger, 1971. A quantitative study of the
    meiofauna of an exposed sandy beach, at Robin's Hood Bay,
    Yorkshire, *J. Mar. Biol. Ass. U. K. 51:* 1-20.

Guille, A. and J. Soyer, 1968. La faune benthique des substrates
    meubles de Banyuls-sur-Mer. Premières donnés qualitatives
    et quantitatives, *Vie Milieu 19:* 323-360.

Harrison, B., 1977. The effect of sand ripples on the small-
    scale distribution of the intertidal meiobenthos with
    particular reference to the harpacticoid copepods.
    Abst. 3rd Intn. Conf. Meiofauna Hamburg, FRG.

Heip, C., 1976. The spatial pattern of *Cyprideis torosa* (Jones,
    1850) (Crustacea: Ostracoda), *J. Mar. Biol. Ass. U. K.
    56:* 179-189.

Heip, C. and W. DeCraemer, 1974. The diversity of nematode
    communities in the southern North Sea, *J. Mar. Biol. Ass.
    U. K. 54:* 251-255.

Heip, C. and P. Engels, 1977. Spatial segregation of copepod
    species from a brackish water habitat, *J. Exp. Mar. Biol.
    Ecol. 26:* 77-96.

Heip, C. and N. Smol, 1975. On the importance of *Protohydra
    leuckarti* as a predator of meiobenthic populations, *10th
    European Symp. Mar. Biol., Ostend, 2:* 285-296.

Hogue, E. W., 1978. Spatial and temporal dynamics of a subtidal
    estuarine gastrotrich assemblage, *Mar. Biol. 49:* 211-222.

Hummon, W. D., J. W. Fleeger, and M. R. Hummon, 1976. Meiofauna-
    macrofauna interactions: I. Sand beach meiofauna affected
    by maturing *Limulus* eggs, *Chesapeake Sci. 17:* 297-299.

Ivester, M. S. and B. C. Coull, 1977. Niche fractionation
    studies of two sympatric species of *Enhydrosoma*,
    Mikrofauna Meeresboden 61: 131-145.

Jansson, B.-O., 1967. The importance of tolerance and preference
    experiments for interpretation of mesopsammon field experi-
    ments, *Helgolander wiss. Meeresunters. 15:* 41-58.

Jansson, B.-O., 1968. Quantitative and experimental studies of the interstitial fauna in four Swedish beaches, *Ophelia 5:* 1-72.

Kikuchi, T., 1966. An ecological study on marine communities of the *Zostera marina* belt in Tomioka Bay, Amakusa, Kyushu, *Publ. Amakusa Mar. Bio. Lab 1:* 3-106.

Krogh, A. and R. Spärck, 1936. On a new bottom sampler for investigation of the microfauna of the sea bottom with remarks on the quantity and significance of the benthonic microfauna, *Kgl. Danske Vidensk. Biol. Meddr 4:* 1-12.

Lasserre, P., J. Renaud-Mornant, and J. Castel, 1975. Metabolic activities of meiofaunal communities in a semi-enclosed lagoon. Possibilities of trophic competition between meiofauna and mugilid fish, *10th European Symp. Mar. Biol., Ostend, 2:* 393-414.

Lee, J. J., J. H. Tietjen, C. Mastropaolo, and H. Rubin, 1977. Food quality and the heterogeneous spatial distribution of meiofauna, *Helgolander wiss. Meeresunters. 30:* 272-282.

Levin, S. A. and R. T. Paine, 1974. Disturbance, patch formation, and community structure, *Proc. Nat. Acad. Sci., U. S., 71:* 2744-2747.

MacArthur, R. H. and E. O. Wilson, 1967. *The Theory of Island Biogeography,* 203 pp., Princeton Univ. Press., Princeton, N. J.

McIntyre, A. D., 1961. Quantitative differences in the fauna of boreal mud associations, *J. Mar. Biol. Ass. U. K. 41:* 599-616.

McIntyre, A. D., 1964. Meiobenthos of sublittoral muds, *J. Mar. Biol. Ass. U. K. 44:* 665-674.

McIntyre, A. D., 1968. The meiofauna and macrofauna of some tropical beaches, *J. Zool., Lond., 156:* 377-392.

McIntyre, A. D., 1969. Ecology of marine meiobenthos, *Biol. Rev. 44:* 245-290.

McIntyre, A. D., 1971. Control factors on meiofauna populations, *Thall. Jugoslavica 7:* 209-215.

McIntyre, A. D. and D. J. Murison, 1973. The meiofauna of a flatfish nursery ground, *J. Mar. Biol. Ass. U. K. 53:* 93-118.

McLachlan, A., 1977. Studies on the psammolittoral meiofauna of Algoa Bay, South Africa II. The distribution, composition and biomass of the meiofauna and macrofauna, *Zool. Afr. 12:* 33-60.

McLachlan, A., P. E. D. Winter, and L. Botha, 1977. Vertical and horizontal distribution of sub-littoral meiofauna in Algoa Bay, South Africa, *Mar. Biol. 40:* 355-364.

Maguire, C. and P. J. S. Boaden, 1975. Energy and evolution in the thiobios: an extrapolation from the marine gastrotrich *Thiodasys sterreri, Cah. Biol. Mar. 16:* 635-646.

Mansi, W., 1958. Slide gel diffusion precipitin test, *Nature 181:* 1289-1290.

Marcotte, B. M., 1977. *The Ecology of Meiobenthic Harpacticoids (Crustacea: Copepoda) in West Lawrencetown, Nova Scotia,* 212 pp., Ph.D. Thesis, Dalhousie Univ.

Mare, M. F., 1942. A study of the marine benthic community with special reference to the micro-organisms, *J. Mar. Biol. Ass. U. K. 25:* 517-554.

Marshall, N., 1970. Food transfer through the lower trophic levels of the benthic environment, In: *Marine Food Chains,* edited by J. H. Steele, 52-66, Oliver & Boyd, Edinburgh.

Mauchline, J., 1971a. The biology of *Neomysis interger* (Crustacea, Mysidacea), *J. Mar. Biol. Ass. U. K. 51:* 347-354.

Mauchline, J., 1971b. The biology of *Praunus flexuosus* and *P. neglectus* (Crustacea, Mysidacea), *J. Mar. Biol. Ass. U. K. 51:* 641-652.

Moore, H. B., 1931. The muds of the Clyde Sea area, III. Chemical and physical conditions; rate and nature of sedimentation and fauna, *J. Mar. Biol. Ass. U. K. 17:* 325-358.

Muus, B. J., 1967. The fauna of Danish estuaries and lagoons. Distribution and ecology of dominating species in the shallow reaches of the mesohaline zone, *Meddr Danm. Fisk.-og Havunders., N.S. 5:* 3-316.

Odum, W. E. and E. T. Heald, 1972. Trophic analyses of an estuarine mangrove community, *Bull. Mar. Sci. 22:* 671-738.

Ott, J. A., 1972. Studies on the diversity of the nematode fauna in intertidal sediments, In: *5th Eur. Symp. Mar. Biol.,* 275-285, Padova.

Pianka, E. R., 1974. *Evolutionary Ecology*, 356 pp., Harper & Row, New York.

Pielou, E. C., 1969. *An Introduction to Mathematical Ecology*, 286 pp., Wiley-Interscience, New York.

Purasjoki, K. J., 1947. Quantitative Untersuchungen uber die Mikrofauna des Meeresbodens in der Umgebung der Zoologischen Station Tvarminne an der Sudkuste Finnlands, *Commentat. Biol. (Helsinki) 9:* 1-24.

Remane, A., 1933. Verteilung und organisation der benthonischen Mikrofauna der Keiler Bucht, *Wiss. Meeresunters (Abt. Kiel) 21:* 161-221.

Remane, A., 1952. Die Besiedlung des Sandbodens im Meere und die Bedeutung der Lebensformtypen für die Ökologie, *Verh. Deutsch. Zool. Ges. 1951:* 327-359.

Renaud-Debyser, J., 1959. Contribution à l'étude de la fauna interstitielle du Basin d'Arcahon, *Proc. 15th Internatl. Congr. Zool., London:* 323-325.

Renaud-Debyser, J., 1963. Recherches écologiques sur la fauna interstitielle des sables, Basin d'Arcachon, île de Bimini, Bahamas, *Vie Milieu,* suppl., *15:* 1-157.

Renaud-Debyser, J. and B. Salvat, 1963. Éléments de prospérité des biotopes des sédiments meubles intertidaux et écologie de leurs populations en microfaune et macrofaune, *Vie Milieu 14:* 463-550.

Renaud-Mornant, J. and Ph. Serene, 1967. Note sur la microfauna de la orientale de la Malaisie, *Cah. Pacifique 11:* 51-73.

Rees, C. B., 1940. A preliminary study of the ecology of a mud flat, *J. Mar. Biol. Ass. U. K. 24:* 195-199.

Rhodes, D. C., R. C. Aller, and M. B. Goldhaber, 1977. The influence of colonizing benthos on physical properties and chemical diagenesis of the estuarine seafloor, In: *Ecology of Marine Benthos,* edited by B. C. Coull, 113-138, Univ. S. Carolina Press, Columbia, S. C.

Richard, G. A., 1978. Seasonal and environmental variations in sediment accretion in a Long Island salt marsh, *Estuaries 1:* 29-35.

Riemann, F., 1966. Die enterstitielle Fauna im Elbe estuar, *Arch. Hydrobiol.,* suppl., *31:* 1-279.

Rieper, M., 1978. Bacteria as food for marine harpacticoid
    copepods, *Mar. Biol. 45:* 337-346.

Schultze, M., 1853. Uber *Chaetonotus* und *Ichtydium* Ehrbg. und
    eine neue Verwandte Gattung *Turbanella, Arch. Anat. Physiol.
    6:* 241-254.

Sibert, J., T. J. Brown, M. C. Healy, B. A. Kask, and R. J.
    Naiman, 1977. Detritus-based food webs: exploitation by
    juvenile chum salmon *(Oncorhynchus keta), Science 196:*
    649-650.

Sikora, J. P., W. B. Sikora, C. W. Erkenbrecher, and B. C. Coull,
    1977. Significance of ATP, carbon, and caloric content of
    meiobenthic nematodes in partitioning benthic biomass,
    Mar. Biol. 44: 7-14.

Sikora, W. B., 1977. *The Ecology of* Palaemonetes pugio *in a
    Southeastern Salt Marsh Ecosystem with Particular Emphasis
    on Production and Trophic Relationships,* 122 pp., Ph.D.
    Thesis, Univ. South Carolina.

Smidt, E. L. B., 1951. Animal production in the Danish Waddensea,
    Meddr Danm. Fisk.-og Havunders. Ser: Fiskeri 11: 1-151.

Stripp, K., 1969. Jahreszeitliche Fluktuationen von Makrofauna
    und Meiofauna in der Helgoländer Bucht, *Verröff. Inst.
    Meeresforsch. Bremerh. 12:* 65-94.

Swedmark, B., 1964. The interstitial fauna of marine sand,
    Biol. Rev. 39: 1-42.

Teal, J. M., 1962. Energy flow in the salt marsh ecosystem of
    Georgia, *Ecology 43:* 614-624.

Teal, J. and W. Wieser, 1966. The distribution and ecology of
    nematodes in a Georgia salt marsh, *Limnol. Oceanog. 11:*
    217-222.

Tenore, K. R., J. H. Tietjen, and J. J. Lee, 1977. Effect of
    meiofauna on incorporation of aged eelgrass, *Zostera marina,*
    detritus by the polychaete *Nephthys incisa, J. Fish. Res.
    Bd Can. 34:* 563-567.

Thiel, H., 1966. Quantitative Untersuchungen uber die Meiofauna
    des Tiefseebodens. (Vorlaufiges Erge buis des "Meteor"-
    Expedition in den Indischen Ozean), *Veröff. Inst.
    Meeresforsch. Bremerh. 2:* 131-147.

Thiel, H., 1975. The size structure of the deep sea benthos, *Int. Revue ges. Hydrobiol. 60:* 575-606.

Thistle, D., 1979. Harpacticoid copepods and biogenic structures: implications for deep-sea diversity maintenance, In: *Ecological Processes in Coastal and Marine Systems,* edited by R. J. Livingston, Plenum, N. Y. (this volume).

Tietjen, J. H., 1966. *The ecology of Estuarine Meiofauna with Particular Reference to the Class Nematoda,* 238 pp., Ph.D. Thesis, Univ. Rhode Island.

Tietjen, J. H., 1967. Observations on the ecology of the marine nematode *Monohystera filicaudata* Allgen 1929, *Trans. Amer. Micros. Soc. 86:* 304-306.

Tietjen, J. H., 1969. The ecology of shallow water meiofauna in two New England estuaries, *Oecologia (Berl.) 2:* 251-291.

Tietjen, J. H. and J. J. Lee, 1972. Life cycles of marine nematodes. Influence of temperature and salinity on the development of *Monohystera denticulata* Timm., *Oecologia (Berl.) 10:* 167-176.

Tietjen, J. H. and J. J. Lee, 1977. Feeding behavior of marine nematodes, In: *Ecology of Marine Benthos,* edited by B. C. Coull, 21-36, Univ. S. Carolina Press, Columbia, S. C.

Vitiello, P., 1968. Variations de la densité du microbenthos sur une aire restreinte, *Rec. Trav. St. Mar. d'Endoume, Bull., 43:* 261-270.

Walter, M. D., 1973. Fressverhalten und Darminhaltsuntersun-chungen bei Sipunculiden, *Helgolander wiss. Meeresunters. 25:* 486-494.

Wiens, J. A., 1977. On competition and variable environments, *Amer. Sci. 65:* 590-597.

Wieser, W., 1953. Die Beziehung zwischen Mundhöhlengestalt, Ernährungsweise und Vorkommen bei freilebenden Marinen Nematoden. Eine okologischmorphologische Studie, *Ark. Zool., Ser. II 4:* 439-484.

Wieser, W., 1959. The effect of grain size on the distribution of small invertebrates inhabiting the beaches of Puget Sound, *Limnol. Oceanog. 4:* 181-194.

Wieser, W., 1960. Benthic studies in Buzzards Bay II. The meiofauna, *Limnol. Oceanog. 5:* 121-137.

Wieser, W., J. Ott, F. Schiemer, and E. Gnaiger, 1974. An ecophysiological study of some meiofauna inhabiting a sandy beach at Bermuda, *Mar. Biol. 26:* 235–249.

Wigley, R. and A. D. McIntyre, 1964. Some quantitative comparisons of offshore meiobenthos and macrobenthos south of Martha's Vineyard, *Limnol. Oceanog. 9:* 485–493.

Woodin, S. A., 1976. Adult-larval interactions in dense infaunal assemblages: patterns of abundance, *J. Mar. Res. 34:* 25–41.

HARPACTICOID COPEPODS AND BIOGENIC STRUCTURES: IMPLICATIONS FOR

DEEP-SEA DIVERSITY MAINTENANCE[1,2]

David Thistle

Florida State University

ABSTRACT

Although several models have been proposed to explain the maintenance of enhanced diversity in the deep sea, the data do not clearly support a particular view. This paper reports a study of the relationship of harpacticoid copepods of the San Diego Trough (1200 m depth) to biogenic environmental structures. Harpacticoid species are shown to be significantly associated with such structures. The associations appear to be weak, but the sign of the correlation coefficient between particular species and individual structural classes is conserved on the average. These results support Jumars' grain-matching model and have implications for several of the other models of diversity maintenance.

INTRODUCTION

Hessler and Sanders (1967) demonstrated that the apparent low diversity of the deep sea (see *Ekman*, 1953; *Marshall*, 1954; *Bruun*, 1957) was a sampling artifact, and that, rather than being a region of few species, the deep sea was among the most diverse marine, soft-bottom areas. This result presented deep-sea workers

---

[1]This paper is derived from a Ph.D. dissertation submitted to the Scripps Institution of Oceanography, University of California, San Diego.

[2]Contribution number 12 from Expedition Quagmire.

with an apparent paradox because the deep sea did not appear to
have characteristics which readily suggested why it should be
more diverse than other marine regions.

Several models have been proposed to resolve the paradox of
high deep-sea diversity. *Sanders* (1968, 1969), and *Slobodkin and
Sanders* (1969) developed the stability-time hypothesis. In this
view, the stability of conditions in the deep sea over long periods
of time permitted species to become biologically accommodated to
each other by evolving highly specialized niches which overlap
minimally. In contrast, *Dayton and Hessler* (1972) suggested that,
given the constancy of the environment and food resources, there
were insufficient independent factors to permit large numbers of
species to have non-overlapping niches. Rather, predation and
disturbance kept potential competitors sufficiently low in
abundance that resources were never limiting. *Grassle and Sanders*
(1973) and *Menge and Sutherland* (1976) have presented arguments
which incorporate aspects of both models.

Jumars (1975a, b) presents another view. Many mud-bottom
organisms create microenvironmental heterogeneity. In physically
stable habitats, such heterogeneity persists at least as long as
the lifetime of the organism and, therefore, is available to other
animals for habitat partitioning. Further, because this heterogeneity is organism-generated and maintained, it has the same
spatial and temporal scales as animal ambits and life spans.
Following *Hutchinson's* (1961) arguments, environmental changes on
these scales are those most likely to minimize competitive exclusion. As a result, Jumars suggests that scale matching between
species and biogenic environmental heterogeneity could permit
high diversity to be maintained in the deep sea.

Jumars (1975b) supported his model by citing evidence that
small-ambit polychaete species as a group were more diverse than
large-ambit species, as would be expected if biogenic microenvironmental heterogeneity regulated community diversity. *Thistle* (1978)
analyzed harpacticoid copepod dispersion patterns and found that
harpacticoid species were discordant in their abundance patterns
in pairs of contiguous 10 x 10 cm samples; and, for some highly
aggregated species, there was evidence that the patch size was less
than 10 cm in diameter. Thistle suggested that these data supported
the grain-matching model because organism-generated environmental
heterogeneity was likely to generate patches with scales on the
order of centimeters in dimension which corresponded to scales of
patchiness observed for harpacticoids.

However, these scale data, at best, indirectly test the grain-
matching model. This paper attempts to establish a more direct
relationship between organism-generated environmental heterogeneity and harpacticoid copepod species.

MATERIALS AND METHODS

Locality

The sample site was located in the San Diego Trough (Fig. 1) at 1218.3 to 1223.8 m depth near the base of the Coronado Escarpment (32° 35.75'N, 117° 29.00 W) away from areas of known turbidity channels. *Thistle* (1978) found that the study site can be characterized as having the physical stability typical of the deep sea: granulometric analyses showed no evidence of recent disturbance by turbidity flows; measurements of temperature, salinity, and dissolved oxygen revealed little variability.

The samples were taken as part of Expedition Quagmire (*Thiel and Hessler*, 1974). The project was designed around the capabilities of the Remote Underwater Manipulator, which took cores *in situ* with great deliberateness yielding samples which were essentially undisturbed. In particular, the shock wave which precedes non-deliberate samplers (*McIntyre*, 1971; *Hessler and Jumars*, 1974; *Jumars*, 1975a, b) was eliminated. Sample locations were determined to within a meter.

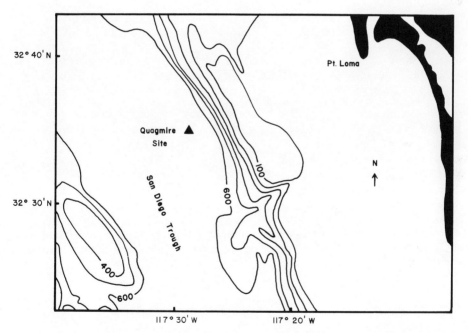

Fig. 1. Chart of sampling area. The filled triangle marks the Quagmire site. Depth contours are in fathoms. Modified from Coast and Geodetic Survey Map N. 5101.

Fifty-eight samples were taken in a stratified random manner
from the triangular study site (Fig. 2) using a modified Ekman
grab (20 x 20 cm). The grab was partitioned internally into four
subcores (10 x 10 cm) which were the units of this study. I
analyzed the harpacticoid fauna from fourteen subcores (pairs
of subcores from six cores and two single subcores). Figure 2
shows the distribution of samples in the study site; Table 1 gives
the intersample distances.

The top 1 cm layer and overlying water for each subcore were
formalin fixed at sea. In the laboratory, each sample was
screened through sieves of 1.00 mm and 0.062 mm mesh and transferred to ethanol. The 0.062 mm fractions containing the
harpacticoids were sorted; adults were identified to species
and counted.

The > 1.00 mm fraction was used to quantify the abundance
of biogenic structures (e.g. tubes, mud balls). Structures were

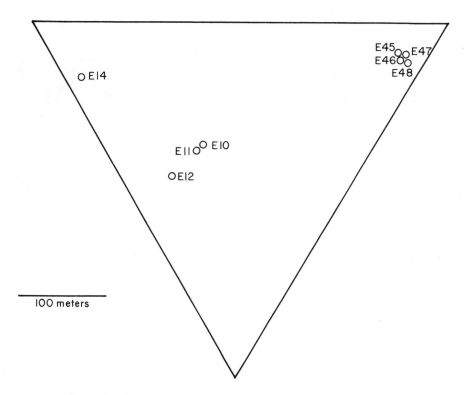

Fig. 2. The Quagmire-site sampling triangle. The Ekman
cores treated in this study are indicated by circles.

TABLE 1.   Distances in meters between Ekman cores.

Sample

	E11	E12	E14	E45	E46	E47	E48
E10	6.0	50.5	152.2	239.9	239.8	239.9	242.1
E11		44.6	152.3	244.5	244.5	244.5	246.7
E12			151.5	284.9	284.9	284.8	286.9
E14				356.6	356.9	357.6	359.9
E45					0.1	2.1	3.4
E46						2.0	3.4
E47							2.2

grouped into seven classes: (1) mud balls formed by the cirratulid polychaete *Tharyx luticastellus* (*Jumars*, 1975b, c);
(2) smaller mud balls made by a congener, *Tharyx monilaris*,
Fig. 3A; (3) all other polychaete tubes; (4) tests of the
agglutinating foraminiferan genus *Orictoderma*, Fig. 3A;
(5) tube-shaped agglutinating Foraminifera, Fig. 3B; (6) bush-
like agglutinating Foraminifera, Fig. 3C; (7) tanaid crustacean
tubes.   All structures or fragments of structures which exceeded
0.5 mm in minimum axial dimension and were retained on a 1.0 mm
sieve were classified.   The maximum orthogonal length and width
were measured except that the second widest dimension was used
for branched forms to **represent more accurately** the diameter of
the volume occupied by the organism.   The shape of each class
was approximated by a sphere (2, 4 of above), a cylinder (3, 5,
7), or a prolate elipsoid (1, 6) and volumes for each class were
calculated (Table 2).

The correlation coefficients cited in *Results* are *Kendall*
(1948) rank correlation coefficients.   The five percent significance level was used unless stated otherwise.

RESULTS

The habitat varies locally in the amount of a given
structural class present (Table 2).   I calculated rank correlation
coefficients using all 14 subcores for each harpacticoid species'

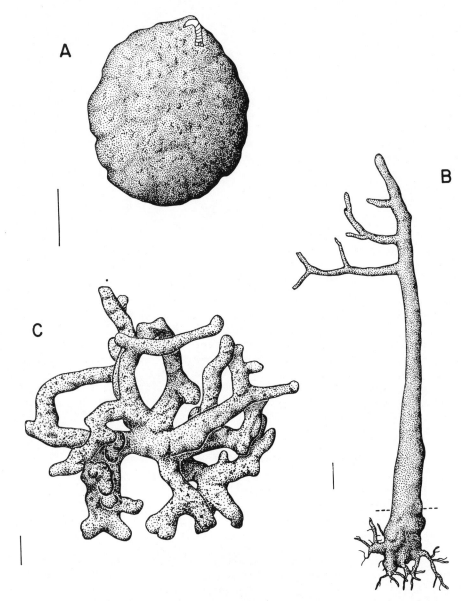

Fig. 3. Quagmire-site biogenic structures. A. The empty
test of the foraminiferan *Orictoderma* sp. which, in this instance,
is occupied by the polychaete *Tharyx monilaris*. B. A tube-shaped
foraminiferan; the dashed line indicates the surface of the sediment. C. A bush-like foraminiferan. The scale lines equal 1.0 mm.

TABLE 2. Volume (mm^3) of biogenic structures by class in 100 cm^2 subcores from the San Diego Trough.

Subcore	Structural Class						
	Tharyx luticastellus	*Tharyx monilaris*	Polychaete tubes	*Orictoderma*	Tube-shaped Foraminifera	Bush-like Foraminifera	Tanaid tubes
E10X	0.00	0.00	55.35	37.00	506.30	257.24	7.03
E11X	0.00	0.00	372.75	169.90	717.05	146.58	44.76
E12W	49394.46	38.97	414.03	152.77	1602.39	31.07	34.41
E12Z	668.52	32.78	193.46	21.58	624.47	1243.64	13.19
E14X	5782.00	0.00	267.12	16.34	1270.43	76.10	0.00
E14Y	6423.81	0.00	271.74	0.00	894.61	791.92	0.77
E45X	0.00	203.15	408.91	674.85	1655.34	354.10	2.11
E45Z	3279.22	426.02	543.78	126.24	2375.86	1247.59	31.80
E46Y	13391.23	0.00	770.46	220.16	3638.50	1019.85	23.22
E46Z	3420.61	48.97	156.76	190.17	1976.05	18.75	29.52
E47W	0.00	84.08	257.17	703.14	1684.00	357.07	0.00
E47Z	307.96	28.10	354.33	720.04	771.87	74.79	29.25
E48Y	763.63	113.84	210.48	117.77	1359.26	114.44	7.23
E48Z	10719.57	47.38	508.01	178.36	2629.75	160.13	7.10

abundance with the volume of each structural class. Species which occurred at only one station were omitted because there is no information about species covariance with structure under those circumstances. Of the 868 coefficients calculated, 88 were significant (two-tailed test), significantly more than would be expected by chance alone (chi square 1 d.f. = 48.2455, p less than 0.0005). These significant correlations are distributed among 58 species. Species differ in the class of structure with which they are correlated; among those species correlated with the same class of structure, some are negatively and some positively correlated (Table 3).

To estimate the strength of individual species-structure correlations, I divided the data into two equal stratified-random subsets. In subset A, I calculated all possible rank correlation coefficients between species and structural classes. Among those species which occurred in more than one subcore in both subset A and subset B, forty-eight significant correlations were observed in subset A (alpha less than or equal to 0.10, two-tailed test). These forty-eight correlations were used to predict specific species-structure relationships for testing in subset B. Of these 48 predicted relationships, only two were significant at the 10% level. Table 4 is a contingency table

TABLE 3.   Summary of the significant correlations between harpacticoid species and biogenic structural classes found in the San Diego Trough.

Structural Class	Number of Significant Positive Correlations	Number of Significant Negative Correlations
Tharyx luticastellus	17	4
Tharyx monilaris	6	5
Polychaete tubes	8	2
Orictoderma	7	7
Tube-shaped Foraminifera	9	7
Bush-like Foraminifera	1	6
Tanaid tubes	3	6

TABLE 4.   Number of correlation coefficients predicted to be
            significant or nonsignificant and the number observed
            to be significant or nonsignificant in data subset B.

		OBSERVED	
		Significant	Nonsignificant
	Significant	2	46
PREDICTED			
	Nonsignificant	37	494

which was used to test whether specific correlations predicted
to be significant in subset B actually were significant more
frequently than expected by chance. They were not (chi square
1 d.f. = 0.169). In summary, when 14 subcores are used to
calculate correlation coefficients, a species-structure correlation is detected; when 7 subcores are used, one is not.

The question of whether weak species-structure interactions
exist can be pursued further using this two-data subset approach.
If there is a weak but real association of a species with a
structural class, the signs of the correlations should be the
same in the two subsets. If there is no real association, then
the signs of the correlations should, on the average, differ as
often as they are the same. Table 5 gives the results of a test
of this hypothesis. The significant total chi square means that
the null hypothesis of no association between species and
structural classes can be rejected (there is no evidence of
significant heterogeneity in the results for the various
structural classes, chi square heterogeneity 6 d.f. = 7.7038,
p greater than 0.30). The chi square values for each structural
class show that *Tharyx monilaris*, *Tharyx luticastellus*, tube-
shaped Foraminifera, and tanaid tubes are responsible for the
effect.

In each of the six pairs of subcores, more individuals
occur in the more highly structured subcore (binomial p = 0.032,
two-tailed test). This effect results in large part because of
the behavior of the individually aggregated species. In each
pair of replicate subcores, the index of dispersion detects
aggregated species. For those species one subcore contains
many more individuals than the other. Table 6 shows that
these species are more abundant in the more highly structured
subcore (binomial p = 0.0025, two-tailed, *a posteriori* test).

TABLE 5.   The chi-square values resulting from testing the
           observed versus expected proportion of matching to
           mismatching in signs of correlation coefficients
           between harpacticoid species abundances and struc-
           tural class volumes in the two data subsets.

Structural Class	Chi Square	Degrees of Freedom	Probability
Tharyx luticastellus	5.1282	1	<0.025
Tharyx monilaris	6.6952	1	<0.010
Polychaete tubes	0.0127	1	>0.050
Orictoderma	0.1084	1	>0.050
Tube-shaped Foraminifera	4.3784	1	<0.050
Bush-like Foraminifera	1.2821	1	>0.050
Tanaid tubes	3.4000	1	>0.050
Total	21.0060	7	<0.005

However, this result cannot be extended to a general relationship
between harpacticoid abundance and habitat-structure abundance
(rank correlation coefficient = 0.0).

DISCUSSION AND CONCLUSIONS

    Jumars (1975a, b) in his grain-matching model of deep-sea
diversity maintenance argues that biogenic structures can create
small-scale environmental heterogeneity. Because this heterogeneity has space and time scales which correspond to the
spatial and temporal scales of species' ambits and life spans,
it creates a patch structure which permits large numbers of
similar species to co-occur. The model predicts that the
abundance of species of taxa which show enhanced diversity
in the deep sea should vary nonrandomly with variation in abundance of biogenic structures. Jumars presents data showing the
impact of mud balls made by a cirratulid polychaete on species
of the family Paraonidae (*Jumars*, 1975b) and the substantial
difference in the species composition of a bathyal polychaete

TABLE 6. The effect of biogenic structure on individually
          aggregated species at the centimeter scale. The
          table shows the number of individually aggregated
          species in a pair of subcores which have their
          greater abundance in either the more highly struc-
          tured or the less highly structured subcore.

Core	Subcore with Greater Structural Volume	Subcore with Lesser Structural Volume
E12	6	0
E14	1	0
E45	1	0
E46	4	0
E47	2	2
E48	3	1
Total	17	3

assemblage associated with the remains of a glass sponge (*Jumars*,
1974).

     This paper further tests the grain-matching model. *Thistle*
(1978) showed the enhanced deep-sea diversity of harpacticoid
copepods in the San Diego Trough. The present results document
a statistically significant association of harpacticoid species'
abundances with seven classes of biogenic structure as predicted
by the model. However, the associations between individual
species and particular classes of biogenic structure are weak
on the average; significant correlations in one data subset do
not predict significant correlations in a second subset. This
apparent weakness could arise because of a poor measure of structure or because of an inappropriate sample scale. Given that
the 10 x 10 cm sampler is large relative to the size of the
biogenic structures, other sources of variance are likely to
have been included which could obscure a strong species-structure
correlation. Whatever the true strength of such associations,
they are almost certainly real because the direction of the
correlation is maintained between data sets more often than one
would expect by chance. Further, the most aggregated species

between pairs of subcores have been shown to be more abundant in
the more highly structured subcore of a pair. These results
support the grain-matching model of deep-sea diversity maintenance
because they show that the species of a diverse deep-sea taxon
are perceiving biogenic structures as sources of environmental
heterogeneity.

The species-structure interaction results have implications
for certain of the other models of deep-sea diversity maintenance.
If the correlation of species with biogenic structures results
from competitive microhabitat partitioning, then *Sanders'* (1968)
stability-time hypothesis is supported. Moreover, it weakens
Dayton and Hessler's (1972) criticism of Sanders' model because
food resources need provide fewer niches if a portion of the
niche partitioning occurs in terms of microhabitat specialization.

Dayton and Hessler (1972) emphasize food specialization to
the neglect of other potential niche axes. *Jumars* (1975b)
criticizes this imbalance in their view using evidence of
habitat specialization in some deep-sea polychaetes. If the
correlation of harpacticoid species with biogenic structures
indicates habitat separation among potential competitors, then
their model is weakened because it predicts that such partitioning should not occur.

Menge and Sutherland's (1976) model of diversity maintenance
by predation on low trophic levels has been supported by *Rex*
(1976, 1977). In terms of the harpacticoid results, if the
covariance between harpacticoid species and biogenic structures
indicates competitive microhabitat partitioning among harpacticoids, then their model would be weakened because it suggests
that harpacticoids, as low trophic level species, should be
primarily under predator control.

The results suggest a possible explanation for the higher
harpacticoid diversity in the deep sea than in other environments.
Harpacticoid species appear to display microhabitat partitioning
in terms of biogenic environmental structures. These structures
are relatively delicate mud aggregations. In the stable physical
conditions of the deep sea, such structures persist long enough
to serve as a niche axis for harpacticoids. As *Jumars* (1976)
has argued, in the higher energy situation of less stable,
shallower environments, such structures are likely to be too
short lived to serve in this way. Without this niche dimension,
it seems reasonable to expect that these communities would be
able to accommodate fewer species. This consequence of the
physical stability appears to be one reason that the deep-sea
benthos has higher diversity than that observed in shallow-water
soft bottoms.

ACKNOWLEDGMENTS

Dr. Robert R. Hessler provided support and counsel during my tenure in his laboratory. The samples were taken during Expedition Quagmire, R. R. Hessler principal investigator; B. R. Burnett, K. Fauchald, P. A. Jumars, H. Theil, G. D. Wilson, the members of the RUM group, and the crew of R/P ORB aided in the sampling. B. R. Burnett inked Figure 3. E. L. Venrick aided in computer programming. B. C. Coull provided taxonomic counsel. B. B. Bernstein, P. A. Jumars and O. S. Tendall helped identify taxa of habitat structures. Computer work was supported by a Grant-in-Aid-of-Research from the Society of the Sigma Xi. I have benefited from discussions with W. H. Berger, J. T. Enright, A. Fleminger, R. R. Hessler, P. A. Jumars, K. Fauchald, J. A. McGowan, J. F. Siebenaller and C. Wills. B. C. Coull, P. A. Jumars, and J. M. Lawrence commented on an earlier version of this manuscript. I would like to thank these people for their help.

REFERENCES

Bruun, A. F., 1957. Deep-sea and abyssal depths, *Geol. Soc. America Mem. 67:* 641-672.

Dayton, P. K. and R. R. Hessler, 1972. Role of biological disturbance in maintaining diversity in the deep sea, *Deep-Sea Res. 19:* 199-208.

Ekman, S., 1953. *Zoogeography of the Sea,* 417 pp., Sidgwick and Jackson, London.

Grassle, G. F. and H. L. Sanders, 1973. Life histories and the role of disturbance, *Deep-Sea Res. 20:* 643-659.

Hessler, R. R. and P. A. Jumars, 1974. Abyssal community analysis from replicate box cores in the central North Pacific, *Deep-Sea Res. 21:* 185-209.

Hessler, R. R. and H. L. Sanders, 1967. Faunal diversity in the deep sea, *Deep-Sea Res. 14:* 65-78.

Hutchinson, G. E., 1961. The paradox of the plankton, *Am. Naturalist 95:* 137-145.

Jumars, P. A., 1974. *Dispersion Patterns and Species Diversity of Macrobenthos in Two Bathyal Communities,* 204 pp., Dissertation, University of California, San Diego.

Jumars, P. A., 1975a. Methods for measurement of community
    structure in deep-sea macrobenthos, *Mar. Biol. 30:* 245-252.

Jumars, P. A., 1975b. Environmental grain and polychaete species'
    diversity in a bathyal benthic community, *Mar. Biol. 30:*
    253-266.

Jumars, P. A., 1975c. Target species for deep-sea studies in
    ecology, genetics, and physiology, *Zool. J. Linn. Soc. 57:*
    341-348.

Jumars, P. A., 1976. Deep-sea species diversity: does it have
    a characteristic scale?, *J. Mar. Res. 34:* 217-246.

Kendall, M. G., 1948. *Rank Correlation Methods,* 160 pp.,
    Griffin, London.

Marshall, N. B., 1954. *Aspects of Deep-sea Biology,* 380 pp.,
    Hutchinson, London.

McIntyre, A. D., 1971. Observations on the status of subtidal
    meiofauna research, *Smithson. Contr. Zool. 76:* 149-154.

Menge, B. A. and J. P. Sutherland, 1976. Species diversity
    gradients: synthesis of the roles of predation, competi-
    tion, and temporal heterogeneity, *Am. Naturalist 110:*
    351-369.

Rex, M. A., 1976. Biological accommodation in the deep-sea
    benthos: comparative evidence on the importance of
    predation and productivity, *Deep-Sea Res. 23:* 975-987.

Rex, M. A., 1977. Zonation in deep-sea gastropods: the
    importance of biological interactions to rates of
    zonation, In: *11th Europ. Symp. Mar. Biol.,* edited by
    B. F. Keegan, P. O. Ceidigh, and P. J. S. Boaden, 521-
    530, Pergamon, Oxford.

Sanders, H. L., 1968. Marine benthic diversity: a comparative
    study, *Am. Naturalist 102:* 243-282.

Sanders, H. L., 1969. Benthic marine diversity and the
    stability-time hypothesis, *Brookhaven Symp. Biol. 22:*
    71-80.

Slobodkin, L. B. and H. L. Sanders, 1969. On the contribution
    of environmental predictability to species diversity,
    Brookhaven Symp. Biol. 22: 82-93.

Thiel, H. and R. R. Hessler, 1974. Ferngesteuertes Unterwasser-
fahrzeug erforscht Tiefseeboden, *Umschau in Wiss. und
Techn. 74:* 451-453.

Thistle, D., 1978. Harpacticoid dispersion patterns: implications for deep-sea diversity maintenance, *J. Mar. Res.
36:* 377-397.

PREDATION, COMPETITIVE EXCLUSION, AND DIVERSITY IN THE SOFT-

SEDIMENT BENTHIC COMMUNITIES OF ESTUARIES AND LAGOONS

Charles H. Peterson

University of North Carolina at Chapel Hill

ABSTRACT

A review of experiments in which large, epibenthic preda-tors are excluded from soft-sediment marine benthic communities in unvegetated portions of estuaries and lagoons and a compari-son of unvegetated areas with nearby grassbeds, where predators on the infauna are less effective, demonstrate that such soft-sediment systems, when freed from predation, usually exhibit 1) an increase in total density, 2) an increase in species richness, and 3) no tendency toward competitive exclusion by some dominant species. The currently accepted model of community organization, developed from experimental work in marine rocky intertidal communities, would predict that significant simpli-fication of the community should occur as a consequence of intense competition in such a system where density had increased sub-stantially following the removal of predators.

Four general types of explanation are developed to account for this anomalous behavior of the soft-sediment benthic communities. First, the experimental exclusions of predators may have been carried out for an insufficient length of time. Second, interference competition, which produces rapid mortality on hard substrates, may be much less common in soft substrates because crushing of individuals is made difficult by the lack of adhesive contact in soft substrates and because the three-dimensionality of soft sediments permits some segregation along another resource axis. Interference in the form of over-growth is also less common in soft sediments because the competitors best suited to utilize this strategy, colonial forms, are excluded by the mobility of the sediments and because

*the infaunal mode of existence prevents colonization onto
"secondary space," the surface of other organisms. Third,
intense negative interactions between densely established adults
and potentially colonizing larvae may be sufficient in the absence
of predation to maintain the community density below carrying
capacity and thereby prevent competitive exclusion. Fourth, if
interference competition is rare among soft-sediment benthic
organisms exploitation may be the dominant mode of competition
in infaunal systems. Sluggish, cold-blooded marine invertebrates are tolerant of the stress of low food concentration
because they have low basal metabolic rates and an ability to
restrict growth. This low energy requirement may make competitive
exclusion through exploitation competition an ineffective or
extremely slow process.*

*Thus, although the process of interference competition as
mediated by predation and physical disturbance evidently does not
govern the structure of soft-sediment marine benthic communities
in estuaries and lagoons, intense biological interactions,
particularly selective predation and disturbance by epibenthic
predators and predation by benthic organisms on colonizing
larvae, retain an important organizing role. Interspecific
competition also plays a significant role, but only through the
preemption of resources by established individuals: competitive
exclusion is apparently not achieved by invaders' causing the
mortality of established adults. However, additional observations on the natural history of infaunal organisms are needed to
confirm the significance of adult-larval interactions and the
ineffectiveness of adult-adult competitive interactions.*

INTRODUCTION

Because many marine benthic invertebrates can be successfully manipulated under natural field conditions, research on
marine macrobenthic communities during the last two decades
has included numerous experimental tests of hypotheses designed
to explain how these communities are organized. For the marine
benthic systems on rocky intertidal shorelines, results of such
manipulative experimentation consistently conform to a single
organizational pattern involving the interplay among competitive
interactions and agents of disturbance, both biological and
physical. As first rigorously demonstrated by *Connell* (1961a),
intense competition for attachment space, a limited resource in
the rocky intertidal zone, leads toward the exclusion of inferior
competitors by the competitively dominant species and toward a
virtual spatial monopoly by this single best competitor. Disturbance, in either one of two general forms, can serve to
deflect the course of competition. Predation, a form of biological
disturbance, analyzed first by *Paine* (1966), as well as physical

disturbance (*Dayton*, 1971), can prevent competitive exclusion by maintaining competing species at relatively low densities such that interactions are less intense and by periodically providing available free resources (open space in the rocky intertidal) such that available larvae can always colonize the system.

Although many of the important details of this pattern remain to be defined (for example, just what characteristics of the disturbance process are both necessary and sufficient to produce persistent co-existence among the competing occupiers of intertidal space?), subsequent experimentation in numerous, very different rocky shorelines has confirmed the ubiquitous nature of these organizational processes (see the review by *Menge and Sutherland*, 1976). Consequently, this organizational pattern has become a paradigm of rocky intertidal community organization. Furthermore, the paradigm appears to have even broader significance in that it has been shown that either bio-logical or physical disturbance can play this same role in maintaining the diversity of competing species in several other very different systems: the zooplankton of lakes (*Brooks and Dodson*, 1965; *Dodson*, 1970; *Sprules*, 1972), vegetation of grass-lands under the grazing of ungulates (*Harper*, 1969), algae on intertidal rocks (*Lubchenco*, 1978), trees in rain-forests (*Connell*, 1978) and corals on reefs (*Connell*, 1978). Although past theoretical work demonstrated that predation that did not encompass a switching mechanism was highly unlikely to maintain high diversity among competitors in a single-celled microcosm (*May*, 1973), recent models have shown that such density-independent predation can be expected to mediate the outcome of competition in an open system composed of many cells, interconnected by migration (*Slatkin*, 1974; *Caswell*, 1978).

The apparent generality of this paradigm of community organization, originally developed from experimental work on marine benthic communities inhabiting hard substrates, prompts me to attempt to apply the paradigm to those marine benthic systems found in soft sediments. In particular, I pose these questions: does this paradigm provide an accurate description of the major forces structuring marine benthic communities in soft sediments and is it, therefore, a useful predictor of events that will occur following the natural or experimental elimination of mobile, epibenthic predators from soft-sediment systems? To answer these questions, I first review the avail-able experimental data on the effects of predator removal from soft-sediment systems. I then follow this review with an inter-pretation of these results, which involves an analysis of how the importance and the nature of adult-adult and adult-larval interactions would be expected to differ between a hard and a soft substrate.

THE CONSEQUENCES OF EXCLUDING
EPIBENTHIC PREDATORS FROM SOFT SEDIMENTS

Most, but not all, of the experimental removals of predators
from marine benthic communities involve the construction of
exclusion cages. The use of cages to exclude predators from an
otherwise natural field system in order to determine the usual
effects of predation upon that system can be notoriously difficult because of the unnatural effects of the cages themselves
(*Virnstein*, 1978). For example, the cages shade the substrate,
thus altering the light regime and possibly the abundance of
macro- and micro-algae. Cages also change the pattern of
current flow, usually reducing the flow rates in their vicinity
and thereby producing often substantial deposition of sediments
inside the cages (e.g., *McCall*, 1977). Such sedimentation
suffocates many suspension-feeding marine invertebrates, whereas
for many deposit-feeding species it represents an addition of
detrital and microbial food. The surfaces of cages become
fouled with numerous organisms. Fouling by diatoms and other
microalgae can increase local food concentrations. In contrast, if suspension-feeding benthic organisms dominate the
fouling community, benthic animals within the cages may experience
a lower abundance of suspended food particles.

Although designed to exclude relatively large predators,
cages often serve as refuges into which predators can recruit
as planktonic larvae or as small juveniles and thrive free from
their own enemies. This can be a particularly severe problem
in quiet waters, where various predaceous arthropods often
increase to unusually elevated densities under the protection of
the cages (e.g., *Young et al.*, 1976; *Arntz*, 1977; *Virnstein*, 1978).

Partial cages (cage controls) constructed in such a way
as to allow normal access by predators but to utilize nearly as
much caging material as a complete cage, can be used in conjunction with full cages and uncaged controls to separate the
effects of caging from the effects of excluding predators.
However, few studies utilize an adequate set of cage controls.
Even in those cases where adequate cage controls are incorporated
into the experimental design, the existence of significant cage
effects may make impossible the goal of determining the effects
of predation on the normal benthic community unless the cage
effects and the predator effects are additive. Furthermore,
cage controls often fail because mobile predators tend to be
attracted in atypically high density to the partial cages, just
as fish are attracted to an artificial reef.

All of these problems in using caging occur commonly
during experiments done in the soft sediments of estuaries and
lagoons. Sedimentation is usually a worse problem than in the

corresponding experiments done on hard substrates because hard
substrates are not necessarily horizontal and often are in areas
free of large quantities of mobile sediments. Because wave
action is minimal in lagoons and estuaries, the recruitment
into cages by fishes, crabs, and other mobile predators tends
also to be a more severe problem during estuarine experimentation than during experiments on wave-beaten shorelines, where
most of the rocky intertidal research has been performed.
Consequently, because the cage effects may be more severe and
because many experiments fail even to include cage controls,
the results of predator exclusion experiments in the soft-sediment
systems of estuaries and lagoons must be interpreted with special
caution.

Exclusion of Predators from Unvegetated Bottoms

Several experiments have now been completed in which large,
epibenthic predators were successfully excluded from the marine
benthic communities of unvegetated soft sediments in estuaries
or lagoons. All of these experiments have used caging to
achieve exclusion. A review of the results of these studies
(Table 1) demonstrates that without exception the density of
macro-invertebrates was significantly higher inside the cages
than in control areas after some period of time. Results are
of two types: those (9 studies) which show density increasing
inside cages while remaining nearly constant outside and those
(2 studies) which show relatively little change in density
inside cages while control densities decline significantly.
In those studies where density increased inside complete cages,
total density and apparently also biomass of the macrobenthos
reached on average a level of about two to three times that of
the uncaged controls. Every species did not respond to the
same degree following the exclusion of predators. Several
species showed large increases inside the cages, while the
density of many others remained unchanged at control levels.
Although several of the large increases are probably consequences of opportunistic deposit feeders taking advantage of
the organic-rich sedimentation inside the cages, this cage
artifact does not appear to explain all of the results. In
many studies, several of the species which showed dramatic
increases inside the cages are known from independent observations (usually gut contents for fish or laboratory observations
for crabs) to be extremely important prey items in the diets
of the major predators (e.g., *Blegvad*, 1928).

One additional bias is operating to influence the conclusions drawn from any such review of published experimental
results. An unsuccessful experiment is unlikely to be published.
In this instance success might be measured by whether the caging

TABLE 1.   The effects of excluding predators from marine benthic
communities in unvegetated areas within estuaries and
lagoons.

Study	Study Site	Depth	Period of Predator Exclusion
Blegvad (1928)	"Nissum Broad" in the Limfjord, Denmark	shallow subtidal	1 summer
Naqvi (1968)	Alligator Harbor, Florida	from mid-intertidal to mean low water	2 weeks to 6 months
Commito (1976)	Newport Marsh, Newport River estuary, North Carolina	intertidal	1 to 14 months
Commito (1976)	Cross Rock, Newport River estuary, North Carolina	intertidal	1 to 14 months
Virnstein (1977)	York River estuary in Chesapeake Bay, Virginia	shallow subtidal	2½ weeks to 12 months

	Effects on Infaunal:	
Density	Species Richness	Evenness of species Abundances
several fold (?) increase; up to 60X for preferred fish foods, but no change for many poor fish foods	no data	no data, but no suggestion of mortality arising through competition at high densities
2½ to 4X increase	up to 2X increase	a slight increase; no evidence of increasing dominance
2 to 3X increase at peak, then return to control levels	up to 3X increase at peak then return to control levels	no change
up to 3X increase after 2 months, then return to control levels	up to 3X increase after 2 months, then return to control levels	a slight increase
2½X increase	2X increase	a slight decline, but little evidence of dominance created at the expense of mortality of sub-dominants

TABLE 1. Continued

Study	Site	Depth	Period
Reise (1977a, 1977b)	German Bight in Königshafen, W. Germany	intertidal	3 months in two experiments; 4 months in another
Wiltse (1977)	Barnstable Harbor, Massachusetts	intertidal	3 months and 6 months for *Polinices* only
Schneider (1978)	in Plymouth estuary, Massachusetts	intertidal	2 months
Lee (1978)	Gulf of Panama	subtidal	9-10 months
Holland *et al.* (1980)	upper Chesapeake Bay, Maryland	subtidal	12 months
Peterson (1979)	Mugu Lagoon, California	shallow subtidal	12-24 months

Summary: 1) 11 out of 11 studies show higher infaunal densities
              in the absence of predators.
          2) 9 out of 9 studies show higher infaunal species
              richness in the absence of predators.
          3) 2 out of 9 studies show a decline in evenness in
              the absence of predators, but none of the
              studies shows a trend of increasing dominance
              being achieved by competitive mortality of
              subdominants.

Effects on Infaunal:		
Density	Species Richness	Evenness
an increase of from 4X to 23X in different experiments at different times	an increase ranging from slight to 4X	small but non-significant increase, with no evidence of domi-nance developing
1.5X increase	1.3X increase	a small increase with no evidence for competitive exclusion
no change, while controls declined greatly	no data	no change, while evenness increased in controls; no evidence of compe-titive exclusion occurring
little change while controls declined greatly	slow decline while con-trols declined greatly	no consistent difference, with no obvious domi-nance emerging
large increase	slight increase	no data, but no trend of increas-ing dominance or mortality at high density
slight increase	slight increase	little change with no emerging dominance

has any significant effect. For instance, R. W. *Virnstein*
(pers. comm.) has unpublished data from just such a series of
"unsuccessful" exclusion experiments. Consequently, because of
the contributions made by this publication bias and the cage
artifacts, it is impossible to conclude unequivocally that predation is controlling the overall density of the macrobenthos in
soft sediments of lagoons and estuaries; nevertheless, predation
appears often to have an impact.

Although not all of these completed studies include data on
the behavior of species richness (the total number of species in
the system), those that do demonstrate a consistent pattern
(Table 1). After a period of predator exclusion, species richness inside cages was consistently greater than in the controls.
Again some of this difference is a cage artifact resulting largely
from the invasion of opportunistic deposit feeders that colonized
the newly deposited sediments inside the cages. That some of the
difference in species richness appears to be a real effect of the
exclusion of predators is not surprising. Predators seem quite
likely to exclude certain more susceptible species both as a
result of the physical disruption of the soft sediments caused
by their movements and also as a consequence of their trophic
activity. Because of the differential availability of certain
categories of prey, various epibenthic predators in soft-sediment
systems are known to consume epifauna more often than infauna
(*Nelson*, 1978; *Peterson*, 1979) and more shallow-dwelling animals
than deep-dwelling ones (*Woodin*, 1978; *Virnstein*, 1979). In
other words, the exclusion of predators probably does not by
itself allow new species to invade, but rather substantially
increases the survivorship of several of the more susceptible
species such that they become large enough and numerous enough
to be apparent only after predators are excluded.

Such an apparent increase in species richness immediately
following the removal of predators is not surprising and has
been observed in other systems (*e.g.*, *Harper*, 1969). However,
the surprising aspect of the experimental results from marine
soft sediments is that, even after substantial periods of time
free from predation, the species richness fails to show a convincing decline in any of the areas studied, despite the relatively
short life spans of most marine infaunal species relative to epifaunal organisms. Furthermore, no species even begins to show
numerical dominance in any of these studies. In Table 1, I utilized
a measure of how evenly the species are represented to attempt an
objective documentation of whether dominance increases following
the exclusion of predators. In only two instances does evenness
decline (indicating an increase in dominance) following the
removal of predators. However, even in these experiments, no
single species or set of species appears to be in the process of
establishing a true monopoly.

Consequently, this summary of the results of experiments
in which large epibenthic predators were excluded from certain
unvegetated areas within soft-sediment systems demonstrates that,
although the total density of the macrobenthos usually increases
to levels substantially higher than in the controls, the species
richness also increases after the predators have been removed
and remains higher than in the controls even after substantial
periods of time at elevated density. Apparently, adult-adult
competitive interactions even at the relatively high density
achieved inside predator-exclusion cages are incapable of
simplifying the soft-sediment benthic community and producing
a near monopoly of the single best competitor, as would be
predicted from the rocky intertidal paradigm of community
organization. Although the increase in macrobenthic density
inside complete cages and the initial increase in species richness may be largely the result of experimental artifact, the
general failure of these experiments to demonstrate competitively
caused mortality and subsequent community simplification at
elevated densities is a significant result that is free of cage
artifacts and other biases. Relative to the controls, density
was substantially higher inside complete cages thus providing
a test of whether mortality and some competitive exclusion of
the poorest competitors would occur as a function of density
in the soft-sediment system. In contrast to results of experiments in the rocky intertidal, such simplification of the benthic
community did not occur.

Exclusion of Predators from Grassbeds

Those few experiments (*Young et al.*, 1976; *Young and Young*,
1977; *Orth*, 1977; *Reise*, 1977a; *Virnstein*, 1978) in which large
epibenthic predators have been excluded from estuarine or lagoonal
grassbeds tend to suggest that the removal of predators has very
little effect on the macrobenthic community. This result may
be partly a consequence of successful recruitment of predaceous
shrimp and crabs inside the cages (*Young et al.*, 1976), but *Orth*
(1977) and *Reise* (1977a) suggest that the dense root mat typically
formed by marine grasses actually tends to protect infaunal
organisms from the usual physical and biological disturbances
caused by predators.

If one accepts this hypothesis that benthic organisms that
live in the sediments are at least partly protected from the
activity of their predators by the stabilizing influence of
dense root mats, one can then compare the benthic infaunal
communities of marine grassbeds to the infaunal communities in
the nearby unvegetated bottom in order to determine the natural
impact of predators on the infauna. In other words, to some
degree one can consider the infauna in densely vegetated areas

to represent a system that is naturally free of the effects of
large epibenthic predators.

       Available data (*Warme*, 1971; *Thayer et al.*, 1975; *Orth*,
1977; *Reise*, 1977a; *Peterson*, unpublished data for Bogue Sound,
N. C.) comparing the infauna of grassbeds to that in the surrounding unvegetated sediments indicate that in grassbeds, the
infauna has much higher density and biomass (perhaps by an
order of magnitude), yet far greater species richness with no
obvious tendency toward monopolization by dominant species.
Consequently, to the degree that one can view the comparison
of densely vegetated to unvegetated areas as the outcome of a
natural predator exclusion experiment, the data from this comparison support the conclusion drawn from the review of the
manipulative experimental work done in unvegetated sediments.
Much of the difference between the infaunal communities of
vegetated and unvegetated areas is certain to arise from
factors unrelated to the degree of predation (such as higher
productivity, increased physical heterogeneity, etc.). Yet
this comparison at least serves to suggest that competitive
dominance does not seem to be common in the infaunal communities
of grassbeds, where the densities of macrobenthos tend to be
extremely high and thus where competitive exclusion should
operate if the soft-sediment system shares the organizational
paradigm of the rocky intertidal.

       WHY COMPETITIVE EXCLUSION FAILS TO OCCUR IN SOFT-SEDIMENT
BENTHIC COMMUNITIES FOLLOWING THE REMOVAL OF EPIBENTHIC PREDATORS

       Because both the results of experimentally removing pre-
dators and the results of comparing the macrobenthos of grassbeds
to that of unvegetated areas suggest that marine macrobenthic
communities in soft sediments are not organized in ways that are
adequately modeled by the rocky intertidal paradigm, I develop
some possible and probable explanations for the existence of
this fundamental difference. The explanations, based upon
what is known about modes of interactions among species in
soft sediments, are not mutually exclusive and may be operating
simultaneously to produce this somewhat paradoxical failure of
soft-sediment macrobenthic communities to become simplified in
the absence of large epibenthic predators. I develop four
general explanations which may account for this phenomenon:
(1) insufficient time has passed in the absence of predators
for competitive exclusion to occur; (2) soft sediments provide
a much reduced opportunity for interference mechanisms to be
employed; (3) the extreme importance of adult-larval interactions
in soft sediments maintains densities below carrying capacity
even in the absence of predators; and (4) under food limitation
marine invertebrates exhibit extreme plasticity in growth and

reproduction and possess low energy needs, both of which help to insure continued high survivorship.

Insufficient Time

Most of the experiments done to exclude epibenthic predators were carried out over relatively short periods of time, usually 3 to 6 months (Table 1). Some exclusions lasted for a longer period, 12 months for *Virnstein's* (1977) longest treatment, nearly 10 months in *Lee's* (1978) experiment, and 24 months for *Peterson's* (1979) exclosures (although only 12 months went without the disturbance of sampling). Even though the majority of these experiments were terminated after relatively short periods, analogous experiments done to exclude predators from the rocky intertidal where generation times of the organisms tend to be longer would typically reveal substantial trends toward competitive exclusion over similar 3-6 month periods (*e.g.*, *Connell*, 1961a; *Dayton*, 1971; *Menge*, 1976). Consequently, it appears that insufficient time is not an important explanation for the failure of competitive exclusion to operate inside cages free from predators, unless the exclusion processes occur at far slower rates in soft sediments than in the rocky intertidal.

Even if cages have successfully reduced predation rates for a sufficiently long period of time for competitive effects to operate, time may still be insufficient for exclusion to begin to appear if the effects of seasonality are very strong. That is, environmental fluctuations through the seasons may be of sufficient magnitude to produce widespread mortality and/or replacement of the benthic macro-invertebrates before competitive interactions have an opportunity to simplify the community. Indeed, many marine infaunal systems include a large proportion of very short-lived, seasonally abundant organisms (*e.g.*, *Commito*, 1976; *McCall*, 1977). While this effect of seasonality may explain some of the paradoxical results of the caging experiments in soft sediments, it probably is not the major cause of the differences between the benthic systems of hard and soft substrates. Many of the hard substrate systems reveal strong seasonality (*e.g.*, *Connell*, 1961a; *Menge*, 1976), yet during favorable seasons competitive exclusion occurs convincingly in the absence of predators.

Reduced Opportunity for Interference Mechanisms

I suspect that the real explanation for the failure of competitive exclusion to occur following the removal of epibenthic predators from soft substrate systems probably involves the modes of interaction among the benthic infauna. Developing this explanation requires an examination of how exclusion through

competitive interactions is achieved on marine hard substrates
and a subsequent comparison of hard and soft substrates to see
how well these same modes of interactions might serve to simplify
benthic systems in soft sediments.

In marine benthic systems on hard substrates, competition
occurs on and for a firm two-dimensional surface, resulting in
the mortality of subdominant species and the ultimate monopolization of this limited surface space by the best competitor.
Two major mechanisms of interference competition serve to
produce this competitive mortality. First, the growth of
individuals attached to the primary substrate can produce
mortality by crushing, prying, and pushing other organisms off
the surface, well illustrated by the interaction between *Balanus
balanoides* and *Chthamalus stellatus*, barnacles on the rocks at
Millport, Scotland (*Connell*, 1961a, b). The mechanism involves
direct physical interference and its success in producing
mortality is closely tied to the adhesive contact with the
two-dimensional surface. The second major mechanism of competitive mortality on hard substrates involves overgrowth and
the subsequent suffocation or starvation of the overgrown
organisms. This process is exhibited by various sea mussels
which preferentially settle on "secondary substrates" (the
surface of barnacles and other organisms that are attached to
"primary" rock surfaces) and form a suffocating blanket over
these doomed underlying organisms, as illustrated by *Mytilus
californianus* in competition with barnacles, chitons, and other
organisms on the Pacific northwest coast (*Paine*, 1966; *Dayton*,
1971). Overgrowth and subsequent mortality of underlying
individuals can also be achieved on hard surfaces by the extended
growth of colonial forms, such as sponges and bryozoans like
Schizoporella unicornis at Beaufort, North Carolina (*Sutherland*,
1977).

It is quite conceivable that space could be an ultimately
limiting resource in the soft-sediment benthic communities of
shallow-water estuaries and lagoons. Field and laboratory
data from studies by *Woodin* (1974), *Whitlatch* (1976), and
Peterson (1977) suggest that space limitation is even reasonably
likely to operate in such systems. Assuming that surface space
may be potentially limiting in the soft-sediment systems of
shallow water, it becomes appropriate to ask whether any form
of interference competition might be expected to produce
competitive exclusion in soft sediments, analogous to the way
interference competition for space serves to simplify the benthos on hard substrates. The crushing, undercutting, and
evicting of animals through the increased growth of individuals seems far less likely to be an effective means of causing
mortality in soft substrates. First, there is no firm basis of
attachment from which to push and apply the force that might

produce such an outcome. Second, there are three dimensions in which organisms may be distributed in soft sediments. That is, there is a third resource axis along which organisms may segregate and thus help to avoid competitive interference. (There may actually be a little studied and largely ignored third dimension to systems on hard substrates also, as exemplified by the microfauna which inhabit the mussel byssus after mussels have monopolized all primary space.) Several soft-sediment marine benthic communities appear to possess definite vertical stratification of the abundant species (*e.g.*, *Levinton and Bambach*, 1975; *Levinton*, 1977; *Peterson*, 1977), suggesting that this means of avoiding interference competition at the sediment surface may indeed be utilized by infaunal organisms, at least in some cases. This mechanism is presumably effective only where burrows extend the area of sediment-water contact or where the cross-sectional area of organisms is sufficiently greater than the cross-sectional area of siphons, etc., so that space at depth becomes limiting before the two-dimensional surface space is filled.

The second common mode of interference competition on hard surfaces, overgrowth by colonial forms or by individuals recruited to secondary substrates, also seems less likely to be important in organizing soft-sediment benthic systems. Most of the benthic organisms in soft sediments are found below the sediment surface and consequently do not provide a secondary surface to be colonized and overgrown. Furthermore, the extreme mobility of the soft-sediments tends to exclude the colonial life form from this environment. It is the colonial life form that is best suited to this strategy of overgrowth on hard substrates (*Jackson*, 1977). Thus, competitive interference through overgrowth is not ordinarily possible in soft sediments because colonial invertebrates cannot survive in their attempt to overgrow the unstable sediment surface. In summary, both of the interference mechanisms which commonly produce mortality during competition for space on rocky shorelines are unlikely to be very important factors in soft sediments. In addition, no other mode of interference competition appears to play a major role in causing mortality of infaunal organisms. This ineffectiveness of interference competition should lead to a very different organizational structure in soft-substrate benthic systems.

Extreme Importance of Adult-Larval Interactions

Even if crushing of individuals is made difficult by the softness of the loose sediment and by the three-dimensionality of a soft substrate, if overgrowth by colonial forms is prevented by the mobility of the loose sediments, and if colonization of secondary surfaces is rendered unlikely because of the infaunal

habits of most soft-sediment organisms, competitive exclusion
might still be achieved by a species whose larvae colonize the
surface zone and survive in sufficient density to produce
mortality of other species. Such mortality could result either
from the inability of deeper organisms to penetrate the layer
of surface animals and thereby to make necessary contact with
the water column or from severe depletion of food supplies.
However, in order that these mechanisms be effective, invading
larvae must survive in sufficient numbers to reduce the abundance of the limiting resource to a level where the carrying
capacity is exceeded for some of the resident species. Such
high larval survivorship must be achieved despite the existence
of strong "cannibalism" by established adults upon invading
larvae. These negative interactions will become more significant as the density of established organisms increases,
leading to the hypothesis that even after mobile epibenthic
predators are excluded from the soft-sediment systems of
estuaries and lagoons, the resultant high intensity of adult-
larval interactions may serve to maintain densities below
carrying capacity for most organisms. This would provide
another mechanism by which established species would tend to
persist in the soft-sediment system even after the system was
freed from the control of predators (Fig. 1).

Although the mechanisms discussed in the previous section
presume that competition occurs for limited surface space in
the absence of predation, the hypothesis that adult-larval
interactions might maintain densities below the carrying
capacity for most species could be effective independent of
the type of limiting resource (food or space). Nevertheless,
to the degree that colonization occurs through post-larval
immigration rather than larval recruitment and to the degree
that certain larvae possess behavioral or structural adaptations
to enable them to avoid mortality caused by adults, this hypothesis will cease to be an important determinant of low
infaunal densities.

Adult-larval interactions are at least occasionally important
determinants of community structure in the marine benthic communities on hard substrates, as well illustrated by *Sutherland's*
(1977) experiments in which recruitment was often heavy on
empty settlement plates, suspended subtidally, while little
or no recruitment occurred at the same time on adjacent plates
which were covered by an established *Schizoporella* colony.
However, such interactions on hard surfaces generally fail to
maintain densities below carrying capacity; space becomes completely covered despite the existence of the interactions.
Consequently, if adult-larval interactions are to maintain
relatively low densities in marine soft sediments following the

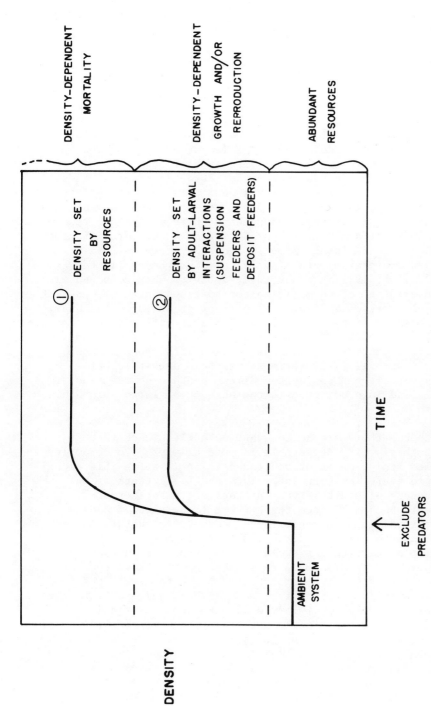

Fig. 1. The effects of removing predators from infaunal systems: 1) where adult-larval interactions are ineffective and 2) where adult-larval interactions are important. In the ambient system, density is set by a combination of predation by mobile, epibenthic predators and adult-larval interactions.

removal of predators, such interactions must be more effective at inhibiting density increase on soft substrates than on hard surfaces.

In general, successful recruits to any marine benthic community, on hard or in soft substrates, must pass through several filters of two major types: suspension feeders, which capture the larvae while they are still in the water column, and grazers (or deposit feeders), which consume larvae after settlement has occurred. While the impact of both suspension feeders and grazers upon communities on hard substrates is almost entirely a consequence of their trophic effects, these two types of consumers each play a dual role in soft-sediment systems. First, they play a trophic role: in feeding they can reduce larval survivorship and perhaps thereby maintain relatively low adult densities (thus acting as agents of biological disturbance, *sensu Paine*, 1966). Secondly, they perform a significant role as agents of physical disturbance (in the mode of *Dayton*, 1971) because their activity when combined with water movement tends to mobilize the soft sediments and prevent successful invasion and survival of many organisms. The duality of the role played by established adults in soft-sediment benthic communities leads me to propose the hypothesis that intense adult-larval interactions in conjunction with the resultant sediment instability may serve to maintain densities below carrying capacity in the soft-sediment benthic communities of estuaries and lagoons, even in the absence of large epibenthic predators.

Although there are no adequate data with which to test this hypothesis, it is testable in several ways. For example, controlled manipulation of adult densities is a feasible experimental procedure for both laboratory and field tests of the effectiveness of adult-larval interactions. Furthermore, recent work has emphasized the extreme importance of adult-larval interactions in determining community structure in dense infaunal assemblages (*Woodin*, 1976) such as might be expected to develop after predator exclusion in estuaries and lagoons. This hypothesis is also made reasonable by the prevalence of those studies which have demonstrated how extensively the feeding or physical disruption from established adults affects the growth or survivorship of larvae or other nearby organisms in soft-sediment systems (*Kristensen*, 1957; *Segerstrale*, 1962; *Mills*, 1969; *Rhoades and Young*, 1970; *Myers*, 1977; *Peterson*, 1977; *Brenchley*, 1978). A long history of study of larval settlement and subsequent survival (*e.g.*, see *Thorson*, 1966; *Gray*, 1974) suggests that larval invasion may often be the limiting step in the development of benthic communities in soft substrates.

If the negative interactions between established adults and settling larvae do play a major role in organizing marine benthic communities in the soft sediments of estuaries and lagoons where predators are absent or ineffective, an examination of how the importance of adult-larval interactions might vary along an offshore depth gradient becomes interesting. Convincing data (*Rowe*, 1971; *Mills*, pers. comm.) demonstrate that the biomass and density of benthic organisms steadily decline along a depth gradient on the continental shelf. Assuming that this trend represents a changing energy input which creates a decline in carrying capacity with increasing depth and distance from shore, I pose the question of whether adult-larval interactions can retain their ability to keep density below the carrying capacity when the carrying capacity is greatly reduced.

The effectiveness of adult-larval interactions is dependent upon the absolute abundance of the adults and their mobility. By abundance I mean biomass abundance more than numerical density in that biomass probably more closely predicts the impact of all relevant activity. Consequently, unless mobility increases with water depth and declining carrying capacity, which seems unlikely given that rates of most biological processes decline with decreasing temperature, the adult-larval interactions should lose their capacity to maintain density below carrying capacity at some depth offshore. Thus, if the effectiveness of adult-larval interactions is a major reason why in shallow water the exclusion of predators fails to produce substantial simplification of the soft-sediment benthic community, one might make the tentative prediction that predator exclusion would succeed in permitting the development of competitively produced monocultures at some depth offshore. There are presently no data with which to test this prediction.

Developmental Plasticity of Marine Invertebrates

If, as I have argued, interference mechanisms are unlikely to produce much mortality in soft-sediment systems following the removal of predators, it then seems appropriate to consider how marine invertebrates would respond to prolonged exposure to high density conditions, where food resources are presumably in short supply. Table 2 presents a review of those manipulative field experiments in which the density of at least one species of marine mollusc was varied experimentally and in which direct physical interference does not appear to play a major role. These experiments utilize marine molluscs from both hard and soft substrates. In each experiment, at least one and rarely all of the following responses were measured: 1) emigration, 2) growth, 3) reproductive effort, and 4) mortality.

TABLE 2. Intraspecific effects of varying density of marine benthic invertebrates in controlled, field experiments.*

Species	Treatment	Effects of varying density on:				Source
		Emigration	Reproductive Effort	Growth Rate	Mortality Rate	
Acmea digitalis	up to 2X	+ (large)	+ (small)		0 (except perhaps indirectly)	*Frank* (1965)
Acmea scabra	up to 5X			+ (large)	0	*Sutherland* (1970)
Littorina scutulata	over 2X			+	–	*Behrens* (1971)
Littorina sitkana	over 2X			+	+	*Behrens* (1971)
Acmea digitalis	up to 3X	+ (small)		0		*Breen* (1972)
Littorina irrorata	up to 2X			+ (large)	0	*Stiven and Kuenzler* (1979)
Nerita atramentosa	up to 5X			juveniles + (large) adults, no data	juveniles 0 adults + (large)	*Underwood* (1976)
Patelloida alticosta	up to 2X	0		+ (large)	+ (small) (?)	*Black* (1977)

TABLE 2. Continued.

Species	Density					Reference
Acmea digitalis	up to 6X	0	0	+ (large)	0	*Choat* (1977)
Venerupis semidecussata (lab experiment)	up to 100X			+ (large)	+ (small)	*Ohba* (1956)
Cardium edule	up to 13X			+ (large)	0	*Hancock* (1970)
Venus striatula (lab experiment)	up to 100X				0	*Ansell* (1961)
Protothaca staminea	up to 8X		+ (large)	+ (large)	0	*Peterson* (1979)
Chione undatella	up to 8X		+ (large)	+ (large)	0	*Peterson* (1979)
Geukensia demissa	up to 2X			+ (large)	+ (small)	*Stiven and Kuenzler* (1979)

*Key to symbols: + means an inhibitory effect of high density as would be expected,
  - means the opposite, counterintuitive effect,
  0 means no effect
  a blank entry means no data available to test that effect.

This review illustrates the relative ease with which growth rate is affected by experimental changes in the density of a marine mollusc and the relative difficulty of achieving significantly higher mortality rates as a response to increased density. Reproductive effort may also be more readily influenced by changes in density than is mortality rate, but there are insufficient experimental data to test this notion adequately. Data on emigration are also few and inconclusive. Consequently, the major conclusion suggested by Table 2 is that marine molluscs appear to possess great developmental plasticity which enables them to survive sustained periods of reduced food levels by cutting back on energy put into growth and perhaps also into reproduction. Mortality rates presumably can be increased by establishing sufficiently high densities, but generally survivorship is unaffected over a wide range of increased densities where growth rate is significantly suppressed (see references in Table 1).

Although all of the experiments in Table 2 involve molluscs, it seems unlikely that this general conclusion applies to molluscs alone. The ability to survive resource stress arising from high population density seems likely to be held in common by all poikilotherms and to be particularly evident among sedentary or very sluggish invertebrates that do not expend much energy in moving. Thus, in the absence of effective interference mechanisms, this plasticity of marine invertebrates and general ability to survive periods of resource stress provide another probable explanation for our failure to observe simplification of soft-sediment benthic communities following the removal of epibenthic predators. In other words, exploitation competition is unlikely to create rapid mortality and obvious monocultures despite its probable importance as a selective force producing segregation over evolutionary time. Even if mortality is the eventual result of exploitation competition at high infaunal densities, it may only be achieved over so long a period of time as to exceed the lifetime of the usual experiments. In that case, the rocky intertidal paradigm may still apply to soft-sediment benthic communities, but over a greatly expanded time scale, such as is required by exploitation competition. This is certainly a possibility that cannot be totally dismissed on the basis of the few long-term exclusion experiments that have been completed in soft sediments.

DISCUSSION OF COMMUNITY ORGANIZATION IN ESTUARINE AND LAGOONAL SOFT SEDIMENTS

Although in this paper I have presented a review of experimental evidence as well as a series of arguments to suggest that the generally accepted rocky intertidal paradigm of community

organization does not apply well to the soft-bottom benthic
communities of estuaries and lagoons, I do not thereby mean to
deny the importance of biological factors in the organization
of these systems. The results of field experiments tend to
suggest that predation from and disturbance by large, mobile
epibenthic predators have a significant impact on the density
of organisms in shallow-water soft sediments and on the community
structure, even though some of the apparent impact is surely a
consequence of experimental artifacts. In addition, I concur
with the ideas of others in suggesting that interactions between
adult organisms and invading larvae are extremely important
determinants of density and structure in soft sediments. This
sort of interaction is also biological and clearly has the
potential to organize soft-sediment systems.

While I acknowledge the probable importance of these
trophic interactions, both predation and cannibalism in its
broad sense, I question some of the currently held beliefs
about the role and modes of interspecific competition in
soft-sediment systems. Both distributional (e.g., Bradley and
Cooke, 1959; Sanders et al., 1962; Levinton and Bambach, 1975;
Ronan, 1975; Whitlatch, 1976; Peterson, 1977; Myers, 1977) and
experimental (Woodin, 1974) evidence suggests that interspecific
competition plays a significant role in the organization of many
shallow-water infaunal communities. However, none of these
studies demonstrates the active process of competitive exclusion
in which one species actually eliminates another; each illustrates
instead a status quo, usually some allopatric pattern of spatial
heterogeneity in species distributions. In other words, there
apparently can be active competition for a limited resource, .
usually space, but in a stable environment the competition
favors those relatively long-lived species that are initially
established and permits them to exclude potential invaders.
Thus, where densities are high, the history of occupation of
free space is preserved to a great degree in what one sees of
current soft-sediment benthic communities (as similarly argued
by Johnson, 1973).

Competition might also have played an important role through
successive generations by providing strong selective forces for
habitat partitioning in infaunal communities (as argued by
Levinton, 1977). This is a competing explanation for the
existence of allopatric patterns in the spatial distributions
of the infauna. To the degree that different soft-sediment
systems arise under essentially identical conditions as a
function of chance availability of larvae, competition for
resources is important in ecological time. To the degree that
slight differences in sediment character and in other physical
properties produce predictably different soft-sediment communities, competition over evolutionary time is implicated as a

likely cause of the allopatry. Both of these processes are probably significant in marine soft-sediment communities. However,
distinguishing among them may be difficult. Simple correlation
between community type and physical parameters is insufficient
because the organisms in soft sediments can and do alter their
own physical environments substantially (*e.g.*, *Rhoads et al.*,
1978). What is probably required to distinguish between these
mechanisms is careful observation of the development of
communities on cleared substrata of similar types set out
at different times and followed long enough to test for con-
vergence of community structure.

That history is an important determinant of soft-sediment
community structure is implicit in *Woodin's* (1976) notion that
adult-larval interactions are significant organizational factors
in these communities. Such interactions produce substantial
community inertia by retarding change. When free space is
made available by some sort of disturbance, those larvae available by chance in the plankton can colonize the area and, if
they survive at sufficient density to inhibit later colonization, persist. Subsequent change in the local system is then
retarded by their ability to exclude larvae of potentially
invading species and by their ability to survive the stress
of relatively high density. This process can produce the extreme
patchiness often observed even within a given sediment type in
soft-sediment benthic communities because when free space is made
available through disturbance only a small subset of all benthic
species will have sufficient larvae in the plankton to take
advantage of that opportunity for settlement. This sort of
chance event will lead to extreme patchiness on the scale of the
dominant disturbances. This view of soft-sediment community
organization resembles, therefore, the fouling community studied
by *Sutherland* (1974) more closely than it does the true rocky
intertidal systems.

I suggest here that in the absence of significant predation
this patchiness will persist because of 1) the lack of effective
interference mechanisms to produce subsequent competitive
exclusion of the poorest competitors, 2) the plasticity and
sluggishness of marine invertebrates which permits their survival at high density, and 3) the effectiveness of adult-larval
interactions at high density. A patch will become available
again for colonization only after the resident animals eventually
die, either from physical or biological disturbance or because
of natural physiological limits to their longevity. A body of
data exists which may seem to contradict this suggestion that
initial dense colonists can successfully repel subsequent larval
invasion without the threat of competitive displacement. In
disturbed and in experimentally defaunated sediments, an apparently successional process occurs in which the early, very

abundant opportunistic colonizers are commonly replaced by more long-lived species (*Simon and Dauer*, 1977; *McCall*, 1977; *Rhoads et al.*, 1978). However, the cause of the crash of the opportunists is not at all clear. *Rhoads et al.* (1978) suggest that predation and/or competition are the most likely causes of the mortality. It is also possible that the demise of the opportunists is not at all a consequence of direct biological interactions, but that early physiological aging and a resultant short life span are natural corollaries to an opportunistic life history that includes the ability to reproduce quickly and colonize early. Such opportunists must rely upon a succession of generations in order to persist in any one patch. At some stage, the later successional species may have reached an abundance that excludes further colonization and thereby effectively eliminates the opportunists. That is, the mortality of the opportunists during the crash may be a quite natural event independent of the presence of other species; however, adult-larval interactions may inhibit subsequent reinvasion by the larvae of the opportunists. That such outbreaks of opportunists are lacking or greatly damped in undisturbed communities of established adult organisms (*Rhoads et al.*, 1978) is again suggestive of the importance of the history of preoccupation of marine soft sediments.

These suggestions concerning the organization of infaunal communities are made somewhat tentative by the very real paucity of extensive natural history observations on infaunal organisms. We really know very little about how individuals interact within the sediments; this surely affects our ability to understand the dynamics and organization of infaunal communities. In addition, there is little real knowledge of the effectiveness or the selectivity of adult-larval interactions. Future progress in understanding the organization of soft-sediment systems is strongly dependent upon filling the present void in our knowledge about the basic natural history of adult-adult and adult-larval interactions.

ACKNOWLEDGMENTS

This work was supported in part by research grants from the Biological Oceanography Division of NSF, OCE 77-07939, and from Sea Grant, NOAA Sea Grant Program 04-7-158-44121. The paper benefited from discussions with W. Ambrose, R. Fulton, J. Hunt, P. Hyland, N. Mountford, W. Nelson, H. Porter, S. Shipman, H. Stuart, H. Summerson, J. Sutherland, and A. Williams. The manuscript has been improved by comments from W. Ambrose, J. Hunt, C. Onuf, M. Scott, A. Sih, R. Virnstein, and W. Wilson. I continue to express my appreciation to the U.S. Navy for its preservation of Mugu Lagoon. A portion of this project was

completed at the Marine Science Institute of the University of
California at Santa Barbara.

REFERENCES

Ansell, A. D., 1961. Reproduction, growth and mortality of
    Venus striatula (Da Costa) in Kames Bay, Millport,
    J. Mar. Biol. Ass. U. K. 41: 191-215.

Arntz, W. E., 1977. Results and problems of an "unsuccessful"
    benthos cage predation experiment (Western Baltic), In:
    Biology of Benthic Organisms, edited by B. F. Keegan,
    P. O. Ceidigh, P. J. S. Boaden, 31-44, Pergamon Press, N. Y.

Behrens, S., 1971. *The distribution and abundance of the
    intertidal prosobranchs* Littorina scutulata *(Gould 1899)
    and* L. sitkana *(Philippi 1845),* M.Sc. Thesis, Univ. of
    British Columbia, Vancouver.

Black, R., 1977. Population regulation in the intertidal
    limpet *Patelloida alticostata* (Angas, 1865), *Oecologia
    30:* 9-22.

Blegvad, H., 1928. Quantitative investigations of bottom
    invertebrates in the Limfjord 1910-1927 with special
    reference to the plaice food, *Rep. Dan. Biol. Stn. 34:*
    33-52.

Bradley, W. H. and P. Cooke, 1959. Living and ancient
    populations of the clam *Gemma gemma* in a Maine coast tidal
    flat, *Fishery Bull. Fish. Wildl. Serv. U. S. 58:* 305-334.

Breen, P. A., 1972. Seasonal migration and population regula-
    tion in the limpet *Acmea (Collisella) digitalis, Veliger
    15:* 133-141.

Brenchley, G. A., 1978. Mobility modes of marine infauna:
    distribution and composition of infaunal assemblages,
    Bull. Ecol. Soc. 59: 62.

Brooks, J. L. and S. Dodson, 1965. Predation, body size, and
    composition of plankton, *Science 150:* 28-35.

Caswell, H., 1978. Predator-mediated coexistence: a nonequili-
    brium model, *Amer. Nat. 112:* 127-154.

Choat, J. H., 1977. The influence of sessile organisms on the
    population biology of three species of acmeid limpets,
    J. Exp. Mar. Biol. Ecol. 26: 1-26.

Commito, J. A., 1976. *Predation, Competition, Life-history Strategies, and the Regulation of Estuarine Soft-bottom Community Structure*, Ph.D. Thesis, Duke Univ., Durham.

Connell, J. H., 1961a. The influence of interspecific competition and other factors on the distribution of the barnacle *Chthamalus stellatus*, *Ecology 42:* 710-723.

Connell, J. H., 1961b. Effects of competition, predation by *Thais lapillus*, and other factors on natural populations of the barnacle *Balanus balanoides*, *Ecol. Monogr. 31:* 61-104.

Connell, J. H., 1978. Diversity in tropical rain forests and coral reefs, *Science 199:* 1302-1310.

Dayton, P. K., 1971. Competition, disturbance and community organization: the provision and subsequent utilization of space in a rocky intertidal community, *Ecol. Monogr. 41:* 351-389.

Dodson, S. I., 1970. Complementary feeding niches sustained by size-selective predation, *Limnol. Oceanogr. 15:* 131-137.

Frank, P. W., 1965. The biodemography of an intertidal snail population, *Ecology 46:* 831-844.

Gray, J. S., 1974. Animal-sediment relationships, *Oceanogr. Mar. Biol. Ann. Rev. 12:* 223-261.

Hancock, D. A., 1970. The role of predators and parasites in a fishery for the mollusc *Cardium edule* L., *Proc. Adv. Study Inst. Dynamics Numbers Popul.* (Oosterbeek, 1970): 419-439.

Harper, J. L., 1969. The role of predation in vegetational diversity, *Brookhaven Symp. Biol. 22:* 48-62.

Holland, A. F., N. K. Mountford, J. A. Mihursky, M. Hiegel, and R. K. Kaumeyer, 1980. The influence of large mobile bottom-feeding predators on infaunal abundance in the upper Chesapeake Bay, Ms. in preparation.

Jackson, J. B. C., 1977. Competition on marine hard substrata: the adaptive significance of solitary and colonial strategies, *Amer. Nat. 111:* 743-767.

Johnson, R. G., 1973. Conceptual models of benthic communities, In: *Models in Paleobiology*, edited by T. J. M. Schopf, 148-159, Freeman, Cooper, and Co., San Francisco.

Kristensen, I., 1957. Differences in density and growth in a cockle population in the Dutch Wadden Sea, *Archs. Neerl. Zool.* 12: 351-453.

Lee, H., 1978. *Predation and Opportunism in Tropical Soft-bottom Communities*, Ph.D. Thesis, Univ. of North Carolina, Chapel Hill.

Levinton, J. S., 1977. Ecology of shallow water deposit-feeding communities Quisset Harbor, Massachusetts, In: *Ecology of Marine Benthos*, edited by B. C. Coull, 191-228, Univ. of South Carolina Press, Columbia.

Levinton, J. S. and R. K. Bambach, 1975. A comparative study of Silurian and Recent deposit-feeding bivalve communities, *Paleobiol.* 1: 97-124.

Lubchenco, J., 1978. Plant species diversity in a marine intertidal community: importance of herbivore food preference and algal competitive abilities, *Amer. Nat.* 112: 23-39.

McCall, P. L., 1977. Community patterns and adaptive strategies of the infaunal benthos of Long Island Sound, *J. Mar. Res.* 35: 221-266.

May R. M., 1973. *Stability and Complexity in Model Ecosystems*, 235 pp., Princeton Univ. Press, Princeton, N. J.

Menge, B. A., 1976. Organization of the New England rocky intertidal community: role of predation, competition, and environmental heterogeneity, *Ecol. Monogr.* 46: 355-393.

Menge, B. A. and J. P. Sutherland, 1976. Species diversity gradients: synthesis of the roles of predation, competition, and temporal heterogeneity, *Amer. Nat.* 110: 351-369.

Mills, E. L., 1969. The community concept in marine zoology, with comments on continua and instability in some marine communities: a review, *J. Fish. Res. Bd Canada* 26: 1415-1428.

Myers, A. C., 1977. Sediment processing in a marine subtidal sandy bottom community: II. Biological consequences, *J. Mar. Res.* 35: 633-647.

Naqvi, S. M. Z., 1968. Effects of predation on infaunal invertebrates of Alligator Harbor, Florida, *Gulf Res. Rep.* 2: 313-321.

Nelson, W. G., 1978. *The Community Ecology of Seagrass Amphipods: Predation and Community Structure, Life Histories, and Biogeography*, Ph.D. Thesis, Duke Univ., Durham.

Ohba, S., 1956. Effects of population density on mortality and growth in an experimental culture of a bivalve, *Venerupis semidecussata*, *Biol. J. Okayama Univ. 3:* 169-173.

Orth, R. J., 1977. The importance of sediment stability in seagrass communities, In: *Ecology of Marine Benthos*, edited by B. C. Coull, 281-300, Univ. South Carolina Press, Columbia.

Paine, R. T., 1966. Food web complexity and species diversity, *Amer. Nat. 100:* 65-75.

Peterson, C. H., 1977. Competitive organization of the soft-bottom macrobenthic communities of southern California lagoons, *Mar. Biol. 43:* 343-359.

Peterson, C. H., 1979. The effects of predation and intra- and interspecific competition on the population biology of two infaunal, suspension-feeding bivalves, *Protothaca staminea* and *Chione undatella*, Ms. in preparation.

Reise, K., 1977a. Predation pressure and community structure of an intertidal soft-bottom fauna, In: *Biology of Benthic Organisms*, edited by B. F. Keegan, P. O. Ceidigh, and P. J. S. Boaden, 513-519, Pergamon Press, N. Y.

Reise, K., 1977b. Predator exclusion experiments in an intertidal mud flat, *Helgoländer wiss. Meeresunters. 30:* 263-271.

Rhoads, D. C., P. L. McCall, and J. Y. Yingst, 1978. Disturbance and production on the estuarine seafloor, *Amer. Scientist 66:* 577-586.

Rhoads, D. C. and D. K. Young, 1970. The influence of deposit-feeding organisms on sediment stability and community trophic structure, *J. Mar. Res. 28:* 150-178.

Ronan, T. E., Jr., 1975. *Structural and Paleoecological Aspects of a Modern Marine Soft-sediment Community: An Experimental Field Study*, Ph.D. Thesis, Univ. California, Davis.

Rowe, G. T., 1971. Benthic biomass and surface productivity, In: *Fertility of the Sea*, edited by J. D. Costlow, 441-454, Gordon and Breach Sci. Publ. 2, New York.

Sanders, H. L., E. M. Goudsmit, E. L. Mills, and G. E. Hampson, 1962. A study of the intertidal fauna of Barnstable Harbor, Massachusetts, *Limnol. Oceanogr. 7:* 63–79.

Schneider, D., 1978. Equalisation of prey numbers by migratory shorebirds, *Nature 271:* 353–354.

Segerstrale, S. G., 1962. Investigations on Baltic populations of the bivalve *Macoma baltica* (L.). Part II. What are the reasons for the periodic failure of recruitment and the scarcity of *Macoma* in the deeper waters of the inner Baltic?, *Commentat. Biol. 24:* 1–26.

Simon, J. L. and D. M. Dauer, 1977. Reestablishment of a benthic community following natural defaunation, In: *Ecology of Marine Benthos*, edited by B. C. Coull, 139–154, Univ. South Carolina Press, Columbia.

Slatkin, M., 1974. Competition and regional coexistence, *Ecology 55:* 128–134.

Sprules, W. G., 1972. Effects of size-selective predation and food competition on high altitude zooplankton communities, *Ecology 53:* 375–386.

Stiven, A. E. and E. J. Kuenzler, 1979. The response of two salt marsh molluscs, *Littorina irrorata* and *Geukensia demissa*, to field manipulations of density and *Spartina* litter, *Ecol. Monogr. 49:* (in press).

Sutherland, J. P., 1970. Dynamics of high and low populations of the limpet, *Acmea scabra* (Gould), *Ecol. Monogr. 40:* 169–188.

Sutherland, J. P., 1974. Multiple stable points in natural communities, *Amer. Nat. 108:* 859–873.

Sutherland, J. P., 1977. The effect of *Schizoporella* (Ectoprocta) removal on the fouling community at Beaufort, North Carolina, USA, In: *Ecology of Marine Benthos*, edited by B. C. Coull, 155–176, Univ. South Carolina Press, Columbia.

Thayer, G. W., S. M. Adams, and M. W. LaCroix, 1975. Structural and functional aspects of a recently established *Zostera marina* community, In: *Estuarine Research, Vol. I. Chemistry, Biology and the Estuarine System*, 518–540, Academic Press, N. Y.

Thorson, G., 1966. Some factors influencing the recruitment and
    establishment of marine benthic communities, *Netherlands
    J. Sea Res. 3:* 267–293.

Underwood, A. J., 1976. Food competition between age–classes
    in the intertidal neritacean *Nerita atramentosa* Reeve
    (Gastropoda: Prosobranchia), *J. Exp. Mar. Biol. Ecol.
    23:* 145–154.

Virnstein, R. W., 1977. The importance of predation by crabs
    and fishes on benthic infauna in Chesapeake Bay, *Ecology
    58:* 1199–1217.

Virnstein, R. W., 1978. Predator caging experiments in soft
    sediments: caution advised, In: *Estuarine Interactions,*
    edited by M. L. Wiley, 261–273, Academic Press, New York.

Virnstein, R. W., 1979. Predation on estuarine infauna:
    response patterns of component species, *Estuaries 2:*
    (in press).

Warme, J. E., 1971. Paleoecological aspects of a modern
    coastal lagoon, *Univ. Calif. Pubs. Geol. Sci. 87:* 1–131.

Whitlatch, R. B., 1976. *Seasonality, Species Diversity and
    Patterns of Resource Utilization in a Deposit-feeding
    Community,* Ph.D. Thesis, Univ. of Chicago, Chicago.

Wiltse, W. I., 1977. Effects of predation by *Polinices
    duplicatus* (Gastropoda: Naticidae) on a sand-flat
    community, *Biol. Bull. 153:* 450–451.

Woodin, S. A., 1974. Polychaete abundance patterns in a marine
    soft-sediment environment: the importance of biological
    interactions, *Ecol. Monogr. 44:* 171–187.

Woodin, S. A., 1976. Adult-larval interactions in dense infaunal
    assemblages: patterns of abundance, *J. Mar. Res. 34:* 25–41.

Woodin, S. A., 1978. Refuges, disturbance and community
    structure: a marine soft-bottom example, *Ecology 59:*
    274–284.

Young, D. K., M. A. Buzas, and M. W. Young, 1976. Species
    densities of macrobenthos associated with seagrasses: a
    field experimental study of predation, *J. Mar. Res. 34:*
    577–592.

Young, D. K. and M. W. Young, 1977. Community structure of
    the macrobenthos associated with seagrass of the Indian
    River estuary, Florida, In: *Ecology of Marine Benthos,*
    edited by B. C. Coull, 359-382, Univ. of South Carolina
    Press, Columbia.

THE COMMUNITY STRUCTURE OF CORAL-ASSOCIATED DECAPOD CRUSTACEANS IN VARIABLE ENVIRONMENTS

Lawrence G. Abele

Florida State University

ABSTRACT

The decapod crustacean community associated with the branching coral Pocillopora damicornis *is described from two localities on the Pacific coast of Panama. The corals are of the same approximate geologic age but the physical conditions at one locality fluctuate unpredictably while conditions at the second locality are relatively constant. The species-area relationship does not differ between the two regions while the individuals-area regression of the constant environment has a higher intercept but the same slope. The dominance diversity curves for the two regions are similar in shape but dominance is greater in the constant environment. A generalist is dominant in the fluctuating environment while a specialist is dominant in the constant environment. There is no evidence for species packing as similar-sized coral heads have the same number of species in both regions and the species composition of individual coral heads suggests a random colonization process. There is no difference in the frequency of co-occurrence of congeneric species between the two regions; in both regions more congeners coexist than would be expected by chance. Comparison of species richness between the two areas based on either equal sample size or number of individuals demonstrates a greater total species richness in the fluctuating environment. It is suggested that a natural physical disturbance, upwelling, reduces dominance and increases species richness in the fluctuating environment.*

265

INTRODUCTION

Communities which occur under different environmental conditions are thought to be structured by different processes. In
relatively constant environments communities are thought to be
structured by biological processes and to be composed of a
relatively large number of stenotopic species. In contrast,
communities which occur in fluctuating environments are thought
to be structured by physical processes and to be composed of
relatively few eurytopic species (*Sanders*, 1968; *MacArthur*, 1969).
Most communities, however, are affected to some degree by both
physical and biological process (*e.g. Dayton*, 1971; *Glynn*, 1976).

This report examines the effects of area, environmental
stability, and disturbance on the community structure of decapod
crustaceans associated with the branching coral *Pocillopora
damicornis* Linnaeus.

METHODS AND MATERIALS

Collections of the *Pocillopora* coral heads and their associated macrofauna were made in two regions off the Pacific coast
of Panama (Table 1). Coral heads were either enclosed in a
plastic bag and removed, or broken off into a bucket and taken
to the surface where all macroorganisms were removed. Only live
coral heads were collected and examined. The macrofauna of
these coral heads consists almost entirely of decapod crustaceans:
96 percent of the individuals and 89 percent of the species in
the Pearl Islands (*Abele and Patton*, 1976) and 80 percent of the
individuals and 76 percent of the species at Uva Island.

A volume measurement (length x width x height) was used as
an estimate of the size of the coral heads. The dead base and
occasional dead center of large corals were not measured, nor
were animals collected from these portions. The dry weight of
the coral is highly correlated with this volume measurement
(r = 0.88). Surface area is related to volume by $SA \sim KV^{2/3}$.

The expected number of decapod species for a given number of
individuals, $E(S_n)$, was calculated by the method of *Hurlbert*
(1971) using the variance calculation of *Heck et al.* (1976). The
same statistical methods were used to calculate the expected
number of species per genus, $E(G_s)$, for different numbers of
species. All programs were run on the Florida State University
Computing Center CDC 6400.

TABLE 1. Comparative data on the *Pocillopora* coral–decapod
crustacean community

Region	Pearl Islands, Panama	Uva Island, Panama
Age (years)	4500 ± 65	2500 to 5585
Annual sea-surface temperature range ($^{\circ}$C)	17–29	28–29
Annual sea-surface salinity range (ppt)	22–36	30–35
Sample size	35	35*
Number of species	55	37 ± 2
Dominance		
D_1	16	32
D_2	26	46
Number of individuals	1107	1107*

*Randomly drawn from a sample of 119 coral heads, including
4724 individuals and 50 species.

Dominance was measured by DI:

$$DI = \frac{m_1}{N} \; x \; 100$$

where m is the number of decapod individuals of the most abundant species and N is the total number of decapod individuals. For

$$DI_2 = \frac{m_1 + m_2}{N} \; x \; 100$$

m_2 is the number of individuals of the second most abundant species.

The Study Areas

The samples from Pearl Islands, Gulf of Panama, consisted of 35 coral heads taken from about three meters depth during June and July, 1973, from relatively unstructured patch reefs dominated by *Pocillopora damicornis* (*cf. Glynn and Stewart*, 1973; *Glynn et al.*, 1972). The patch reefs are about 4550 ± 65 years old based on growth rate and carbon-14 dating (*Glynn*, pers. comm.). Temperature, salinity, and turbidity vary greatly in the area. Heavy rains during part of the year result in salinity changes of at least 10 parts per thousand. During the dry season high winds create turbid conditions and upwelling which can lower the temperature by as much as 12°C in 24 hours (*Glynn*, 1972). The water temperature during upwelling regularly drops to 20°C, a point at which the growth rate of *P. damicornis* declines markedly or ceases, sometimes resulting in loss of soft parts from the tips of branches (*Glynn and Stewart*, 1973). In addition, aseasonal upwelling can occur (*Graham and Abele*, 1973), increasing the unpredictable aspects of the environment.

The samples from Uva Island reef (described in detail by *Glynn*, 1976) in the Gulf of Chiriqui, consisted of 119 coral heads taken from about three to ten meters depth during January, April, June, and August, 1973. The reefs at Uva Island are structured and the samples were taken at random from the fore and rear crest and flank. *Pocillipora damicornis* is the dominant species on the crest but is present over the entire reef (*Glynn*, 1976). The reef is about 2500 to 5585 years old based on growth rate data and carbon-14 dating (*Glynn*, pers. comm.). The yearly temperature variation is minor, usually about 2°C while the salinity may vary 2 to 5 parts per thousand during the rainy season. Portions of the shallow reef crest may be exposed with increased environmental variation during extreme low tides

(*Glynn*, 1976), but the deeper flanks remain unaffected. However, there is little variation in species composition between the crest and flank portions of the reef.

RESULTS

Species-Area Relationship

Numbers of decapod crustacean species associated with each of the coral heads are plotted against coral head size in Fig. 1. There is no significant difference between the regression lines in either slope or intercept by analysis of covariance (F 1/151 < 1) or Student's t-test (t = 1.37 for slopes, t = 1.34 for intercepts). This result is particularly important because it supports the hypothesis that the number of decapod species associated with the coral depends on the areal extent and

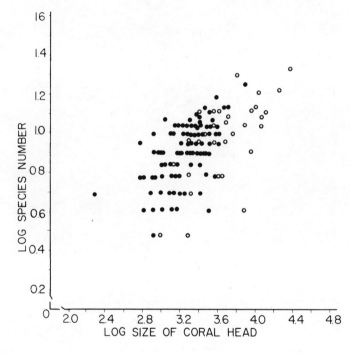

Fig. 1. Log-log plot of the number of coral-associated decapod crustacean species against the size of the coral head. Closed circles denote Uva Island and enclosed stars the Pearl Islands. For the Pearl Islands, log y = 0.356 log x - 0.316, r = .64, P < .001, N = 35; for the Uva Island, log y = 0.269 log x + 0.010, r = .48, P < .001, N = 119. There is no significant difference between the regression lines.

structural characteristics of the coral. There appears to be no
obvious difference in the way species perceive the coral head as
a habitat in the two areas.

Individuals-Area Relationship

Total numbers of individuals of all decapod species associated
with each of the coral heads is plotted against coral head size in
Fig. 2. Comparison of these regressions is somewhat equivocal.
There is no difference between the slopes by Student's t-test
(t < 1). The analysis of covariance model, assuming equal slopes,
reveals a significant difference between the intercepts (Uva >
Pearl, F 1/151 = 32, p < 0.001). However, if the intercepts are
tested using the numerical slope values derived for each region

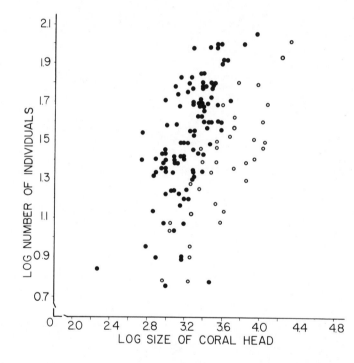

Fig. 2. Log-log plot of the number of individuals of coral-
associated decapod crustacean species against the size of the
coral head. Closed circles denote Uva Island and enclosed stars
the Pearl Islands. For the Pearl Islands, log y = 0.618 log x
- 0.841, r = .77, P < .001, N = 35; for Uva Island, log y = 0.652
log x - 0.675, r = .58, P < .001; N = 119. There is a significant difference between the intercepts but not between the slopes.

(rather than assuming some mean value as in analysis of covariance) there is no significant difference between the intercepts. Visual examination of Fig. 2 supports the analysis of covariance result; there is an obvious tendency for the Uva Island samples to lie above those from the Pearl Islands. In fact, there are about 25 percent more individuals per equal-sized coral head at Uva Island than at the Pearl Islands.

Not all species of decapods at Uva Island and the Pearl Islands increased in numbers of individuals with increasing coral head size. In both areas, different species respond in different ways to increasing coral head size. For example, *Alpheus lottini* and *Synalpheus charon* occurred in single male-female pairs in both areas. The xanthid crab *Trapezia ferruginea* occurred in the Pearl Islands in single male-female pairs on coral heads smaller than 4000 cm^3 while two pairs occurred more commonly on larger corals (*Abele and Patton*, 1976). At Uva Island, however, multiple pairs commonly occurred on coral heads of all but the smallest size classes (*Glynn*, 1976; *Finney and Abele*, in prep.). The palaemonid shrimps *Harpiliopsis depressa*, *H. spinigera* and *Fennera chacei* were more abundant per coral head at Uva Island than in the Pearl Islands. Most species, however, were not sufficiently abundant to allow comparison between the two regions.

Relative Abundance and Dominance

The relative abundances of the decapod crustaceans from the two sites are plotted on a dominance-diversity graph in Fig. 3. The curves are generally similar in shape although that for Uva Island tends to be steeper. This similarity is due to the greater numerical dominance of individual species at Uva Island (DI_1 = 32, DI_2 = 46) than at the Pearl Islands (DI_1 = 16, DI_2 = 26). Thus in the constant environment the most abundant species accounts for 32 percent of all individuals while the first and second most abundant species account for just under half of all the individuals. In contrast, the most abundant species in the fluctuating environment of the Pearl Islands accounts for only 16 percent of the individuals and the first and second most abundant species for only 26 percent of all of the individuals.

The most abundant species in the Pearl Islands is a porcellanid crab, *Pisidia magdalenensis*, which may be considered a generalist species. The species occurs in live corals, dead corals, in the rocky intertidal area, and among rubble (*Abele*, 1976a, *Gore and Abele*, 1976). Analysis of the stomach contents indicates that the crab is a generalized filter feeder taking a wide variety of small particles. The second most abundant species at the Pearl Islands is the palaemonid shrimp *Harpiliopsis depressa* which accounted for about 10 percent of the individuals.

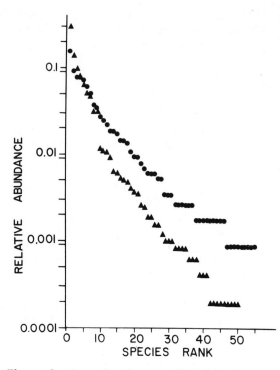

Fig. 3. The relative abundance of individuals of coral-associated decapod crustacean species plotted against their rank from most to least abundant. Closed circles denote the Pearl Island and triangles the Uva Island fauna. N = 1107, S = 55 for the Pearl Islands and N = 4724, S = 50 for Uva Island.

The shrimp is an obligate commensal of pocilloporid corals and occurs throughout the Eastern, Central, and Indo-West Pacific. At Uva Island, the most abundant species is the xanthid crab *Trapezia ferruginea* which may be considered a specialist. The crab is an obligate associate of pocilloporid corals and feeds on coral mucus (*Knudsen*, 1967; *Patton*, 1974). The species is wide ranging throughout the Eastern, Central, and Indo-West Pacific. In the Gulf of Panama the crabs usually occur in one or two male-female pairs, but they occur in multiple pairs in the Gulf of Chiriqui. In the latter region they are important in defending the coral against predation by *Acanthaster* starfish (*Glynn*, 1976). The second most abundant species at Uva Island is another specialist, a small palaemonid shrimp, *Fennera chacei*, which is an obligate associate of pocilloporid corals and has a range similar to that of *T. ferruginea*. At Uva Island it accounts for about 14 percent of the individuals compared to about 8 percent in the Pearl Islands.

At both sites most remaining species are rare. In the Pearl Islands nine species are represented by single individuals and 34 (about 61 percent) species are represented by fewer than 10 individuals. At Uva Island with a very much larger sample size, nine species are represented by single individuals. To make the comparison more reasonable, we note that at the Pearl Islands 61 percent of the species are each represented by less than one percent of the total individuals while at Uva Island 64 percent of the species are so represented. Not only are most of the species at each site rare but they also have a low frequency of occurrence on individual coral heads. In the Pearl Islands 16 of the species were found on single coral heads and 42 of the species occurred on nine or fewer coral heads. At Uva Island nine of the species occurred on single coral heads while 35 species occurred on nine or fewer coral heads.

Comparative Species Composition

The decapod crustacean species which occurred at each site are listed in Table 2. The species composition can be compared in a variety of ways. The Jaccard index omits consideration of negative matches and is $a/(a + u)$ where u = the mismatches and a = the number in common; it thus varies between 0 and 1. For the present comparison of Uva with the Pearl Islands there are 32 species from Uva in common with the Pearl Islands fauna. The Jaccard index is thus $32/(18 + 24 + 32)$ or 0.43. Another way to examine the data is to note that only 18 of the 50 Uva Island species do not occur in the Pearl Islands while 24 of the 55 Pearl Islands species are not found at Uva Island. Thus the Uva Island fauna has 64 percent (32/50) similarity to the Pearl Islands while the Pearl Islands has a 58 percent (32/55) similarity to Uva Island.

Comparing the species composition in the two areas reveals some differences in the biology of the two faunas. In the Pearl Islands *Harpiliopsis depressa, Trapezia ferruginea, Fennera chacei, Alpheus lottini, Harpiliopsis spinigera, Hapalocarcinus marsupialis* and *Synalpheus charon* all appear to be obligate associates of *Pocillopora*. Other species in the Pearl Islands may also be coral specialists (e.g. *Mithrax pygmaeus, Synalpheus digueti, Pagurus cf. lepidus* and *Palaemonella cf. asymmetrica*) but not enough data are available. Thus, at least seven of the 55 species appear to be specialists. At Uva Island the same seven species occur plus *Domecia hispida, Trapezia digitalis, Pomognathus corallinus, Carpilodes cinctimanus* and *Daira americana* which also seem to be coral specialists. However, the above species are represented by one or few individuals so that they may occur at the Pearl Islands but not yet have been collected, especially since all but *T. digitalis* and *D. americana* have been

TABLE 2. Comparison of species composition, numbers of individuals, relative abundance and frequency of occurrence of decapod crustaceans associated with *Pocillopora damicornis* at two localities in Panama. For the Pearl Islands n = 35, for Uva Island n = 119.

Species	Number of individuals		Relative abundance		Frequency of Occurrence (%)	
	Pearl	Uva	Pearl	Uva	Pearl	Uva
Trapezia ferruginea	91	1493	.0822	.3160	100	100
Fennera chacei	89	673	.0804	.1427	57.1	70.6
Teleophrys cristulipes	22	508	.0198	.1077	34.3	79.8
Harpiliopsis depressa	109	396	.0984	.0839	88.6	90.7
Pagurus cf. lepidus	26	318	.0234	.0674	34.3	55.5
Alpheus lottini	83	254	.0749	.0538	91.4	94.9
Petrolisthes haigae	71	240	.0641	.0509	65.7	50.4
Micropanope xantusii	0	159	0	.0337	0	50.4
Harpiliopsis spinigera	17	158	.0153	.0335	11.4	20.1
Thoe sulcata panamense	0	59	0	.0125	0	26.8
Synalpheus cf. mexicanus	2	56	.0018	.0118	5.7	25.2
Synalpheus digueti	29	52	.0261	.0110	37.1	23.5
Mithrax pygmaeus	31	45	.0280	.0095	37.1	10.9
Petrolisthes edwardsii	6	32	.0054	.0067	5.7	6.7
Thor cf. maldivensis	7	30	.0063	.0063	5.7	13.4
Petrolisthes agassizzi	40	27	.0361	.0057	40	5.8
Synalpheus charon	4	25	.0036	.0053	5.7	11.7
Palaemonella cf. asymmetrica	58	24	.0523	.0050	54.3	13.4
Pachycheles biocellatus	2	20	.0018	.0042	2.8	8.4
Platypodia rotundata	0	18	0	.0038	0	10.1
Trizopagurus magnificus	0	17	0	.0036	0	5.8
Actaea sulcata	0	13	0	.0027	0	7.5
Domecia hispida	0	12	0	.0025	0	8.4
Petrolisthes glasselli	0	9	0	.0019	0	4.2
Calcinus obscurus	3	9	.0027	.0019	2.8	3.4

TABLE 2. Continued

	Number of individuals		Relative abundance		Frequency of Occurrence (%)	
	Pearl	Uva	Pearl	Uva	Pearl	Uva
Alpheus paracrinitus	0	8	0	.0016	0	5.0
Majidae sp. 1	0	8	0	.0016	0	2.5
Herbstia tumida	3	6	.0027	.0012	8.6	3.4
Stenorhynchus debilis	11	5	.0099	.0010	14.3	4.2
Alpheus sp. 2	2	5	.0018	.0010	5.7	3.4
Actaea dovii	0	5	0	.0010	0	4.2
Xanthidae sp. 2	3	4	.0027	.0008	2.8	1.7
Thyrolambrus erosus	0	4	0	.0008	0	2.5
Stenocionops sp. 1	0	4	0	.0008	0	3.4
Pilumnus reticulatus	4	4	.0036	.0008	11.4	3.4
Cycloxanthus vittatus	22	3	.0198	.0006	25.7	1.7
Majidae sp. 2	0	3	0	.0006	0	2.5
Trapezia digitalis	0	3	0	.0006	0	1.7
Synalpheus sp. 2	2	2	.0018	.0004	5.7	0.8
Xanthidae sp. 1	1	2	.0009	.0004	2.8	1.7
Typton sp. 1	1	2	.0009	.0004	2.8	1.7
Pomognathus corallinus	0	1	0	.0002	0	0.8
Pelia pacifica	2	1	.0018	.0002	5.7	0.8
Hapalocarcinus marsupialis	7	1	.0063	.0002	2.8	0.8
Palaemonidae ?	0	1	0	.0002	0	0.8
Uhlias ellipticus	1	1	.0009	.0002	2.8	0.8
Paguristes sp.	0	1	0	.0002	0	0.8
Petrolisthes polymitus	2	1	.0018	.0002	5.7	0.8
Carpilodes cinctimanus	0	1	0	.0002	0	0.8
Daira americana	0	1	0	.0002	0	0.8
Pisidia magdalenensis	180	0	.1626	0	77.1	0
Heteractaea lunata	43	0	.0388	0	60	0

TABLE 2. Continued

	Number of individuals		Relative abundance		Frequency of Occurrence (%)	
	Pearl	Uva	Pearl	Uva	Pearl	Uva
Thor amboinensis	17	0	.0153	0	17.1	0
Synalpheus sp. 5	16	0	.0144	0	8.6	0
Lysmata californica	13	0	.0117	0	5.7	0
Cataleptodius sp. 1	11	0	.0099	0	8.6	0
Pilumnus sp. 1	9	0	.0081	0	5.7	0
Lophocanthus lamellipes	8	0	.0072	0	8.6	0
Alpheus sp. 6	7	0	.0063	0	8.6	0
Alpheus panamensis	6	0	.0054	0	11.4	0
Gnathophyllum panamense	4	0	.0036	0	8.6	0
Petrolisthes galathinus	3	0	.0027	0	5.7	0
Pilumnus stimpsoni	3	0	.0027	0	2.8	0
Megalobrachium smithi	3	0	.0027	0	2.8	0
Synalpheus sp. 1	21	0	.0198	0	17.1	0
Megalobrachium erosum	2	0	.0018	0	2.8	0
Synalpheus sanlucasei	2	0	.0018	0	5.7	0
Megalobrachium tuberculipes	2	0	.0018	0	5.7	0
Ulloaia perpusilla	1	0	.0009	0	2.8	0
Pagurus sp. 2	1	0	.0009	0	2.8	0
Synalpheus biunguiculatus	1	0	.0009	0	2.8	0
Pilumnus sp. 2	1	0	.0009	0	2.8	0
Cycloxanthus bocki	1	0	.0009	0	2.8	0
Pachycheles vicarius	1	0	.0009	0	2.8	0
	55(1107)	(50(4724)				

collected at Taboguilla Island only a short distance from the Pearl
Islands (*Abele*, 1976a).

Despite the general similarities noted above, there are some
striking differences in species composition between the two sites.
The porcellanid crab *Pisidia magdalenensis* was the single most
abundant species in the Pearl Islands, being represented by 180
individuals. Yet not a single individual was collected among
the 4724 individuals examined from 119 coral heads collected over
an eight month period at Uva Island. A similar situation exists
for *Heteractaea lunata, Thor cf. amboinensis*, several species of
Synalpheus, Lysmata californica, Cataleptodius sp., *Pilumnus* sp.,
Lophoxanthus lamellipes, Alpheus panamensis and *Gnathophyllum
panamense*. All of the above species were represented in the
Pearl Islands collections and have been taken from *Pocillopora*
in areas adjacent to the Pearl Islands (*Abele*, 1976a, and
unpublished data) yet were not collected at Uva Island. It is
not that these species do not occur in the Gulf of Chiriqui
since they are found throughout the Panamic Province. All of
these species could be considered habitat generalists as they
occur in a variety of habitats which include pocilloporid corals
(*Abele*, 1976a).

In contrast to the absence of generalized species from Uva
Island the abundance of species that appear to be obligate
associates of pocilloporid corals is higher at Uva Island.
Based on a random sample of 29 coral heads from Uva taken at
the same time of the year as those from the Pearl Islands, the
numbers of individuals found at the two locations were, for
Trapezia ferruginea, 91 (Pearl) and 147 (Uva; 91 and 558 if
juveniles at the coral base are included, for *Fennera chacei*
89 and 243, and for *Harpiliopsis spinigera* 15 and 59. Not only
were coral specialists more abundant at Uva Island than in the
Pearl Islands but their frequency of occurrence on individual
coral heads (Table 2) was higher as well.

Species/Genus Ratios

The number of coexisting congeneric species has often been
considered a community parameter of interest, particularly with
respect to island faunas (*Grant*, 1966; *Simberloff*, 1970). Competition has often been inferred when the observed species/genus
ratios are less than the expected, but these low ratios are open
to alternative explanations (*Simberloff*, 1970). The observed
species/genus ratios for coral heads of various species numbers
are plotted in Fig. 4, where the theoretical distribution and
one standard deviation are given if congeners were drawn at
random. The Pearl Island samples are plotted and compared with
a sample of 29 coral heads from Uva Island. The random theoretical

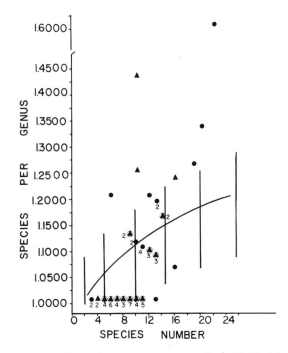

Fig. 4. Comparison between expected (solid line) and observed species per genus ratios for the number of decapod crustacean species found on various-sized coral heads. One standard deviation is noted for the expected ratio. Closed circles denote the Pearl Islands, triangles denote Uva Island and encircled triangles denote both. Numbers adjacent to symbols indicate the number of observations at that point. The confidence intervals for low numbers of species are asymmetrical since the lower limit for an S/G ratio is 1.0.

distribution is based on only those species taken from pocilloporid corals. In the Pearl Islands 11 ratios are above the line and 24 are below it; no congeners coexisted on 16 of the coral heads. However, only one point is more than one standard deviation below the line, while 5 points are more than one standard deviation above the line. At Uva Island five ratios are above the line and 24 are below it; no congeneric species coexisted on 22 of the coral heads. None of the Uva samples is more than one standard deviation below the line, while three points are more than one standard deviation above the line. The ratio for the Pearl Islands fauna is between 1.69 and 1.77 (depending on the generic status of unidentified species) while that for Uva Island is between 1.28 and 1.38 (depending on the generic status of unidentified species). Thus there are fewer congeneric species among the entire fauna of Uva Island and on individual coral heads than at the Pearl Islands.

However, the trend at both sites is for the occurrence of more congeneric species on individual coral heads than expected.

Comparative Species Richness

Any comparison of species richness between communities must be based on equal sample size (*Hurlbert*, 1971) or equal sampling effort so that there is reasonable certainty that most of the species in each area have been sampled. For both the Pearl Islands (*Abele and Patton*, 1976; Fig. 1) and Uva Island the accumulated species curve becomes asymptotic at about 1000 individuals or 30 coral heads (see Fig. 5). Since the sampling effort was so much greater at Uva Island, where 119 coral heads and 4724 individuals were collected as compared to 35 coral heads and 1107 individuals in the Pearl Islands, it is necessary to reduce the Uva Island sample to a level comparable to that of the Pearl Islands. In Fig. 5 cumulative species number is plotted against numbers of individuals. All 35 coral heads from the Pearl Islands are included, and these are compared to 35 coral heads randomly drawn from the Uva Island samples. Thus we can compare species richness for equal numbers of individuals or for equal numbers of coral heads. Comparing equal numbers of coral heads one finds that there are 55 species in the Pearl Islands samples and 39 in the Uva Island samples. Comparing the two

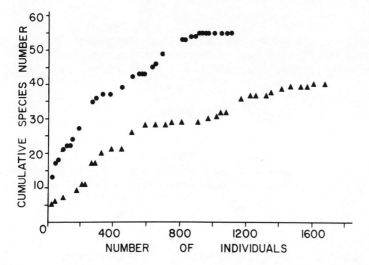

Fig. 5. The relationship between numbers of individuals and cumulative number of coral-associated decapod crustacean species. Closed circles denote the Pearl Islands and triangles Uva Island. Sample size = 35 coral heads for each site.

sites at a sample size of 1000 individuals yields similar results, 55 species from the Pearl Islands and 32 from Uva Island. In addition we can utilize the entire Uva Island sample to calculate an expected number of species and its associated variance for any number of individuals (*Hurlbert*, 1971; *Heck et al.*, 1976). The expected species number for 1107 individuals from Uva Island is 37 ± 2 species. Thus the three methods give the same result: there are significantly fewer species of decapod crustaceans associated with the coral *Pocillopora damicornis* at Uva Island than at the Pearl Islands.

The greater species richness in the Pearl Islands results from a high variation in species composition among coral heads. It is not due to the presence or absence of rare species at either site because, as previously noted, there is about the same proportion of rare species at the two sites. The lower species number at Uva Island as compared to the Pearl Islands does not appear to be due to a random absence of species at the former site. Rather, it appears to result from the absence of a specific group of species which are habitat generalists.

The above remarks, however, apply only to the total number of species found at each site. It may be more meaningful to compare the number of species found on individual coral heads of similar size at the two sites. As already noted, there is no difference in species richness at this level. One finds about the same number of species on similar-sized individual coral heads at the two sites.

Thus it would seem that there is no relationship among total species richness, species packing, and environmental stability. In fact, most species at both sites have a low frequency of occurrence which suggests a random colonization process.

DISCUSSION

The present comparison has focused on two communities with identical physical structure (the coral) and similar geological history. The two communities are in the same biogeographic province, less than 300 km apart and thus share at least a potentially common species pool. Despite the above similarities, the two regions differ considerably in present-day physical characteristics. The Pearl Island region fluctuates both seasonally and aseasonally with little predictability in the magnitude and timing of temperature variation. The Uva Island site is comparatively stable in physical characteristics.

The two regions are similar in terms of the species-area relationship, the relative abundance curves for the two regions are similar, and there are many similarities in their species composition of obligate coral associates and in the relative values of their species/genus ratios. They differ in the individuals-area relationship for total individuals and for the populations of some individual species. There are some obvious differences in species composition of facultative coral associates and in the total number of species in the two areas.

The similarity in species composition between the two regions is not unexpected since the habitat is identical. All of the obligate coral associates found in the Pearl Islands were also found in the Uva collections. Those few obligates that were found at Uva but not in the Pearl samples were represented by only a few individuals and their absence in the Pearl samples is probably due to sampling error since all but two of the species have been collected in the Bay of Panama (*Abele*, 1976a).

The relative abundance curves (or dominance diversity, Fig. 3) are similar in shape although some differences exist in the abundance of the most common or dominant species. Most species at both sites were rare with 61% and 64% of the species accounting for less than 1% of the individuals at the Pearl and Uva sites, respectively. These rare species on coral heads appear to be more common at other localities or in other habitats. For example, the xanthid crab *Daira americana* was represented by a single specimen among the 4724 individuals collected at Uva Island and was absent from the Pearl Islands. Yet this species is very common on live *Pocillopora* coral from the Galapagos Islands (*Garth*, 1946; *Abele*, unpublished). The caridean shrimps *Lysmata californica*, *Gnathophyllum panamense* and *Alpheus panamensis* are relatively uncommon from live corals in the Pearl Islands and were absent from Uva Island. These same species are very common in tidepools of the low intertidal in the Bay of Panama where no coral occurs. It is difficult to assess the ecological relationships, if any, that these species may have in habitats where they are rare or uncommon.

The relationship between the expected and observed numbers of species per genus is shown in Figure 4. An earlier analysis of species to genus ratios by *Abele and Patton* (1976) suggested that fewer congeners than expected occurred on individual coral heads. However, their analysis failed to consider the variance associated with $E(G_s)$. Figure 4 includes the data from Abele and Patton as well as those from the present study and includes $S/(E(G_s))$ and one standard deviation (see *Simberloff*, in press, for a discussion of expected ratios). It is obvious that there is no trend towards fewer congeners on coral heads. Rather, eight of the observations are greater than one standard deviation from the expected while only one is less. If one considers that congeneric species are, on the average, more similar than other

species, then a constant environment does not result in the
coexistence of more similar species. At both sites the majority
of coral heads share the expected number of congeners but both
show the same trend; congeners occur more frequently together
than one would expect if they occurred at random. This is consistent with the observation of *Simberloff* (1970) who found that
more congeners than expected occurred on islands. He attributed
this to the similarities in ecology and dispersal often exhibited
by congeners. There are apparently more congeneric species in
the Pearl Islands pocilloporid fauna than in the Uva
Island fauna; the ratios are 1.69-1.77 for the Pearl Islands
and 1.28-1.38 for the Uva Island fauna. This may be the result
of the higher variation in species composition among coral heads
in the Pearl Islands than at Uva Island (see discussion of species
richness).

One of the most interesting results was the finding that
the species-area relationship was the same in both areas. It
has often been asserted that species are more tightly "packed"
in constant environments, resulting in a large number of coexisting species. If this were true, then the regression line for the
constant Uva site would be above that of the fluctuating Pearl
site because more species would occur on each coral head. However, the regression lines are statistically indistinguishable
and an examination of the plots (Fig. 1) reveals no trend among
the points. This suggests that the coexistence of species in
this habitat is unaffected by the level of environmental variation
described here. Rather, the coexistence of species on individual
coral heads is well predicted by the areal extent of the coral
head. This is an empirical fact which may have at least three
explanations: (1) increased habitat heterogeneity with increased
areal extent, (2) an increase in sampling since more larvae will
come into contact with a larger area, and (3) reduced extinction
rate, with an increase in species richness, due to the possible
larger population sizes of individual species supported by larger
areas (*Connor and McCoy*, in press). There is probably little
change in habitat heterogeneity with an increase in area (*Abele
and Patton*, 1976) so that either increased sampling or decreased
extinction may be the mechanism. At present I favor a sampling
effect because so many of the species are represented by either
a few individuals, single pairs regardless of coral head size,
or show no relationship between numbers of individuals and coral
head size. For example, only 13 of the 55 species from the Pearl
Islands were sufficiently common to analyze and of these the most
abundant, *Pisidia magdalenensis*, showed no relationship between
number of individuals and coral head size. The extinction
hypothesis of island biogeography theory may be relevant for
some of the species but it seems insufficient to account for
the overall pattern.

In contrast to the species-area relationship, the individuals-area relationships of the two sites differ. There was an average of 25% more individuals per similar size coral head in the constant as compared to the fluctuating environment. This result is based on the total numbers of individuals of all species on each coral head. The higher abundance of individuals per coral head at Uva Island was heavily influenced by a dramatic difference in population structure of *Trapezia ferruginea*. In the Pearl Islands one male-female pair or, on large corals, two pairs occurred on each coral. At Uva Island there were as many as 54 individuals on a single coral, and as many as 16 males and females (eight pairs?) occurred together (*Finney and Abele*, in prep.). Other species such as *Synalpheus charon* and *A. lottini* occurred in male-female pairs at both sites. Among those species sufficiently abundant to analyze, *Fennera chacei*, *Harpiliopsis spinigera*, and *Teleophrys cristulipes* showed a slight increase in number of individuals per coral head at Uva Island. It is possible that the increase in density of obligate coral associates is due to increased mucus production associated with a relatively continuous growing season for the coral at Uva Island, as compared to the dry-season reduction in growth in the Pearl Islands (*Glynn and Stewart*, 1973).

The differences in species composition between the two sites are related to the differences in species numbers and are discussed together. As already noted there are significantly fewer species of decapods associated with corals in the constant environment than in the fluctuating environment. I have previously suggested that this difference in species richness is the result of differential extinction rates generated by physical disturbance (*Abele*, 1976b). Space appears to be the limiting resource for many sessile marine organisms and species differ in their ability to obtain and hold space (*Connell*, 1961; *Paine*, 1966; *Dayton*, 1971; *Osman*, 1977). The coral-associated decapods parallel this situation because, through their association with a sessile host, they are functionally sessile. That space is limiting is suggested by both the species-area and individuals-area relationships and by the territorial activities of some species.

In the Pearl Islands upwelling regularly occurs from January through March, periodically lowering the temperature to 20°C, a point at which the growth rate of *P. damicornis* declines markedly or ceases, sometimes resulting in the loss of soft parts from the tips of branches (*Glynn and Stewart*, 1973). Experimental data have demonstrated that *Trapezia ferruginea* feeds on coral mucus and cannot survive on dead corals (*Knudsen*, 1967). Other coral specialists are known to feed on coral mucus (*Patton*, 1974) and are never found on dead corals (*Abele*, 1976a). It seems reasonable to expect some extinctions of the coral mucus-feeding

specialists during upwelling. These extinctions would open up
space for colonization by habitat generalists such as *Pisidia
magdalenensis* and *Heteractaea lunata*. Both species are abundant
on live and dead coral in the Pearl Islands but have not been found
in live corals at Uva Island. Colonization of cold-shocked corals
is likely to be a largely random process, resulting in a large
number of species, each with a low frequency of occurrence, and
a high variation in species composition among corals. This is
the situation in the Pearl Islands where the majority (about 70%)
of the species occur on only a few corals. At Uva Island, only
300 km away, there is no upwelling and comparatively little
variation in physical conditions. Extinction is probably rare
and the dominant coral specialists increase their total population
density and frequency of occurrence on coral heads. In summary,
the number of species per coral head is the same in the two regions
while the total species richness is greater in the fluctuating
environment. The data suggest that environmental constancy *per se*
does not increase species richness.

It should be pointed out that while the Uva Island reef is
not subjected to upwelling it is subject to disturbance much as
any other habitat. The crests of reefs in the Pearl Islands and
at Uva Island are occasionally exposed by extreme mid-day low
tides with considerable coral mortality (*Glynn*, 1976). If the
disturbance hypothesis is valid, one might expect to find more
species on the disturbed reef crest than on the deeper reef flanks.
However there are a few more species on the flank than on the
crest (*Abele*, unpublished). This phenomenon may be due to the
fact that the disturbed corals at Uva Island are surrounded by
other corals and any open space would probably be rapidly
colonized from adjacent corals.

In the Pearl Islands during upwelling all corals are affected
to some degree. In both cases (upwelling and exposure) in the
Pearl Islands, colonization of open space is more likely from
adjacent habitats because there are relatively few unaffected
corals.

While the above comments may seem like an *ad hoc* explanation,
it should be pointed out that the basic premise of the disturbance
hypothesis is amenable to laboratory falsification. It requires
that decapod coral specialists be more susceptible (in some way
affecting survival) to low temperatures than the decapod generalists. In contrast, alternative hypotheses, such as the
stability-time hypothesis of *Sanders* (1968), probably cannot
be falsified (*Peters*, 1976).

ACKNOWLEDGMENTS

The field work was supported by a Smithsonian Tropical Research Institute Fellowship in Tropical Biology. Computer time was provided by a grant from the Florida State University Computing Center. I thank Peter Glynn, S. Faeth, B. Felgenhauer, R. Gore, R. Livingston, and the FSU Ecology Group for constructive criticism.

REFERENCES

Abele, L. G., 1976a. Comparative species composition and relative abundance of decapod crustaceans in marine habitats of Panama, *Mar. Biol. 38:* 263-278.

Abele, L. G., 1976b. Comparative species richness in fluctuating environments: coral-associated decapod crustaceans, *Science 192:* 461-463.

Abele, L. G. and W. K. Patton, 1976. The size of coral heads and the community biology of associated decapod crustaceans, *J. Biogeogr. 3:* 35-47.

Connell, J. H., 1961. The influence of interspecific competition and other factors on the distribution of the barnacle *Chthalamus stellatus, Ecology 42:* 710-723.

Connor, E. F. and E. McCoy, in press. The statistics and biology of the species-area relationship. *Amer. Naturalist.*

Dayton, P. K., 1971. Competition, disturbance, and community. The provision and subsequent utilization of space in a rocky intertidal community, *Ecol. Monogr. 41:* 351-389.

Finney, W. C. and L. G. Abele, in prep. The natural history of the obligate coral symbiont *Trapezia ferruginea.* (Crustacea Decapoda Xanthidae).

Garth, J. S., 1949. Littoral brachyuran fauna of the Galapagos Archipelago, *Allan Hancock Pacific Expeditions 5:* i-iv, 341-522.

Glynn, P., 1972. Ecology of the shallow waters of Panama, *Bull. Biol. Soc. Washington 2:* 13-30.

Glynn, P., 1976. Physical and biological determinants of coral community structure in the Eastern Pacific, *Ecol. Monogr. 46:* 431-456.

Glynn, P. and R. H. Stewart, 1973. Distribution of coral reefs in the Pearl Islands (Gulf of Panama) in relation to thermal conditions, *Limnol. Oceanogr. 18:* 367-379.

Glynn, P., R. H. Stewart, and J. E. McCosker, 1972. Pacific coral reefs of Panama: structure, distribution, and predators, *Geol. Rdsch. 61:* 483-519.

Gore, R. H. and L. G. Abele, 1976. The shallow water porcelain crabs from the Pacific Coast of Panama and adjacent Caribbean Waters (Crustacea: Anomura, Porcellanidae), *Smithsonian Contr. Zool. No. 237:* 1-30.

Graham, J. B. and L. G. Abele, 1973. Fish kill and crab swarming following upwelling and a plankton bloom in Panama Bay, *Smithsonian Center Short-lived Phenomenon. Event 54-73:* 1618-1619.

Grant, P. R., 1966. Ecological compatibility of bird species on islands, *Amer. Naturalist 100:* 451-452.

Heck, K. L., Jr., G. van Belle, and D. Simberloff, 1976. Explicit calculation of the rarefaction diversity measurement and the determination of sufficient sample size, *Ecology 56:* 1459-1461.

Hurlbert, S. H., 1971. The nonconcept of species diversity: a critique and alternative parameters, *Ecology 52:* 577-586.

Knudsen, J. W., 1967. *Trapezia* and *Tetralia* (Decapoda, Brachyura, Xanthidae) as obligate ectoparasites of pocilloporid and acroporid corals, *Pacific Sci. 21:* 51-57.

MacArthur, R. H., 1969. Patterns of communities in the tropics, *Biol. J. Linn. Soc. 1:* 19-30.

Osman, R. W., 1977. The establishment and development of a marine epifaunal community, *Ecol. Monogr. 47:* 37-63.

Paine, R. T., 1966. Food web complexity and species diversity, *Amer. Naturalist 100:* 65-75.

Patton, W. K., 1974. Community structure among the animals inhabiting the coral *Pocillopora damicornis* at Heron Island, Australia, In: *Symbiosis in the Sea,* edited by W. B. Vernberg, 219-243, Univ. S. Carolina Press, Columbia, S. C.

Peters, R. H., 1976. Tautology in evolution and ecology, *Amer. Naturalist 110:* 1-12.

Sanders, H. L., 1968. Marine benthic diversity: a comparative
    study, *Amer. Naturalist 102:* 253-282.

Simberloff, D. S., 1970. Taxonomic diversity of island biotas,
    Evolution 24: 23-47.

Simberloff, D. S., in press. Rarefaction as a distribution-free
    method of expressing and estimating diversity, *Proc.
    Statistical Satellite of Int. Ecol. Congress,* Parma,
    Italy.

V. Climatological Features and
Psysical/Chemical Influence on Biological Systems

AQUATIC PRODUCTIVITY AND WATER QUALITY AT THE UPLAND-ESTUARY INTERFACE IN BARATARIA BASIN, LOUISIANA

Charles S. Hopkinson and John W. Day

Louisiana State University

ABSTRACT

Three fresh- to brackish-water coastal lakes and streams at the upland-estuary interface of Barataria Basin, Louisiana, were studied for a one-year period. Measurements of community production and metabolism, chlorophyll a, and water column nitrogen and phosphorus were used to assess the functional relationship the upland connection has with the estuary. Aquatic community gross production was highest in the lake immediately adjacent to the New Orleans West Bank uplands (696 g C m^{-2} yr^{-1}). A pattern of decreasing community heterotrophy was exhibited in the lakes from upland to the lower estuary. Active chlorophyll a, nitrogen, and phosphorus were also highest in the streams and lake adjacent to the upland area. Chlorophyll a averaged over 55 mg m^{-3} at these locations; remaining stations averaged only 11 mg m^{-3}. Nutrient levels were lowest in lakes, suggesting incorporation into plant biomass and sediments. Dissolved carbon and nutrient components made up an average of 78% of total concentrations.

Spatial distribution of selected water chemistry parameters indicates that artificial drainage and navigation canals in the estuarine and upland area have changed the hydrologic regime. Nutrient loading rates and high production values indicate that the area immediately adjacent to the upland is currently hypereutrophic. Its potential to absorb increasing nutrient loads is in doubt. Altered hydrology is directing nutrient-laden runoff away from Lake Salvador, an area that still has the potential to cleanse upland runoff water. Consequently accelerating eutrophication is to be expected at the new runoff-water discharge location--saline Barataria Bay. The implications of this process in relation to commercial fisheries production are many.

INTRODUCTION

Estuaries and lagoons are unique ecological systems because of their spatial relationship to land and sea. Their structure and function are controlled not only by internal processes but also by connections with the adjacent land and sea. Because of the extremely open nature of coastal systems, it is necessary to consider these connections in order to understand ecological function. Many papers in this volume are concerned with ecological connections between estuarine and shelf systems. Interactions and connections between marshes and bays, between bays and passes, and between passes and the continental shelf have been addressed. A connection which is as yet unexplored, but which is central to a better understanding of coastal areas, is the upland-estuary interface. All three systems--upland, estuary, and shelf--are inter-connected in many complex ways.

Historically, coastal areas have been important as sheltered sites of habitation that provide access to both land and sea. Some of the largest population centers in the United States have developed in or adjacent to estuaries (*e.g.* Boston, Philadelphia, New York, Washington, D. C., and New Orleans). The coastal area has been important not only for transportation but also because it provides natural food resources rich in protein and easy dumping places for waste material. These multiple uses are often incompatible.

Our objective in this paper is to investigate the upland-estuary interface in Barataria Basin, Louisiana. In particular, we are interested in effects of changes at the interface brought about by urban development. This study reports the aquatic production, nutrient loading, and water chemistry of three fresh- to brackish-water coastal lakes that are located at the upland-estuary interface. Water chemistry is helpful in understanding the hydrologic regime in this area and in assessing potential impacts of canals on estuarine eutrophication and fisheries production.

DESCRIPTION OF THE AREA

The study area consists of a connected series of coastal lakes, bayous, and navigation, petroleum, and drainage canals in Louisiana. These are located in the mid-portion of the Barataria drainage basin, a hydrologically unified interdistributary zone bounded by the Mississippi River to the east, Bayou Lafourche on the west, and the Gulf of Mexico to the south (Fig. 1). Five major habitats are included in Barataria Basin: (1) upland habitat located on the natural levee of the Mississippi River, (2) swamp forest at the head of the estuary, (3) fresh marsh, (4) brackish marsh, and (5) saline marsh adjacent to the Gulf of Mexico. Since the leveeing of the Mississippi River, completed around 1920, no overbank flooding has occurred and precipitation (152 cm annually) is the only significant freshwater input to the basin. As it enters the basin, fresh water flows from the headlands and side lands of the basin toward its central axis and drifts toward the Gulf of Mexico mixing with tidally introduced salt water from the Gulf.

The entire Barataria Basin unit is unique for several reasons: (1) it is an extremely productive ecosystem and supports major fisheries, (2) it is an important oil and natural gas producing zone, and consequently has a high concentration of

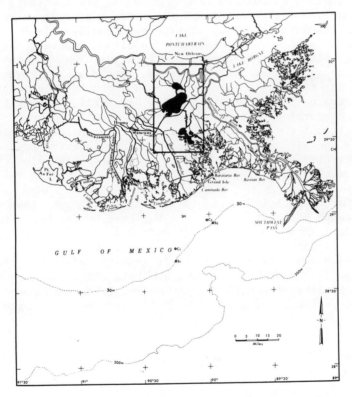

Fig. 1. Location of study area in relation to Barataria Bay and the Louisiana Coast.

petrochemical industries, and (3) the second and fifth largest
ports in the U. S. are adjacent to it.

These human activities have caused environmental degradation.
Recent reports indicate a process of accelerating eutrophication
in Louisiana coastal waters and most dramatically in the upper
freshwater regions of the basins (*Gael and Hopkinson*, 1978;
Craig and Day, 1977; *Van Sickle et al.*, 1976; *Day et al.*, 1977;
Kemp, 1978). These hypereutrophic areas are characterized by
frequent algal blooms, catfish, gar, and shad, fish kills, and
high drainage densities.

Measurements were caried out in three shallow lakes--
Cataouatche, Salvador, and Little--and in connecting waterways
(Fig. 2). The lakes range in size from 3680 to 19000 hectares.
Lake Cataouatche is not in the main drainage axis of the
Barataria Basin. It lies adjacent to the West Bank of New
Orleans to which it is connected via Bayous Verrett and Segnette.
In the natural state, all freshwater runoff entering Lake
Cataouatche would flow into Lake Salvador, but with the dredging
of Bayou Segnette, an unknown amount of water short circuits
Salvador and flows into the Gulf Intracoastal Waterway (GIWW).
The GIWW is regularly dredged; it receives considerable runoff
from New Orleans also. Lake Salvador is also connected to the
GIWW. Water flows into Barataria Bay from the GIWW via two
routes: through the Barataria Waterway, or via Bayous Perot and
Rigolettes to Little Lake. The degree to which Bayou Segnette,
the Gulf Intracoastal Waterway, and the Barataria Waterway have
affected the hydrologic regime is not quantitatively known.
However, personal communications with other scientists working
in the area and with local fishermen indicate that the regime
has been affected and that fish kills, saltwater intrusion, and
higher storm tides have resulted.

Lake Cataouatche is a slightly brackish (0-2 ppt) lake with
a maximum depth of about 2 m. It is bordered by fresh-water
marsh, predominantly bull tongue (*Sagittaria falcata*) and cattail
(*Typha* spp.). Lake Salvador is also slightly brackish (0-5 ppt)
with a maximum depth of about 3 m. It is bordered primarily by
cypress swamps (*Taxodium distichum*) and bull tongue marsh. Little
Lake is a tidally-influenced brackish-water lake (2-10 ppt) with
a maximum depth of about 2 m. The surrounding marsh is primarily
Spartina patens.

METHODS

Studies were conducted from December 1976 through February
1978. Diurnal measurements of dissolved oxygen were taken approximately monthly near the center of each of the three lakes.

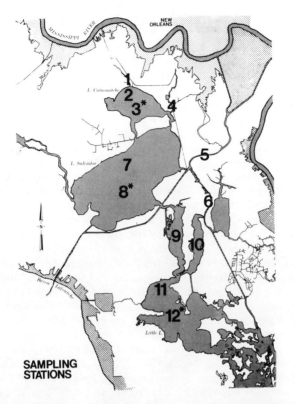

Fig. 2. Sampling site locations. Code: 1-Bayou Verett
(BVER), 2-North Lake Cataouatche (NLCAT), 3-Lake Cataouatche
(LCATS), 4-Bayou Segnette (BSEG), 5-Gulf Intracoastal Waterway
(BIWW), 6-Barataria Waterway (BW), 7-North Lake Salvador (NLSAL),
8-Lake Salvador (LSALS), 9-Bayou Perot (BPER), 10-Bayou Rigolettes
(BRIG), 11-North Little Lake (NLITL), 12-Little Lake (LITLS).
Asterisks indicate sites of production measurements. Cross-
hatched areas are uplands.

Dissolved oxygen (D.O.) was measured with a polarographic
electrode and recorded continuously. Profiles of D.O. and
temperature with depth were made at the beginning and end of
each diurnal monitoring to estimate turbulence. In the summer,
D.O. in Lake Cataouatche was monitored near the surface and
bottom. At other times of the year and in the other lakes it
was measured continuously at the surface only.

Community production and metabolism were calculated
using the method of *Odum* (1956; see also *Hall and Moll*, 1975).
Diffusion of oxygen across the air-water interface was estimated
from published studies of diffusion that were based on reaeration

rates in free water or on reoxygenation of oxygen-free domes
(*Odum and Hoskins*, 1958; *Butler*, 1975; *Gayle*, 1975). The
reported diffusion constant values were set proportional to
wind speed as suggested by *Kanwisher* (1962). Diffusion constants ranged linearly from 0 with 0 m sec^{-1} winds to 0.7 g
$O_2/m^2/hr$ at 100% deficit with 6.8 m sec^{-1} winds. Production
was determined on an areal basis by integrating volumetric
production with depth. In nonstratified conditions this was
done by simply multiplying surface production by depth. During
May, June, July, and August, however, the water column was
stratified and a different procedure was followed. Dissolved
oxygen and temperature-versus-depth curves for the lakes in
summer suggested that some degree of mixing was occurring in
the water column and that the lakes did not separate into two
distinct water masses. In Lake Cataouatche water column production was calculated by multiplying surface production by
half the depth and adding to that bottom production after it
was also multiplied by half the depth of the water column.
Generally, the magnitude of production in the lower half of the
water column was half that at the surface. As D.O. profiles were
similar in the other two lakes, we assumed, in calculating their
productivity, that the same pattern of production held.

At twelve stations (Fig. 2) surface water samples were
taken monthly and filtered on 0.45 μ glass fiber filter paper.
The filter paper was frozen immediately on dry ice for the
determination of chlorophyll. Filtered and unfiltered water
samples were frozen in plastic bottles for nutrient analysis.
Unfiltered samples were frozen in glass for organic carbon
analysis. In May, 1978, an additional 4 stations were sampled
for chlorophyll and nutrient analysis (Fig. 3). Active chlorophyll *a* was analyzed using the acetone extraction method of
Lorenzen (1967). In the lab, carbon samples were filtered in
glass apparatus, and total and dissolved organic carbon was
determined using an infrared carbon analyzer (Oceanography
International, College Station, Tex.). Nutrient levels were
determined as outlined by *Strickland and Parsons* (1968) with
modifications of *Ho and Schneider* (1976). Filtered samples
were used for NH_4-N, NO_3-N, NO_2-N, dissolved Kjeldahl-N and
PO_4-P determinations. Unfiltered samples were used to
analyze for total Kjeldahl-N and total-P.

Water depth, Secchi depth, temperature, wind speed, sea
state, and other physical environmental factors were measured
at each station.

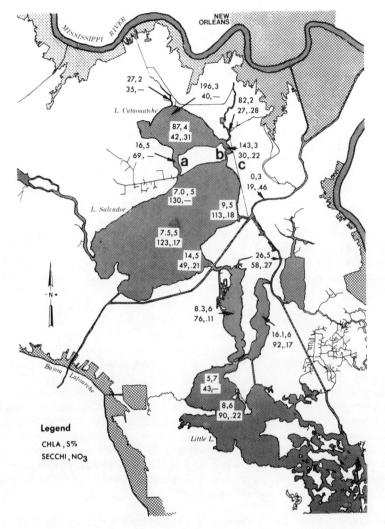

Fig. 3. Selected water chemistry parameters measured in May that aid in the determination of major hydrologic patterns. In each group of four numbers, the upper left refers to chlorophyll a, upper right to salinity, lower left to Secchi depth, and lower right to NO_3- and NO_2-N.

RESULTS AND DISCUSSION

Production

The magnitudes and seasonal patterns of production in the three lakes are similar to those found in other aquatic studies

in Louisiana. Gross production (Fig. 4) during the study ranged
from non-detectable to 15.2 g O_2 m^{-2} day^{-1}. Lake Cataouatche was
by far the most productive lake. At 2222 g O_2 m^{-2} year^{-1} (695 g C
m^{-2} year^{-1}) it was 1.7 times as productive as Little Lake and 2.1
times as productive as Lake Salvador.

Net daytime production (Fig. 4) was calculated as the net
sum of positive and negative rates of change of oxygen during
daylight hours. As with gross production Lake Cataouatche was
the most productive with Little Lake and Lake Salvador following
in that order, 876, 639, and 402 g O_2 m^{-2} year^{-1}, respectively.
The single highest rate was observed in Lake Cataouatche
(6.7 g O_2 m^{-2} day^{-1}). During fall and winter rates were often
below zero, indicating a heterotrophic condition ($P/R<1$).

Nighttime respiration in the water column showed the same
relative degree of activity as gross and net daytime production
(Fig. 4). Respiration at night averaged 1205 g O_2 m^{-2} year^{-1} in
Lake Cataouatche and was 753 and 602 g O_2 m^{-2} year^{-1} for Little
Lake and Lake Salvador, respectively.

The levels of production found in this study fall within
the range of other aquatic areas studied in Barataria Basin
(Table 1). In the headwaters of the basin, *Day et al.* (1977)
measured gross production rates of 888 and 3284 g O_2 m^{-2} year^{-1}
in Bayou Chevreuil and Lac des Allemands, respectively. The low
rate of production in the bayou was presumably due to an extensive
tree cover overhanging and shading the water. Lac des Allemands
receives considerable agricultural runoff and is highly eutrophic.
Lake Cataouatche is comparable to Lac des Allemands in that it is
eutrophic and very productive. In the brackish and saline waters
of Barataria Basin (*Day et al.*, 1973; *Allen*, 1975) production
averaged 1850 g O_2 m^{-2} year^{-1}. The productivity of Little Lake
and Lake Salvador are somewhat lower.

An analysis of net community production (the daily net sum of
all corrected oxygen rates of change) indicates an important attri-
bute of this system. On an annual basis net community production
was negative for all lakes. The lakes are heterotrophic; they
consume more organic matter than is produced. As the lakes are
surrounded by marsh lands and receive allochthonous inputs of
nutrients and organic matter, this situation is as expected. Lake
Cataouatche was the most heterotrophic (-350 g O_2 m^{-2} year^{-1}) and
Little Lake was the least so (-117 g O_2 m^{-2} year^{-1}). Lake
Salvador is heterotrophic at -198 g O_2 m^{-2} year^{-1}. Earlier
studies showed that Lac des Allemands was heterotrophic
(-450 g O_2 m^{-2} year^{-1}, *Day et al.*, 1977) while brackish and
saline stations in the lower basin were either balanced or
slightly autotrophic (0 to +54 g O_2 m^{-2} year^{-1}, *Day et al.*,
1973; *Allen*, 1975). The pattern exhibited is one of decreasing

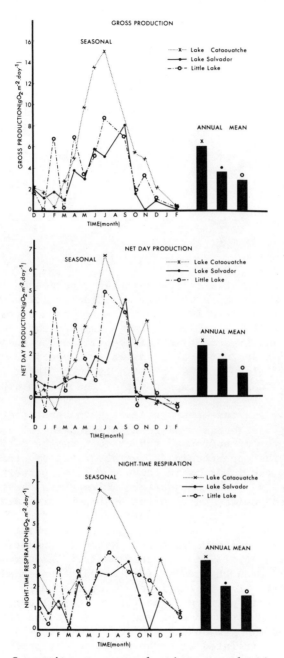

Fig. 4.   Community gross production, net daytime production, and nighttime respiration in three coastal Louisiana lakes.

TABLE 1.   Comparative aquatic productivity in the Barataria Basin.*

Lake	NDP	NR	GP	NCP	Reference
Lake des Allemands			3284	-450	*Day et al.*, 1977
Lake Cataouatche	876	1205	2222	-350	This study
Lake Salvador	402	602	1058	-198	This study
Little Lake	639	753	1307	-117	This study
Brackish-Saline			1850	0 to + 54	*Allen*, 1975 *Day et al.*, 1977

*All values in g O_2 m^{-2} yr^{-1}.

heterotrophy from headwater regions that experience considerable upland runoff to the more brackish and tidally influenced regions (Table 1).   This pattern implies that allochthonous sources of organic material (from upland drainage and wetlands) become increasingly less important in the overall carbon budget of this system in a headwater to tidal water direction.   The land (wetland and highland) to water ratio is much higher in the upper basin and accentuates the importance of the allochthonous input.

This study and the others conducted in Barataria Basin indicate that, on the basis of temporal patterns of production and of the degree of eutrophy, the estuary can be divided into two regions.   The upper basin is characterized by clear seasonal patterns while the lower basin lacks the clear pattern and appears quite oscillatory.   Lake Cataouatche illustrates the former situation where production and respiration levels are relatively low in the winter and rise to peaks during the summer.   This was also the pattern found in Lac des Allemands.   Little Lake, however, lacked this clear seasonal pattern.   Rather, production varies from high to low levels on a monthly basis.   Lake Salvador was intermediate between the two.   *Allen* (1975) found a similar pattern of oscillation in the lower Barataria Basin.   Although the reason for this pattern is not known, we believe that it may be due to light penetration and benthic producers.   In Little Lake the monthly change in Secchi disc depths appears to be strongly correlated with the magnitude of production. When water transparency increased enough to allow light to penetrate to the lake bottom (February and April, Fig. 4), significant production took place.   This same pattern was reported for lower Barataria Bay (*Day et al.*, 1973).   It is

generally only in the brackish and saline regions of Barataria
Basin that the water is shallow enough and clear enough to enable
sufficient light to reach the bottom. Secchi depths were significantly greater in Little Lake than in the other two lakes.

Chlorophyll

The four stations closest to the West Bank of New Orleans
(Bayous Verrett and Segnette and two stations in Lake Cataouatche)
had much higher active chlorophyll a levels than all remaining
stations (Fig. 5). Chlorophyll a at the four locations averaged
above 55 mg m^{-3} and was as high as 230 mg m^{-3} in Bayou Verrett.
The mean chlorophyll a concentration of the other stations was
11 mg m^{-3}. The obvious reason for this difference is the runoff
and nutrient loading from the West Bank uplands into Bayous
Verrett and Segnette which connect to Lake Cataouatche, thus
stimulating phytoplankton growth. This conclusion agrees well
with the high magnitude of gross production in Lake Cataouatche.
With the exception of the Gulf Intracoastal Waterway, none of the
other stations receives any direct nutrient inputs of comparable
magnitude. The Gulf Intracoastal Waterway actually receives more

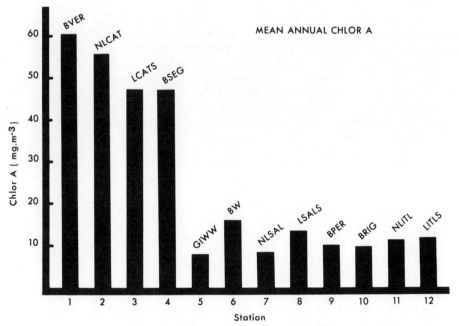

Fig. 5. Average annual active chlorophyll a concentrations
at twelve locations in the Louisiana upland–estuary interface
region (mg m^{-3}).

loading than any other station, but because of barge traffic,
bottom sediments are stirred up, increasing turbidity (average
Secchi disk depth of 16 cm) and causing increased light limitation.
Consequently, although nutrients were available, insolation was
not. Thus, chlorophyll levels in the Intracoastal Waterway
(6.8 mg m^{-3}) were the lowest in the study.

No station showed a clear seasonal trend of chlorophyll a
concentrations, but when monthly means of all stations were combined a significant seasonal pattern emerged. Chlorophyll
concentrations were higher in the summer than during the winter.
In Lac des Allemands, *Butler* (1975) found a seasonal pattern
similar to that of Lakes Cataouatche and Salvador. In contrast,
Allen (1975) and *Happ* (1974) did not find such a pattern in either
the brackish or the saline portions of the Barataria Basin. They
found a pattern similar to that found in Little Lake, where a more
oscillatory pattern was exhibited. Consequently, it appears that
the seasonal pattern found in the fresh headwaters of Louisiana
hydrologic basins is lacking in the more saline reaches. This
distribution pattern coincides with that found for aquatic
production. As chlorophyll a concentrations can change exponentially within hours or days, monthly values are inadequate
to evaluate true cycles of abundance. The data presented here
are only suggestive of seasonal trends.

Chlorophyll a concentrations at the fresher water stations
are similar to concentrations measured previously in freshwater
Lac des Allemands, while those at the more saline stations are
closer to values reported from brackish and saline areas of the
basin. The average concentrations found in the freshwater
stations, Bayous Segnette and Verrett and Lake Cataouatche
(55 mg m^{-3}), are 1.8 times higher than those reported by
Butler (1975) for Lac des Allemands. The average concentration from the remaining stations (11 mg m^{-3}) is similar to that
found by *Happ et al.* (1977) in the most saline portion of
Barataria Basin but somewhat lower than levels ($\overline{x} \cong 24$ mg m^{-3})
reported by *Allen* (1975) in the mid-Barataria Basin brackish
region.

Nutrient Analysis

Generally, highest nutrient levels were found in canals and
waterways draining the Mississippi River natural levee (Bayou
Segnette, Bayou Verrett, Gulf Intracoastal Waterway, and the
Barataria Waterway; see Tables 2 and 3). In contrast, there
were no consistent differences in nutrient levels among lakes,
while marked differences were apparent in productivity and
chlorophyll.

TABLE 2. Stations exhibiting the highest and lowest nutrient concentrations and secchi depths. High Secchi depth indicates clear water. * indicates significant difference from all other stations. TIN = Total inorganic nitrogen, TP = Total phosphorus, GIWW = Gulf Intracoastal Waterway, SEG = Bayou Segnette, VER = Bayou Verrett, CAT = Lake Cataouatche, SAL = Lake Salvador, LL = Little Lake, RIG = Bayou Rigolettes, PER = Bayou Perot, BW = Barataria Waterway.

Component	Highest level			Lowest level	
	1	2	3	1	2
NO$_3$	GIWW*	SEG	VER	CAT	SAL
NH$_4$	SEG*	VER	RIG	SAL	CAT
TIN	GIWW*	SEG	BW	CAT	SAL
TOC	SEG*	VER	CAT	RIG	LL
DOC	SEG*	VER	PER	GIWW	BW
TP	SEG*	GIWW*	LL		
PO$_4$	SEG*	VER	GIWW	CAT	LL
Secchi	RIG	SAL	LL	GIWW*	SEG

TABLE 3. Annual means of six water chemistry parameters measured in the brackish region of Barataria Basin.

Sampling location	Salinity (ppt)	Secchi (cm)	TOC mg ℓ^{-1}	DOC mg ℓ^{-1}	Total Kjeldahl-N mg ℓ^{-1}	Dissolved Kjeldahl-N mg ℓ^{-1}
Bayou Verrett	0.7	46.5	20.1	17.7	1.7	1.3
Bayou Segnette	0.8	38.2	25.1	22.1	2.0	1.5
Gulf Intracoastal Waterway	1.4	16.6	13.4	8.8	1.6	1.1
North Lake Cataouatche	1.1	45.7	---	---	---	---
Center Lake Cataouatche	1.2	48.2	16.8	12.2	1.5	0.8
North Lake Salvador	1.6	53.0	---	---	---	---
Center Lake Salvador	2.4	58.3	13.7	9.2	1.4	0.8
Barataria Waterway	2.9	40.9	13.4	9.2	1.7	1.0
Bayou Perot	3.4	49.0	14.7	12.3	1.2	0.9
Bayou Rigolettes	3.4	64.5	12.5	9.4	1.3	0.9
North Little Lake	3.9	38.8	---	---	---	---
Center Little Lake	4.8	54.6	13.2	10.3	1.4	0.9

Note: --- denotes information not collected.

Secchi depths, total and dissolved organic carbon (TOC, DOC), and total and dissolved Kjeldahl-N exhibited no consistent seasonal pattern. Rather, these parameters varied widely from month to month. With the exception of one or two stations (Table 2) there were no statistically significant differences in the values of these parameters from station to station. The exceptions were that Secchi depths in GIWW were lower than elsewhere and TOC and DOC were higher in Bayou Segnette than elsewhere. Low Secchi depths in GIWW are indicative of sufficient turbulence caused by barge traffic to suspend benthic sediments. That the highest organic carbon values were found in Bayou Segnette (Bayou Verrett secondarily) indicate that the uplands in the region are an important source of organic material. Wetlands adjacent to Bayou Segnette constitute a small area and probably do not contribute major amounts of organic carbon. Dissolved carbon and nitrogen components were consistently found in greater concentrations than particulate components. DOC:TOC ratios averaged 3.5:1 at all stations. The two bayous draining the West Bank had the highest ratios (7.3:1). The ratios at all stations are higher than those found in Chesapeake Bay (1:1 to 2:1) as reported in *Happ et al.* (1977). Such ratios may reflect the low tidal energy range found in Louisiana. To be transported detrital particles must be fractionated to progressively smaller sizes as tidal flushing energies decrease.

The ratio of reactive phosphorus to total phorphorus at various stations reveals several processes that occur in this system (Fig. 6). Note that aside from Bayou Segnette and the GIWW there are no essential differences in the concentration of total phosphorus from station to station. However, the reactive phosphorus-total phosphorus ratio is much lower in the lakes (and Bayou Perot which closely resembles a lake) than at other stations. As the ratio is lowest in Lake Cataouatche which is also the most productive we can conclude that available P is being removed from the water column and incorporated into plant biomass at a rate proportional to production. (A significant negative correlation was found between reactive P and gross production.) The GIWW also has a low reactive to total P ratio, but as mentioned above this is because of the presence of suspended sediments which have a high proportion of unreactive P (*Patrick and Khalid*, 1974). In contrast Bayous Segnette and Verrett exhibited high ratios of reactive to total P. Such ratios are indicative of urban storm water runoff (*Omernik*, 1977), and again illustrate the importance of upland-estuarine fluxes.

Inorganic nitrogen concentrations (Fig. 7) reveal much the same system attributes as did phosphorus but not as clearly. NO_3-N and NO_2-N concentrations are lowest in the lakes and Bayou Perot, a pattern similar to that found for reactive P.

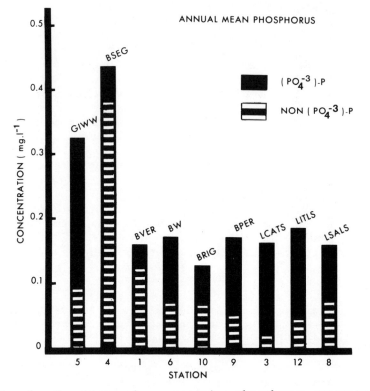

Fig. 6. Reactive and non-reactive phosphorus concentrations
on an annual basis (mg ℓ^{-1}).

In addition, the highest nitrogen levels were found in Bayou
Segnette and GIWW. Nitrogen concentrations were inversely
proportional to gross production (r = 0.57, df. 21: NO_3^-- and
NO_2^--N to gross production).

Hydrology

Although we have no direct measurements of water movement,
the results of this study and others indicate that there has
been a significant change in the hydrology of the area (Fig. 8).
Prior to alteration by human activity, water flowed into Lake
Cataouatche from Bayous Verrett and Segnette and via sheet flow
from surrounding wetlands. Water then moved through East and
West Couba Island Bayous into Lake Salvador which also received
upper basin drainage from Bayou des Allemands. Movement was
then south via Bayous Barataria-Rigolettes or Perot to Little
Lake and thence to Barataria Bay.

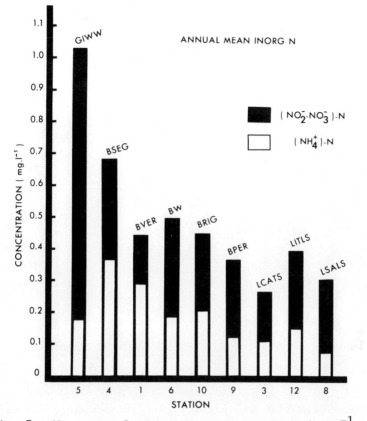

Fig. 7. Mean annual nitrogen concentrations (mg ℓ^{-1}).

At present, we believe that most water entering Lake
Cataouatche does not flow into Lake Salvador, but rather flows
via the Bayou Segnette waterway to the Barataria Waterway and
thence into upper Barataria Bay. Very little seems to flow into
Bayou Rigolettes, formerly a major route. In addition, most
upland runoff currently flows directly into water bodies,
bypassing sheet flow. Water from Lake Salvador may also flow
to Barataria Bay down the Barataria Waterway or via Bayou Perot
and Little Lake.

High productivity and chlorophyll levels in the Lake
Cataouatche-Bayou Segnette area are evidence of the effects
of upland runoff. The waterways draining the West Bank also
had higher nutrient levels and more turbid waters (Table 2).
During May, 1977, a number of additional stations were sampled.
Results for chlorophyll a, salinity, Secchi, and NO$_3$-N illustrate
the effects of hydrologic changes (Fig. 3). The four parameters

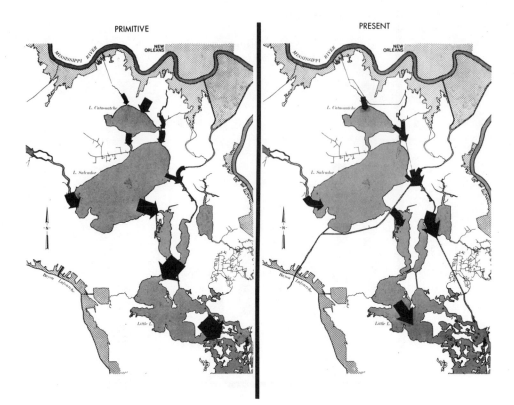

Fig. 8. Primitive and present patterns of water movement in the middle Barataria Basin.

indicate that most Lake Cataouatche water does not pass through West Couba Island Bayou ("a" in Fig. 3). It appears that the freshwater gradient in Lake Cataouatche is from west to east. In fact, saltier water from Lake Salvador appears to gain access to Lake Cataouatche via West Couba Island Bayou. Secchi disk depth and NO_3-N concentrations also indicate an easterly direction. (It should be stated that although the Gulf Intracoastal Waterway has some of the highest and lowest values for the parameters considered, it is not a potential sink or source because there is little net water flow in it.) From eastern Lake Cataouatche concentration gradients indicate that there is little flow to Lake Salvador through Eastern Couba Island Bayou ("b" in Fig. 3). In particular, Secchi depths at East Couba Island Bayou and Lake Salvador are very dissimilar. Rather, water probably flows down Bayou Segnette and thence down Barataria Waterway. High nutrient concentrations in East Couba Island Bayou are probably due to considerable mixing of local water masses in the immediate vicinity. Lake Salvador nutrient concentrations

and Secchi depths are quite dissimilar to those in Lake
Cataouatche and Barataria Waterway. Salinity, chlorophyll,
and nutrient concentrations suggest that Lake Salvador is part
of a regime consisting of Little Lake and Bayou Perot.

   Seaton (1979) conducted an extensive two-year chemical
survey in this area. Her results support our observations.
In addition she found that a station near the junction of the
Barataria Waterway and upper Barataria Bay was more enriched
than either Little Lake or Bayou Rigolettes. This fact supports
the idea that the Barataria Waterway receives most of the drainage from Lake Cataouatche and the GIWW.

   The net result of these hydrological changes is that there
has been an uncoupling of a once unified hydrological system and
a destabilization of the hydrological regime. The major water
flow in the basin has shifted away from Little Lake to the
Barataria Waterway. Upland runoff which originally flowed
slowly as a sheet into water bodies now enters as pulses closely
associated with precipitation episodes. Upbasin-downbasin movement of water was probably originally dampened by shallow sinuous
bayous and gradual mixing in lakes and bays. Now, straight,
hydraulically efficient canals allow much more rapid water
movement. As a result changes in salinity and water level are
more rapid and extreme (*Seaton,* 1979, and personal communication
with residents and fisherman).

Nutrient Loading

   The upland area adjacent to Lake Cataouatche is functionally
important to the Barataria Basin estuary because of discharge of
water and nutrients to the wetlands. Land use and nutrient
runoff from the area have been estimated by *Hopkinson* (1979).
The uplands draining into this area comprise the heavily urbanized
West Bank of New Orleans. The uplands are 24680 hectares in size,
40 percent of which is woodland or openland. Of the remainder, 11
percent is agricultural, 60 percent is strip and residential, 12
percent is commercial or institutional, and 16 percent is industrial. Based on the quantities of nitrogen and phosphorus expected
(calculated by *Hopkinson* (1979) for the entire Barataria Basin) to be
available from these various land use types in storm water runoff
the following annual nutrient output is expected: nitrogen--9.35
x 10^8 g; phosphorus--1.34 x 10^8 g. All runoff from the area is
pumped into the swamp but pumpage records have not been compiled
(*Jefferson Parish Department of Drainage and Sewage,* pers. comm.).
Therefore, runoff is estimated. If it is assumed that 75 percent
of rainfall in urban areas and 15 percent of rainfall in forested
and agricultural areas runs off, then annual discharge into the
wetlands would amount to 2.36 x 10^8 m^3 yr^{-1}. In relation to

nutrient output this quantity is equivalent to average nutrient concentrations of 3.91 mg N ℓ^{-1} and 0.57 mg P ℓ^{-1} in the runoff water. The majority of upland runoff enters the study area via Bayous Verrett and Segnette and GIWW. Amounts of nitrogen, phosphorus and water runoff reaching these waterbodies, based on the drainage basin size of each, are given in Table 4.

All discharge from Bayou Verrett flows into Lake Cataouatche. Areal nutrient loading of Lake Cataouatche from Bayou Verrett is 3.5 g N m^{-2} yr^{-1} and 0.5 g P m^{-2} yr^{-1}. This level of loading alone indicates that Lake Cataouatche is hypereutrophic (*Shannon and Brezonik*, 1972, *Craig and Day*, 1977). The drainage patterns of Bayou Segnette and GIWW have not been determined but we do know that some portion of the former discharges into Lake Cataouatche also. Consequently total areal loading to Lake Cataouatche is higher than that calculated from Bayou Verrett. An additional undetermined source of loading is the wetland surrounding Lake Cataouatche.

CONCLUSIONS

This study has revealed several important processes and attributes of the brackish region of Barataria Basin. Lake Cataouatche and Lake Salvador are currently interface systems valuable to man, serving as efficient nutrient processing traps. High nutrient loads from the upper des Allemands region of Barataria Basin (*Butler*, 1975) and from the West Bank of New Orleans are being processed by Lake Salvador and Lake Cataouatche, respectively. The relatively low trophic state of Lake Salvador indicates that it will have the capacity to continue processing nutrient loads in the future. Lake Cataouatche, however, is already highly eutrophic and its ability to handle further the increased nutrient loads which will undoubtedly occur with increasing development of the West Bank is in doubt. Lake Cataouatche already experiences severe fish kills following

TABLE 4. Estimated runoff of nitrogen, phosphorus, and water to three bodies of water near Barataria Basin.

Water body	Nitrogen (g)	Phosphorus (g)	Water (m^3)
Bayou Verrett	1.3 x 10^8	1.88 x 10^7	3.3 x 10^7
Bayou Segnette	1.39 x 10^8	2.04 x 10^7	3.5 x 10^7
Gulf Intracoastal Waterway	6.6 x 10^8	9.5 x 10^7	1.7 x 10^8

heavy storm water runoff from the West Bank. Fish kills
indicate the area is losing its value as a prime nursery ground
for important Louisiana commercial fisheries. A project which
should be undertaken in the future is investigation of the
status of these lake sediments with respect to nutrient satura-
tion. The dynamics of sediment-water nutrient relationships
need to be investigated in order to determine whether these
lakes will be able to continue to trap nutrients.

The present study also demonstrates the impact of canals on
local hydrology. It appears that two canals, Bayou Segnette and
Barataria Waterway, have drastically modified the original hydrologic regime of the area. Whereas in the past all water from
the upper portion (mainly freshwater area) of the basin passed
through Lake Salvador, only that portion originating in the
des Allemands area does at present. Because of these canals
saltwater intrusion is occurring further north in the basin,
a great deal of West Bank storm water runoff is being shunted
directly to Barataria Bay, and Lake Salvador no longer serves
as a nutrient trap or processor for storm water runoff.

The most obvious weakness encountered in this study is the
lack of hydrological knowledge of the region. Water flow is the
integrating factor in wetlands. It controls pathways of carbon,
nutrients, and sediments. In Barataria Basin, and in all wetland
areas there has been an extreme paucity of research dealing with
hydrology. Future research will have to address more satis-
factorily water flux and the nutrient and organic dynamics driven
by water flow.

In this paper we have dealt with ecological connections at
the upland-estuary interface. It is obvious that alterations at
this interface can have direct impacts such as nutrient runoff
and eutrophication. However, there can also be indirect changes
distant from the site of direct impact, such as changes in
hydrology. An adequate understanding of ecological connections
demands analysis of those indirect, cumulative impacts.

Interactions taking place at the upland-estuary interface
can also affect coastal waters. For example, poor water quality
and destruction of nursery habitat can affect fisheries in
coastal waters. Hydrological changes may change salinity distribution over the estuary and in coastal waters. Therefore, an
understanding of coastal-estuarine interactions requires that
the entire spectrum from uplands to shelf waters be considered.

REFERENCES

Allen, R. L., 1975. *Aquatic Primary Productivity in Various Marsh Environments in Louisiana*, Master's thesis, Louisiana State University.

Butler, T. J., 1975. *Aquatic Metabolism and Nutrient Flux in a South Louisiana Swamp and Lake System*, Master's thesis, Louisiana State University.

Craig, N. J. and J. W. Day (eds), 1977. Cumulative impact studies in the Louisiana coastal zone: Eutrophication, Land Loss, 157 pp., Final Report to Louisiana State Planning Office, Center for Wetland Resources, Louisiana State University, Baton Rouge, La.

Day, J. W., W. G. Smith, P. R. Wagner, and W. C. Stowe, 1973. Community structure and carbon budget of a salt marsh and shallow bay estuarine system in Louisiana, 80 pp., Center for Wetland Resources, Baton Rouge, La. 70803. Sea Grant Publ. No. LSU-SG-72-04.

Day, J. W., T. Butler, and W. H. Conner, 1977. Productivity and nutrient export studies in a cypress swamp and lake system in Louisiana, *Estuarine Processes 2:* 255-269.

Gael, B. T. and C. S. Hopkinson, 1978. Drainage density and trophic state in Barataria Basin, Louisiana, In: *Third Coastal Marsh and Estuary Management Symposium*, edited by J. Day, D. Culley, A. Mumphrey, and R. Turner, Louisiana State Univ., Baton Rouge, in press.

Gayle, T., 1975. *Systems Models for Understanding Eutrophication in Lake Okeechobee*, Master's thesis, Univ. of Florida.

Hall, C. A. S., and R. Moll, 1975. Methods of assessing aquatic primary production, In: *Primary Productivity of the Biosphere*, edited by L. Leith and R. H. Whittaker, 20-43, Springer-Verlag, New York.

Happ, G., 1974. *The Distribution and Seasonal Concentration of Organic Carbon in a Louisiana Estuary*, Master's thesis, Louisiana State University.

Happ, G., J. G. Gosselink, and J. W. Day, 1977. The seasonal distribution of organic carbon in a Louisiana estuary, *Estuarine and Coastal Marine Science 5:* 695-705.

Ho, C. L. and S. Schneider, 1976. Water and sediment chemistry, In: *Louisiana Offshore Oil Port: Environmental Baseline Study, Vol. II, Sec. 2,* edited by J. G. Gosselink, R. R. Miller, M. Hood, and C. M. Bahr, Jr., *Environmental Assessment of a Louisiana Offshore Oil Port, Vol. II, Sec. II;* Technical Appendices VI-VIII, LOOP, Inc., New Orleans, La.

Hopkinson, C. S., 1979. *The relation of man and nature in Barataria Basin, La.,* Ph.D. Diss., Louisiana State University.

Kanwisher, J., 1962. On the exchange of gases between the atmosphere and the sea, *Deep-Sea Research 10:* 195-207.

Kemp, G. P., 1978. *Agriculture Runoff and Nutrient Dynamics of a Swamp Forest in Louisiana,* Master's thesis, Louisiana State University.

Lorenzen, C. J., 1967. Determination of chlorophyll and pheo-pigments: spectrophotometric equations, *Limnol. and Oceanogr. 12:* 343-346.

Odum, H. T., 1956. Primary production in flowing waters, *Limnol. and Oceanog. 1:* 102-117.

Odum, H. T. and C. M. Hoskins, 1958. Comparative studies of the metabolism of marine waters, *Publ. Inst. Mar. Sci. Univ. Tex. 5:* 16-46.

Omernik, J., 1977. Non-point source steam nutrient level relationships: a nationwide study, EPA-600/3-77-105.

Patrick, W. H., Jr. and R. A. Khalid, 1974. Phosphate release and sorption by soils and sediments: effect of aerobic and anaerobic conditions, *Science 186:* 53-55.

Seaton, A. M., 1979. *Nutrient Chemistry in the Barataria Basin--a Multivariate Approach,* Master's thesis, Louisiana State University.

Shannon, E. E. and P. L. Brezonik, 1972. Relationships between lake trophic state and nitrogen and phosphorus loading rates, *Environmental Science and Technology 6:* 719-725.

Strickland, J. and T. Parsons, 1968. A practical handbook of seawater analysis, *Fish. Res. Bd Can. Bull. 167:* 311 pp.

Van Sickle, V. R., B. B. Barrett, L. J. Gulick, and T. B. Ford,
    1976. *Barataria Basin: Salinity Changes and Oyster
    Distribution*, 22 pp., Louisiana State University Center
    for Wetland Resources, Baton Route, La., Sea Grant Publ.
    No. LSU-T-76-002.

LONG-TERM CLIMATOLOGICAL CYCLES AND POPULATION CHANGES IN A RIVER-DOMINATED ESTUARINE SYSTEM

Duane A. Meeter, Robert J. Livingston, and

Glenn C. Woodsum

Florida State University

ABSTRACT

River flow and rainfall patterns in the Apalachicola drainage system were analyzed by time-series methods, and a preliminary comparison was made with commercial harvests and trawl-tow collections of organisms taken during a long-term sampling program. Spectral analysis revealed long-term (5-7 year) cycles in river flow ranging over 25% of the mean flow figures. Cross-spectral analysis indicated that these cycles were somewhat correlated with local (Florida) rainfall but were strongly related to upstream (Georgia) rainfall. A two-parameter model gave a satisfactory fit to the monthly series of annual river flow changes. Dam construction and filling did not appear to be related to long-term cyclic patterns, but weekly cycles in river flow, evident during periods of low water, appeared after dam installations were made. A preliminary comparison of annual river flows with commercial harvests showed strong correlations with oyster and crab catches in associated coastal areas. River flow was weakly correlated with shrimp and crab numbers obtained from a long-term sampling program. There were indications that while long-term river fluctuations were closely associated with commercial landings, such correlations should be carefully scrutinized for possible influence of economic and sociological conditions. Cyclic biological changes driven by key climatological factors may be complicated by highly individualistic species strategies which tend to mask direct phase relationships. However, long-term periodic changes in climatological features of drainage systems need to be considered if the biological

*variability of such areas is to be explained. There are indica-
tions that such cycles may differentially influence population
changes at various levels of biological organization.*

INTRODUCTION

Wiens (1977) points out that virtually all mathematical
treatments of competitive interactions assume functional resource
limitation and equilibrium of the system in question. Long-term
(*i.e.* > 2 years) environmental variation is usually ignored and
impact arising from periodic "ecological crunches" based on
climatological functions is rarely acknowledged as a possible
controlling factor. Long-term climatic fluctuations have been
identified in various areas, and extended responses of individual
populations have been associated with changes in temperature
(*Sutcliffe et al.*, 1977), river flow (*Sutcliffe*, 1972), and
rainfall (*Gunter and Hildebrand*, 1954). The importance of
long-term cyclic variation to an understanding of marine
(coastal) systems has been pointed out by several authors
(*Longhurst et al.*, 1972; *Jensen*, 1976) even though most of
the marine research effort in recent times has been applied
to relatively short-term problems.

The Apalachicola Bay system (Fig. 1) is a shallow, barrier-
island estuary. It is located at the terminus of the Apalachicola
River, which originates at the confluence of the Flint and
Chattahoochee Rivers, which drain considerable portions of
Georgia and Alabama. Lake Seminole, formed by the Jim Woodruff
Dam, serves as the head-waters of the Apalachicola system.
Apalachicola River flow has been shown to be a key determinant
of the physico-chemical environment of the Apalachicola estuary,
and is a predictor of various biological functions of this system
(*Livingston et al.*, 1978; *Meeter and Livingston*, 1978). Because
of the importance of the upland drainage system to receiving
coastal areas of northeast Florida, a study was carried out to
determine the potential for use of "time series" analysis of
key climatic features such as rainfall and river flow as pre-
dictors of variation in several important coastal populations.

THE DATA

River flow data (Blountstown, Florida) consists of mean daily
flow by month (696 observations) and discrete daily readings
(21,170 observations) provided by the U. S. Army Corps of
Engineers (Mobile, Alabama) for the years 1920-1977. Local
rainfall information was provided by the National Oceanic and
Atmospheric Administration (Environmental Data Service,
Apalachicola, Florida) and consisted of 492 monthly totals

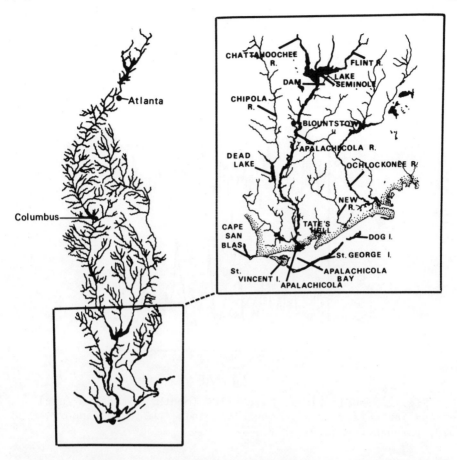

Fig. 1. Map of the Apalachicola drainage system showing
locations of rainfall and river flow monitoring stations (Atlanta,
Georgia; Columbus, Georgia; Blountstown, Florida; Apalachicola,
Florida) and the position of the Apalachicola Bay system (Franklin
County, Florida) relative to the Apalachicola River and the Tate's
Hell Swamp (East Bay rainfall tower).

for the years 1937-1977. Monthly Columbus, Georgia, rainfall
information (696 observations) was provided by the National
Oceanic and Atmospheric Administration (National Climatic Center,
Asheville, N. C.). Blountstown is located on the Apalachicola
River about 129 km (80 miles) upstream from Apalachicola Bay and
48 km (30 miles) south of Jim Woodruff Dam (Fig. 1); Columbus is
located on the Chattahoochee River about 257 km (160 miles) north
of Jim Woodruff Dam (Fig. 1). Monthly river flow is seasonal
(Fig. 2), with considerable variation evident in peak flow rates.
The rainfall data show similar patterns, and obviously contain a

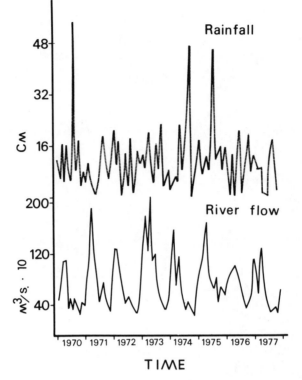

Fig. 2. Apalachicola River flow (monthly means; m³/sec) and Apalachicola rainfall (monthly totals; cm) from January, 1970, through February, 1978.

source of variation with higher frequency. The monthly means of river flow, Apalachicola rainfall, and Georgia rainfall (Fig. 3) confirm the existence of distinctive seasonal patterns. River flow consists primarily of a late winter-spring peak, while Apalachicola rainfall is characterized by a minor spring peak and a major one in late summer (July-September). In Georgia, the rainfall pattern consists of two comparable peaks (March and July). The river flow peak is timed with the major and minor spring rainfall peaks in Georgia and Florida. The fact that there is no summer peak in river flow, despite respective peaks in rainfall, may be related to seasonal, differential evapo-transpiration cycles in the associated wetlands. The modifying effect of wetlands vegetation on major hydrological cycles deserves attention. However, the standard deviations for all variables are large enough to allow considerable differences from year to year. A plot of the twelve monthly means against the corresponding standard deviations in the case of river flow and against variances in the case of the two rainfall variables revealed a linear relationship.

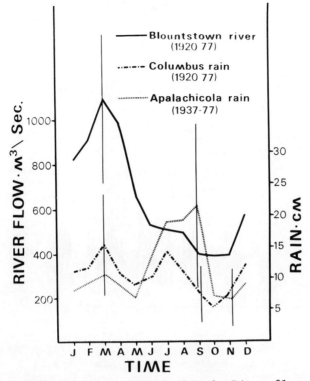

Fig. 3. Monthly means of Apalachicola River flow and
rainfall with selected standard deviations.

This relationship suggests that the statistically appropriate
metrics, in which the standard deviation will be approximately
constant, are log river flow and square root of rainfall
(*Kempthorne*, 1952).

A plot of 36-month moving averages of river flow and rainfall (Fig. 4) suggests that, in addition to the obvious annual
variability, longer-term cycles may be present. A moving average
replaces each observation by the mean of the adjacent observations,
and tends to "smooth" the data, revealing low-frequency cycles.
In this case, certain long-term cycles are indicated. However,
since averaging in this way is known to introduce dependence
even into series of independent observations, a more objective
method of analyzing cycles should be used for verification.

SPECTRAL ANALYSIS

All spectral and cross-spectral analyses used here came from
the SPSS (Statistical Package for the Social Sciences) subprogram
SPECTRAL (1977), and used 120 lags with a Hamming window. Prior

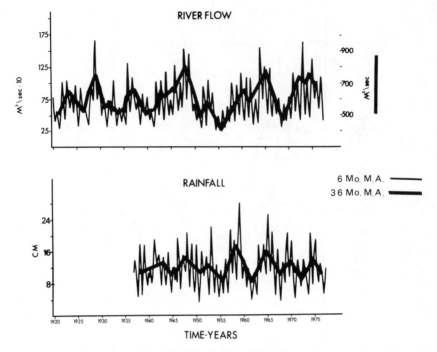

Fig. 4. Six-month and 36-month moving averages of river flow
(m³/sec; 1920-1977) and Apalachicola rainfall (cm; 1937-1977).

to analysis, river flow and rainfall data were transformed by
logarithm and square root, respectively. Comparisons of the
spectra of the transformed and untransformed data revealed that
they were only different away from the major peaks; the transformed spectrum was lower and smoother there. After transformation,
the mean and a linear trend (essentially none) were subtracted from
the data and the resulting observations tapered 20% in the manner
suggested by Bloomfield (1976). That is, the first and last 10%
of the observations were multiplied by a cosine function on the
interval (π, 2π) and on the interval (0, π), respectively. This
process has a mild purifying effect on the spectrum similar to
the effect of the transformations. Confidence intervals for the
power, coherence, and phase spectra were computed as suggested
by *Jenkins and Watts* (1968).

     Spectral analysis is a means of expressing a sample of points
as the weighted sum of periodic functions, and of statistically
analyzing the result. If, when N observations are taken at unit
intervals, sines and cosines of frequency 1/N, 2/N, . . . , 1/2
(of period N, N/2, N/3, . . . , 2) are chosen as the periodic
functions, the functions are orthogonal and uncorrelated. This
choice simplifies the computations and the analysis.

A plot of the sum of the squares of the two coefficients at each of the N/2 frequencies (periods) is called the periodogram. It constitutes an analysis of variance with respect to the period or frequency of the cycles into which the series has been decomposed (two degrees of freedom at each frequency). The sample (power) spectrum is formed by replacing periodogram values with weighted averages of values at adjacent frequencies (smoothing). This process has the effect of decreasing the variance of the estimate of the true spectrum at the cost of biasing upwards (downwards) estimates of low (high) spectrum values.

A sample spectrum for log river flow for the years 1920-1977 is shown in Fig. 5. The height of the spectrum is in logarithmic units. The approximate confidence interval shown can be centered on the sample spectrum at any frequency; for example, with approximately 95% confidence, the true spectral density at period twelve months can be said to be between 2.1 and 10.3. A greater amount of smoothing would decrease the length of the confidence interval but would broaden and de-emphasize the peaks and valleys of the spectrum. There is considerable annual variability. The peaks at periods of six and four months represent harmonics of the annual cycle, as suggested by Fig. 3. A periodic function

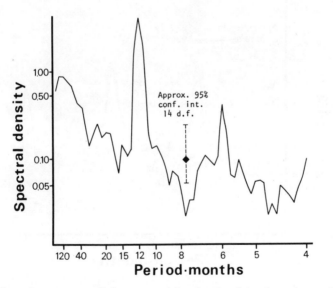

Fig. 5.   Spectrum of log monthly Apalachicola river flow (1920-1977).

of period f which is not sinusoidal will have non-zero spectrum
at some of the frequencies 2f, 3f, ... Equivalently, if the
period of the cycle is p = 1/f, the spectrum will be non-zero at
some of the periods p/2, p/3, ...

A significant feature of this spectrum is the variability at
very low frequencies. The peak is located somewhere in the region
of 60 to 240 months, or from five to twenty years. There are
indications (Fig. 4) that the cycles may be on the order of 6
to 7 years, but there appear to be variations in these cycles of
longer duration (*e.g.* abnormally low river flow around 1954). The
variance at the lowest part of the spectrum is also a reflection
of the change in amplitude of the 6- to 7-year cycles. However,
when the approximate confidence intervals are centered on the
long-term peak, the spectral analysis indicates that there are
definite long-term cycles in the river flow variable.

Spectral analyses of the local (Apalachicola) rainfall and of
that in Columbus, Georgia, were performed. The Apalachicola rainfall spectrum (Fig. 6) resembles that of the river flow. For
example, there is strong evidence of a seasonal cycle and cycles
at multiples of the annual frequency. From Fig. 3, which shows
two peaks in the monthly means (six months apart), it is
evident that the variability at a period of six months is not
merely a harmonic of the annual cycle. Furthermore, there is

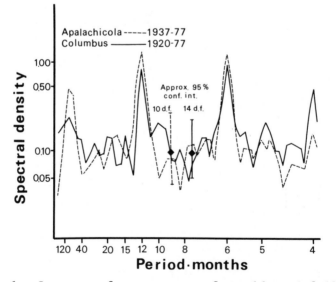

Fig. 6. Spectrum of square root of monthly rainfall data from
Apalachicola, Florida, (1937-1977) and Columbus, Georgia, (1920-
1977).

evidence of long-term cycles in the peak at a period of 80 months.
The Columbus rainfall spectrum shows a slightly different pattern.
Possibly because of the complicated nature of the annual variability
(peaks four months apart in March and July, plus a shoulder or peak
in December; Fig. 3), there are spectral peaks at periods of twelve,
six and four months, plus weaker ones at ten and five months. The
evidence for long-term cycles exists, with a peak again at 80
months, but is weaker than that for the Apalachicola rainfall data.

CROSS-SPECTRAL ANALYSIS

A cross-spectral analysis was run to determine the possible
relationship between the rainfall and river flow spectra (Fig. 7).
The coherence spectrum estimates, at given frequencies, the
correlation between the sinusoidal cycles of two contemporaneous
series. The local rainfall, measured at a point 130 km downstream
from the river flow station at Blountstown, shows high correlation
with the latter at annual, biannual and quarterly periods, as well
as over the long-term (peaks at 60 and 80 months). Confidence
intervals for points on the coherence spectrum are of equal
length, given the nonlinear scale for coherence used here
(*Jenkins and Watts*, p. 379).

The phase spectrum estimates the tendency of cycles in the
two series at a particular frequency (period) to lead or lag

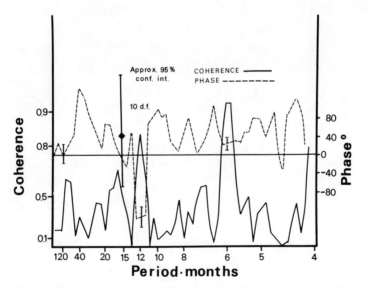

Fig. 7. Cross-spectrum of log Apalachicola river flow and
square root Apalachicola, Florida, rainfall (1937-1977).

each other. The estimate is most precise when the coherence is
high; when the true coherence at a given frequency is zero, the
phase estimate is randomly distributed on the interval $\pm$ 180°.
The three phase confidence intervals shown in Fig. 7 are of
different lengths depending on the strength of the estimated
coherence. The phase estimate for the annual cycles, (-108°,
-158°), is consistent with independent estimates (Fig. 3); the
peak months of the two series are about five months apart, which
equals about 150°. The phase estimate for the 60- to 80-month
cycles is consistent with a phase shift of zero degrees (i.e.,
long-term wet and dry periods in local rainfall tend to occur at
the time as periods of high and low river flow upstream).

     The results of a cross-spectral analysis of Columbus rain-
fall with river flow (Fig. 8) show a uniformly higher coherence
than the above. The Georgia rainfall is more highly associated
with Apalachicola River flow than is the coastal Apalachicola
rainfall. Because the coherence is higher, the estimated phase
spectrum fluctuates less. The estimate of the phase at twelve
months indicates that Georgia rainfall follows river flow by
about 20° (20 days), seemingly a logical contradiction. However,
this estimate is of the phase shift in a sinusoid of period twelve
months, and the spectrum of Georgia rainfall (Fig. 6) indicates
that this variable has a complex annual pattern, requiring
sinusoids of period six, four and possibly other months for
representation. The phase estimate in the region of long-term

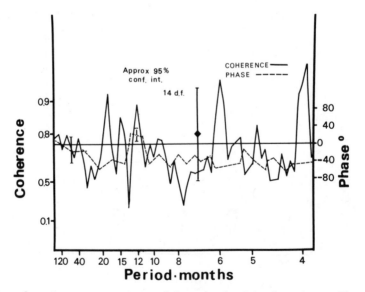

Fig. 8. Cross-spectrum of log Apalachicola river flow and
square root Columbus, Georgia, rainfall (1920-1977).

cycles is consistent with zero degrees shift; thus the long-term Columbus rainfall and Blountstown river flow patterns are in phase.

The cross-spectral analysis of Apalachicola rainfall with Columbus rainfall is not shown; the coherence spectrum was very similar to that of Apalachicola rainfall with river flow (Fig. 7), which is a reflection of the generally high coherence between river flow and Columbus rainfall. The estimated phase spectrum confidence interval for long-term (60-80 month) cycles includes zero, supporting the idea that Apalachicola and Columbus rainfall go through long-term wet and dry periods simultaneously.

TIME-DOMAIN ANALYSIS

Time-based (as opposed to frequency-based) methods for the analysis of time series have been developed (*Box and Jenkins*, 1976). The advantages of these methods over spectral analysis are that smaller sample sizes (*e.g.*, 50 points and above) are needed, and that a function model is fitted which can be used for forecasting. However, it is not as easy to detect long-term cycles as with spectral analysis. The Box-Jenkins approach assumes that the data, possibly after differencing, form a stationary (probabilistic properties unaffected by shifts in time) time series. For example, working with first differences, one would be assuming that the successive slopes of the series were stationary, but that the level of the series was not. The successive observations are assumed to be related by autoregressive (linear combination of past observations) and moving average (linear combination of past random shocks) terms. For example, for time series data $y_0, y_1, y_2, \ldots,$

if

$$z_t = y_t - y_{t-1} \quad \text{(first differences taken)}$$

then

$$z_t - \phi_1 z_{t-1} = e_t$$

and

$$z_t = e_t - \theta_1 e_{t-1}, \quad t = 2, 3, \ldots$$

are respectively first order autoregressive and first order moving average models in the first differences where the e_t are random shocks with zero mean and constant variance.

Tentative identification of the model to be fitted is made
from the autocorrelation function (correlation of observations
k months apart, at lags $k = 1, 2, \ldots$). The autocorrelation
function for log river flow (1957–1977; Fig. 9) reveals cyclical
patterns which decline only slowly at longer lags. The post–
1957 years were used because this period followed the installation
of the Jim Woodruff dam. The rhythmic nature of the function
indicated the need for seasonal (in this case annual) differencing
($z_t = y_t - y_{t-12}$) so that z_t represents changes from the previous
year. The autocorrelation function of the resulting z_t showed
exponentially declining correlation at lags $k = 1, 2, \ldots$,
plus a large negative correlation at lag twelve. This fact
suggested fitting a first order autoregressive with a seasonal
moving average term. The resulting fit was

$$z_t - 0.55z_{t-1} = e_t - 0.86e_{t-12} \; ,$$

or

$$y_t - y_{t-12} - 0.55(y_{t-1} - y_{t-13}) = e_t - 0.86e_{t-12},$$

$$t = 14, 15, \ldots, 239,$$

with a residual standard error of 0.138, with 226 d.f. This
result means that the original log river flow observations, with
a mean of 4.33, are being fit to within 3%. If the model is
adequate, the residual deviations from the fitted model should
behave like a random series from a single population. The

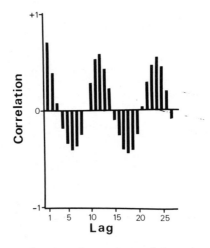

Fig. 9. Autocorrelation function of log Apalachicola river
flow (1958–1977).

autocorrelation function out to 73 lags of the residuals showed
no discernible pattern; the approximate chi-squared statistic
for the presence of autocorrelated residuals was 56.6 with 71
d.f. (p > 0.5). A runs test on the 226 residuals gave 107 runs
from 112 negative and 114 positive residuals, a result less than
one standard deviation from the mean. There was no indication of
autocorrelation in the residuals at lags 60-73, even though the
spectral analysis showed long-term cycles of this length. This
may be because we lack the full power of all 696 observations,
or because the longer-term effect is reduced when annual
differences are used.

Using the methods of *Box and Jenkins* (1976), this model can
be used for forecasting river flow (including confidence intervals).
The one-step-ahead forecast at time t would be of the form

$$\hat{z}_t(1) = 0.55z_t - 0.86(z_{t-11} - \hat{z}_{t-12}(1)),$$

where

$$z_t = y_t - y_{t-12}$$

and $\hat{z}_t(1)$ is the one-step-ahead forecast at time t.

POTENTIAL IMPACT OF IMPOUNDMENTS

A series of dams has been built along the tri-river system.
Data on some of the principal impoundments (Jim Woodruff, Walter
F. George, West Point and Buford Dams) were obtained from their
reservoir regulation manuals and from area-capacity curves
supplied by the Army Corps of Engineers, Mobile, Alabama. The
average daily flow of the Apalachicola River (1920-1977) is
1.67×10^9 m³ per month (22,740 cfs). By comparison of this
figure to the dam capacities and the accounts of their filling,
it was estimated that such activities accounted for no more than
5 to 10% of the Blountstown river flow in any month in which
filling was taking place. After adjustment of the river flow
data by "adding back" the water used to fill each dam (using
allocation proportional to the river flow that month), a
spectral analysis of the adjusted log river flow figures was
not appreciably different from the original spectrum, although
the low frequency peak (6-20 years) was slightly better defined.
Separate spectral analyses on two portions of the unadjusted
data (1920-1945 and 1958-1977) did not differ appreciably from
one another.

A low-pass filter (*Bloomfield*, 1976) replaces each observation by a weighted average of adjacent observations; a moving

average is a crude low-pass filter. Here, the weights were chosen
to remove all cycles with periods shorter than two years (Fig. 10).
The low point on the curve occurs at the end of 1955. All water
impounded by the dams was removed after this point, except for
1.04×10^8 m^3 (taken from May, 1954, through June 7, 1954) of
which almost all was released between June 23 and July 7, 1954,
and 1.20×10^8 m^3 removed from March, 1955, through April 15,
1955. Statistical restoration of this water seems not to have
reduced the magnitude of the low point but did accelerate subsequent recovery. Figure 10 illustrates the irregularity in
the amplitude and period of the long-term cycles as did the
36-month moving averages of the raw data (Fig. 4).

Although the long-term fluctuations of Apalachicola River
flow did not appear to be seriously altered by upland dam construction, there are indications that periods of unusually heavy
power use during times of low river flow affect the short-term
pattern of river flow, sometimes causing rapid increases of as
much as 3 m in river level below dams (U. S. Army Corps of
Engineers, Mobile, Alabama). In addition, comparison of spectral
analyses of *daily* river flow rates before and after construction
of the dams reveals the addition of variability with a period of
seven days, presumably reflecting weekly power consumption
patterns. These changes are visible on plots of the summer
low-water season. At least two of the dams temporarily impound
2.77×10^8 m^3 of water between 15 April and 31 May each year.
Further analysis should include the effects of these periodic
impoundment level changes on the river system, with particular
attention to monthly water elevations at each impoundment and the

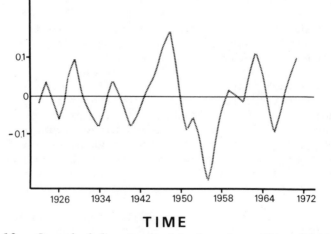

TIME

Fig. 10. Smoothed log Apalachicola river flow (1920-1977).

potential importance of evaporation and groundwater leakage
to flow patterns.

CORRELATION OF CLIMATIC VARIABLES WITH COASTAL BIOTA

Franklin County commercial landings (kg, wet weight) of blue
crabs, penaeid shrimp, and oysters taken annually from 1957-1977
were compared to annual river flow variation (Fig. 11). As a
preliminary step, a correlation analysis of these variables and
total annual rainfall in Columbus, Georgia, and Apalachicola,
Florida, was performed (Table 1). Apalachicola River flow was
highly correlated with rainfall in Columbus, Georgia, rather
than with Apalachicola rainfall. Penaeid shrimp were not correlated (p > 0.05) with any of the other variables although there
was a weak negative association with oysters and a weak positive
correlation with river flow. Blue crabs were postively correlated
with river flow and negatively correlated with oysters, which, in
turn, showed negative correlations with river flow and Apalachicola
rainfall patterns. To explore these associations, we calculated
the partial correlation of oysters with river flow, partialling

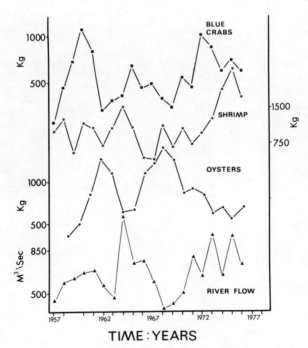

TIME:YEARS

Fig. 11. Annual total catches (kg) of blue crabs, penaeid
shrimp, and oysters, and annual means (m^3/sec) of Apalachicola
River flow from 1957 through 1976.

TABLE 1. Correlations of data on annual levels of Apalachicola River flow, Georgia rainfall, Florida rainfall, and commercial landings (Franklin County, Florida), according to records of the Florida Department of Natural Resources) of dominant invertebrate species (1957–1977), including analysis of numerical and log-transformed data.

	River flow	Rainfall*	Rainfall**	Penaeid shrimp	Blue crabs	Oysters
				LOG_{10}		
River flow	--	0.779 ($p<0.00001$)	0.180 ($p<0.21$)	0.279 ($p<0.11$)	0.51 ($p<0.01$)	-0.578 ($p<0.006$)
Rainfall*	0.788 ($p<0.00001$)	--		-0.0003 ($p<0.50$)	0.219 ($p<0.17$)	-0.391 ($p<0.051$)
Rainfall**	0.186 ($p<0.210$)		--	-0.151 ($p<0.26$)	0.079 ($p<0.36$)	-0.570 ($p<0.005$)
Penaeid shrimp (*Penaeus* spp.)	0.339 ($p<0.067$)	0.088 ($p<0.352$)	-0.139 ($p<0.274$)	--	0.189 ($p<0.21$)	0.293 ($p<0.11$)
Blue crabs (*Callinectes sapidus*)	0.393 ($p<0.038$)	0.145 ($p<0.266$)	0.069 ($p<0.383$)	0.158 ($p<0.247$)	--	-0.702 ($p<0.0005$)
Oysters (*Crassostrea virginica*)	-0.604 ($p<0.003$)	-0.387 ($p<0.051$)	-0.520 ($p<0.011$)	-0.358 ($p<0.066$)	-0.705 ($p<0.00001$)	--

UNTRANSFORMED

*Square-root transformation; Columbus, Georgia
**Square-root transformation; Apalachicola, Florida

out blue crab data. This process is equivalent to correlation
of two sets of residuals: those from regressing oysters on
crabs and residuals from regressing river flow on crabs. The
resulting correlation coefficient was -0.47 (still highly
significant). However, the correlation between crabs and
river flow, partialling out the oyster data, decreased to 0.10.
Thus, annual river flow would explain very little of the annual
variation in crab harvest over and above that explained by oyster
harvest. This fact shows that although there are clear annual
associations of the key physico-chemical functions with catches
of dominant (commercially valuable) organisms in the Apalachicola
estuary, the existence of unidentified .variables such as catch
effort, catch price and fishing preferences remains a strong
possibility.

To qualify these results further, we perform an analysis
on blue crabs and penaeid shrimp taken as part of a long-term
sampling program in the Apalachicola estuary (*Livingston et al.*,
1976b). A correlation analysis was performed of monthly data
(numbers, dry weight biomass) for blue crabs and penaeid shrimp,
rainfall (Apalachicola and East Bay Tower; Columbus, Georgia),
and Apalachicola River flow (Table 2). The high correlations of
Columbus, Georgia, rainfall and Apalachicola River flow were
again significant as were correlations between the East Bay
Tower and Apalachicola rainfall. There were significant negative
correlations of river flow with both numbers and biomass of penaeid
shrimp, while numbers of shrimp were weakly correlated with East
Bay rainfall. Blue crab numbers were weakly correlated with river
flow while blue crab numbers and biomass were weakly correlated
with East Bay rainfall.

These data indicate that short-term changes of shrimp and
crab populations vary considerably on a seasonal basis and that
such populations are not as closely tied to major climatic trends
as the annual catch totals would indicate. These facts mean that
understanding of individual species strategies and recruitment
patterns is an important part of the analysis of long-term population trends. Several such studies are now underway by members
of our research team. In addition, because of serial dependence,
transfer function models (*Box and Jenkins*, 1976) for the relation
of river flow and rainfall to changes in crab, oyster, and shrimp
populations would be preferable to the simple correlation analysis
used here.

It is becoming evident that individual species strategies
contribute to the correlation of commercial fisheries landings
with certain climatic features of the system. Oyster growth is
related to salinity (optimal growth, 15.0-22.5 ppt; *Chanley*, 1957)
and, during periods of high salinity, oyster beds in Apalachicola
Bay are preyed upon by various organisms that are less tolerant

TABLE 2.   Correlations of data on monthly levels of Apalachicola
River flow, Georgia rainfall, Apalachicola rainfall,
East Bay rainfall and otter trawl catches of penaeid
shrimp and blue crabs at 10 stations in the
Apalachicola estuary from March, 1972, through
February, 1977, including analysis of numerical
and log-transformed data.

LOG_{10}

	Shrimp numbers	Shrimp weight	Crab numbers	Crab weight
Shrimp numbers	--	0.910 (p<0.00001)	-0.015 (p<0.45)	0.306 (p<0.005)
Shrimp weight	-0.610 (p<0.00001)	--	0.051 (p<0.335)	0.351 (p<0.001)
Crab numbers	0.030 (p<0.398)	0.073 (p<0.271)	--	0.601 (p<0.00001)
Crab weight	-0.004 (p<0.488)	0.151 (p<0.102)	0.720 (p<0.00001)	--
River flow	-0.370 (p<0.0007)	-0.421 (p<0.00011)	0.230 (p<0.026)	0.138 (p<0.124)
Rainfall*	0.160 (p<0.09)	-0.098 (p<0.21)	-0.033 (p<0.393)	-0.026 (p<0.414)
Rainfall**	-0.080 (p<0.250)	-0.229 (p<0.029)	0.100 (p<0.201)	0.131 (p<0.136)
Rainfall***	0.167 (p<0.050)	-0.051 (p<0.335)	0.215 (p<0.035)	0.237 (p<0.022)

UNTRANSFORMED

*Square-root transformation; Apalachicola, Florida
**Square-root transformation; Columbus, Georgia
***Square root transformation; East Bay Tower

$$LOG_{10}$$

River flow	Rainfall*	Rainfall**	Rainfall***	
-0.654 (p<0.00001)	0.026 (p<0.413)	-0.128 (p<0.142)	0.106 (p<0.187)	Shrimp numbers
-0.544 (p<0.00001)	-0.137 (p<0.126)	-0.125 (p<0.149)	-0.024 (p<0.420)	Shrimp weight
0.236 (p<0.023)	0.009 (p<0.47)	0.095 (p<0.214)	0.162 (p<0.086)	Crab numbers
0.078 (p<0.259)	-0.008 (p<0.472)	0.022 (p<0.426)	0.193 (p<0.052)	Crab Weight
--	0.069 (p<0.281)	0.366 (p<0.0008)	0.066 (p<0.291)	River flow
-0.011 (p<0.464)	--	0.436 (p<0.00007)	0.846 (p<0.00001)	Rainfall*
0.820 (p<0.00001)	0.340 (p<0.002)	--	0.485 (p<0.00001)	Rainfall**
-0.021 (p<0.431)	0.820 (p<0.00001)	0.416 (p<0.00014)	--	Rainfall***

UNTRANSFORMED

of reduced salinity (*Menzel et al.*, 1966). Extended periods of
low salinity can have an adverse effect on oyster growth (*Butler*,
1969). This fact could explain the annual trends of oyster catches
and river flow patterns. Young blue crabs have the highest catch
in low salinity, indicating a strong preference of juveniles for
such conditions (*Copeland and Bechtel*, 1974), especially during
periods of high river flow. On the other hand, white shrimp
(*P. setiferus*), the dominant penaeid in the Apalachicola estuary,
show no such relationship to salinity, being prevalent at a range
of low salinities (*Copeland and Bechtel*, 1974). *Barrett and
Gillespie* (1973) report that white shrimp production increases
during years of the lowest summer river discharges while *Gunter*
(1950) indicates that white shrimp tend to have an optimum
salinity between 10.0 and 14.9 ppt. All of these species thus
respond to combinations of various factors such as productivity
and trophic functions, temperature, salinity, water quality
conditions, and biological interactions (predator-prey relationships, competition), and their strategies may vary from one system
to the next. In the Apalachicola estuary, *Livingston et al.* (1976a)
found that juvenile penaeid shrimp numbers usually peak during
summer-fall periods of increased local rainfall while blue crabs
have major peaks during winter-early spring months (periods of
high river flow) and secondary peaks in late summer. *Livingston
et al.* (1976a) found that river flow, rainfall, and seasonal
changes in temperature account for a temporal succession of
trophic phenomena that, together with particular, time-labile
microhabitat distributions, tended to account for the spatial/tem-
poral occurrence of biological associations in the system. It is
thus evident that, in addition to response to local rainfall
patterns, the Apalachicola estuary is driven by precipitation
in the upper portions of the tri-river basin (*i.e.*, Georgia),
and any factor which would tend to disrupt the flow rate and
periodicity of river discharge could have an important effect
on the biological productivity of the system.

The relationship of commercial fisheries landings and
environmental factors deserves further attention. *Koo* (1970)
found fairly regular patterns of commercial catches of striped
bass (*Morone saxatilis*) which followed periods approximating 6
years. The striped bass catches reflected density-independent
factors which seemed to be operative in determining the 6-year
cycles, and the author concluded that such cycles, if extant,
could have an important influence on the design of sampling
programs and the performance of impact assessments. Long-term
climatic cycles do have a major role in the impact of a given
disturbance (*Livingston and Duncan*, 1979). However, recent
studies with commercial coastal striped bass data (*Van Winkle
et al.*, 1979) indicate that while statistically significant
periodicities of 6-8 years and 20 years are common to data
from most states and regions, they are neither pronounced nor

simple, and causal factors are difficult to determine. Based on
their analysis, the authors concluded that monitoring programs
and impact assessments need not take into account long-term
periodicities of the appearance of dominant year classes of
commercial populations.

Various investigators have assumed that, overall, commercial
fisheries data provide a "reasonable" measure of such long-term
fluctuations. Unfortunately, other factors are often not included
in such an assessment. Anthropogenic impact on the subject
estuaries has been virtually ignored, as have been potential
offshore interactions and population controls. Changes in tax
laws and technological developments in fishing methodology are
difficult to assess. Certain socio-economic factors are known
to divert harvest activity in such a way that landings do not
reflect true abundance. This is especially true when there is
competition for jobs or the same commercial fishermen are involved
in multiple fishing efforts. The relatively high negative correlations between blue crab and oyster catches in our study could
reflect such interactions. These factors are difficult to assess
in a systematic fashion but may have an important influence on
the pattern of commercial catches.

Without a long-term series of scientifically reliable data,
the importance of established cycles in key climatic features
remains in doubt, and the relationship of density-dependent and
density-independent control of population and community variability
cannot be assessed. This problem, together with an almost total lack
of reliable information concerning inshore-offshore interactions,
increases the possibility of serious misunderstandings of long-
term changes at the systems level. Our data indicate that
long-term (supraannual) biological cycles are quite possible and,
indeed, probable. If they in fact exist, the usual short-term
environmental sampling programs which are currently popular
could be invalidated. With this in mind, we intend to use our
multi-disciplinary, long-term sampling program in two north
Florida bay systems (Apalachicola Bay, Apalachee Bay) to test
specific hypotheses related to the existence of long-term cyclic
biological phenomena in coastal areas. If this analysis is combined with further studies of long-term changes in commercial
fisheries, it is possible that some estimate can be made of
background variability in assemblages of coastal populations.

CONCLUSIONS

1) Apalachicola River flow, perhaps the most important
single variable affecting the environmental fluctuations of the
Apalachicola estuary, goes through long-term cycles whose amplitude and duration are irregular, but which seem to last between

5 and 7 years. The amplitude of the cycle is considerable; the annual average flow rate of 540.71 m³/sec could easily range from 475 to 625 m³/sec in a "typical" cycle.

2) Long-term cycles in Apalachicola River flow are somewhat related to coastal (Florida) rainfall, but are strongly related to Georgia rainfall. Presumably, the monthly pattern is not the same because of differences in runoff conditions between spring and summer, and because of seasonal changes in the respective weather systems. Coastal processes along the northern Gulf coast of Florida are thus influenced by climatic variation in upper portions of the drainage system.

3) While damming upriver systems does not seem to have substantially altered long-term Apalachicola River flow patterns, there are considerable changes in short-term flows during periods of low flow which, together with other hydrological features of the system, deserve further analysis.

4) Annual river flow is related to commercial oyster and crab harvests in associated Gulf coastal systems. Such relationships depend on individual species strategies relative to short- and long-term climatic changes, micro-habitat and productivity fluctuations, associated biological relationships, and, possibly, socio-economic factors. The use of commercial fisheries data in such assessments should be qualified by in-depth analysis of intensive sampling programs over long periods before strong generalizations are made. However, the 5-7-year climatic cycles could be an important determinant of long-term functions of biological response, and this possibility deserves further attention if monitoring programs and impact assessments are to have credibility.

ACKNOWLEDGMENTS

Portions of the data collection were funded by NOAA Office of Sea Grant, Department of Commerce, under Grant No. 04-3-158-43. Data analysis was supported by EPA Program Element No. 1 BA025 under Grant No. R-803339. Matching funds came from the people of Franklin County, Florida.

LITERATURE CITED

Barrett, B. B. and M. C. Gillespie, 1973. Primary factors which influence commercial shrimp production in coastal Louisiana, *La. Wildl. and Fish. Comm., Tech. Bull. 9:* 3-11.

Bloomfield, P., 1976. *Fourier Analysis of Time Series: an introduction*, 258 pp., John Wiley & Sons, New York.

Box, G. E. P. and G. M. Jenkins, 1976. *Time Series Analysis: forecasting and control*, 575 pp., Holden-Day, Inc.

Butler, P. A., 1969. Monitoring pesticide pollution, *BioScience 19:* 889-896.

Chanley, P. E., 1957. Survival of some juvenile bivalves in water of low salinity, *Proc. Nat. Shellf. Assoc. 48:* 52-65.

Copeland, B. J. and T. J. Bechtel, 1974. Some environmental limits of six Gulf coast estuarine organisms, *Contrib. Mar. Sci. 18:* 169-204.

Gunter, G., 1950. Seasonal population changes and distributions as related to salinity of certain invertebrates of the Texas coast, including the commercial shrimp, *Publ. Inst. Mar. Sci., Texas 1:* 7-52.

Gunter, G. and H. H. Hildebrand, 1954. The relation of total rainfall of the state and catch of the marine shrimp (*Penaeus setiferus*) in Texas waters, *Bull. Mar. Sci. Gulf Caribb. 4:* 95-103.

Jenkins, G. M. and D. G. Watts, 1968. *Spectral Analysis and Its Applications*, 525 pp., Holden-Day, Inc., San Francisco.

Jensen, A. L., 1976. Time series analysis and forecasting of Atlantic menhaden catch, *Chesapeake Sci. 17:* 305-307.

Kempthorne, O., 1952. *The Design and Analysis of Experiments*, 628 pp., John Wiley & Sons, Inc.

Koo, T. S. Y., 1970. The striped bass fishery in the Atlantic states, *Chesapeake Sci. 11:* 73-93.

Livingston, R. J., C. R. Cripe, R. A. Laughlin, F. G. Lewis, III, 1976a. Avoidance responses of estuarine organisms to storm water runoff and pulp mill effluents, *Estuarine Processes 1:* 313-331.

Livingston, R. J. and J. L. Duncan, 1979. Short- and long-term effects of forestry management on water quality and the epibenthic biota of a north Florida estuary. In: *Ecological Processes in Coastal and Marine Systems*, R. J. Livingston, ed., Plenum, N. Y. (this volume).

Livingston, R. J., R. L. Iverson, and D. C. White, 1976b. Energy
    relationships and the productivity of Apalachicola Bay, Final
    Report, 437 pp., Florida Sea Grant Program (Gainesville,
    Florida).

Livingston, R. J., N. P. Thompson, and D. A. Meeter, 1978.
    Long-term variation of organochlorine residues and assemblages
    of epibenthic organisms in a shallow north Florida (USA)
    estuary, *Mar. Biol. 46:* 355-372.

Longhurst, A., M. Colebrook, J. Gulland, R. Le Brasseur,
    C. Lorenzen, and P. Smith, 1972. The instability of ocean
    populations. *New Sci. 54:* 500-502.

Meeter, D. A. and R. J. Livingston, 1978. Statistical methods
    applied to a four-year multivariate study of a Florida
    estuarine system, In: *Biological Data in Water Pollution
    Assessment: quantitative and statistical analyses,* K. L.
    Dickson, J. Cairns, Jr., and R. J. Livingston, eds. 53-67,
    ASTM STP 652.

Menzel, R. W., N. C. Hulings, and R. R. Hathaway, 1966. Oyster
    abundance in Apalachicola Bay, Florida, in relation to
    biotic associations influenced by salinity and other
    factors, *Gulf Res. Rept. 2:* 73-96.

Statistical Package for the Social Sciences: 2nd ed. 1975.
    675 pp., McGraw-Hill Book Co., New York.

Sutcliffe, W. H., Jr., 1972. Some relations of land drainage,
    nutrients, particulate material, and fish catch in two
    eastern Canadian bays, *J. Fish. Res. Board Can. 29:* 357-362.

Sutcliffe, W. J., Jr., K. Drinkwater, and B. S. Muir, 1977.
    Correlations of fish catch and environmental factors in the
    Gulf of Maine, *J. Fish. Res. Board Can. 34:* 19-30.

Van Winkle, W., B. L. Kirk, and B. W. Rust, 1979. Periodicities
    in Atlantic coast striped bass (*Morone saxatilis*) commercial
    fisheries data, *J. Fish. Res. Board Can. 34:* 19-30.

Wiens, J. A., 1977. On competition and variability environments,
    Amer. Sci. 65: 590-597.

CLIMATOLOGICAL CONTROL OF A NORTH FLORIDA COASTAL SYSTEM AND

IMPACT DUE TO UPLAND FORESTRY MANAGEMENT

Robert J. Livingston and James L. Duncan

Florida State University

ABSTRACT

The assessment of the influence of upland runoff on coastal systems should be carried out within the context of long-term (i.e. supraannual) climatic fluctuation. Forestry operations such as roadbuilding, ditching, and draining tend to alter natural drainage and channelize runoff from local precipitation into receiving water bodies. Long-term (6-year) studies in the Apalachicola estuary indicated that such altered drainage, together with extensive clearcutting, caused temporary flashing of runoff which, in addition to causing rapid changes in salinity, was responsible for periodic increases in nutrient levels and water color and decreases in dissolved oxygen and pH in upper portions of the bay. Periodic incursion of low quality runoff occurred during summer-fall periods of intense precipitation and high productivity of the estuarine nursery. Water quality effects tended to be of short duration and were associated with the timed interaction of the clearcutting events and long-term rainfall patterns. Such effects were temporally modified by revegetation of clearcut areas. The biological effects depended on the timing of input and the specific response of individual populations. Reductions of species richness and abundance were characteristic of shocked upper bay areas, and were considered to be a consequence of exacerbation of the natural instability of the physical environment. The significance of the repeated though temporary reduction of water quality was understandable only within the context of specific natural shocks to the system such as low winter temperatures and periodic river flooding. Natural variation of estuarine interactions was equivalent to the impact due to anthropogenic manipulation of upland drainage areas. It can be inferred that wetlands vegetation in upland drainage areas tends to stabilize pulsed movements of water through such systems, thereby modifying the hydrologic regime in receiving water bodies.

INTRODUCTION

Although forestry management (including roadbuilding, draining, clearcutting, replanting, and fertilization) is common in drainage systems along the Gulf of Mexico, there is little published information concerning associated impact on downstream aquatic systems. Watershed alterations due to timbering have been noted in various areas (*Tebo*, 1955; *Hewlett and Hibbert*, 1961; *Swank and Douglass*, 1974). Long-term studies in the Hubbard Brook Experimental Forest (New Hampshire) determined the effects of clearing activities on water quality and mass flow conditions (*Bormann et al.*, 1968; *Likens et al.*, 1969, 1970; *Smith et al.*, 1968; *Hornbeck et al.*, 1970; *Pierce et al.*, 1970). These studies indicated that clearcutting caused increases in the flow rates, ions, and nutrients of runoff water. Decreases in pH were observed. In various instances, water quality in receiving streams was affected by forestry activities, such as clearcutting.

In addition to the direct impact of various forestry operations, there are often other contributing factors to such effects. These include rainfall patterns, vegetation and soil types, land slope, and specific watershed characteristics. Forest cover tends to control lateral water movement by means of interception, infiltration (basin recharge), and evapotranspiration (*Sokolovskii*, 1968), although such processes are modified by antecedent soil water capacity as a function of the intensity and distribution of previous rainfall. Reductions of water yield in forested systems can arise because of high infiltration of litter/humus (*Lull and Reinhart*, 1972) and evapotranspiration (*Ziemer*, 1964; *Harr et al.*, 1975). Clearcutting can cause increased runoff (*Satterlund*, 1975; *Hewlett and Helvey*, 1970; *Lull and Reinhart*, 1972) by reducing the storage ability of the affected soils (*Hornbeck*, 1973). Deforestation can also cause increased loads of suspended solids and changes in the dissolved oxygen regime of receiving aquatic systems (*Packer*, 1965; *Dickenson et al.*, 1967; *Lull and Reinhart*, 1972). In addition, roadbuilding and associated ditching/draining activities can contribute to runoff and erosion by increased compaction (*i.e.*, reduced permeability), flow channelization, and expansion of the immediate drainage area.

Estuarine habitats are spatially and temporally variable with respect to key forcing functions (*Copeland and Bechtel*, 1974; *Cherry et al.*, 1975; *Livingston et al.*, 1978). Episodic influxes of stormwater runoff are a natural part of physically dynamic nearshore Gulf coastal systems. Intermittent storms cause rapid

changes in salinity, dissolved oxygen, and pH in receiving coastal
systems (*Livingston*, 1978). Although there is some evidence that
various populations nurserying in inshore waters are adapted to
such changes, there is relatively little information regarding
short- and long-term system response to such repeated shocks.
Aquatic organisms can be periodically exposed to low pH, which
then may cause adverse biological response (*Lloyd and Jordan*, 1964;
Kwain, 1975). Such effects can also lead to direct or indirect
changes in population structure, reproductive effectiveness, and
adult distribution (*Powers*, 1941; *Collins*, 1952). Lethal limits at
pH 5.0 and under have been experimentally established for fishes
(*Bishai*, 1960; *Jones*, 1964). However, there are problems in the
direct application of laboratory data to field situations
(*Livingston et al.*, 1976; *Laughlin et al.*, 1978). Species-
specific responses to various combinations of rapidly changing
environmental factors are a possible cause of these problems.
Despite numerous studies concerning physical and chemical changes
in coastal systems, there is little detailed information available
concerning the relationship of short-term changes in estuarine
habitat and long-term patterns of biological response.

From 1970-75, about 8500 hectares of the Tate's Hell Swamp
were clearcut by local forestry interests (Fig. 1). Previous
studies (*Livingston*, 1978) indicated that such clearing, together
with roadbuilding and draining (Fig. 2), tended to increase the
drainage area and local water yield during periods of precipitation. Runoff from upland swamps, characterized by low pH and
relatively high color, turbidity, and nutrients, was channelized
by networks of ditches into upper East Bay via West and East
Bayous (stations 5b and 5c: Fig. 1). Preliminary surveys
indicated that this input occurred in the form of pulsed surges of
runoff during winter/summer periods of high local rainfall. Based
on this preliminary information, a study was initiated to determine
the effects of forestry management in the Tate's Hell Swamp on
water quality and the biota of the Apalachicola Estuary. The East
Bay system was considered useful for this study since it is
relatively free of other forms of pollution and is off-limits to
commercial netfishing. In addition, long-term changes in the
biological assemblages of different East Bay drainage areas (both
affected and unaffected by forestry practices) were compared to
biotic trends in the Apalachicola estuary as a whole.

MATERIALS AND METHODS

The design of the field sampling program was based on a dual
approach. An intensive, short-term (3-year) survey in upper
East Bay (stations 4a, 5b, 5c; Fig. 1) was superimposed over a
long-term (6-year) comparison of physico-chemical features and
biological associations in the bay system. The extensive forestry

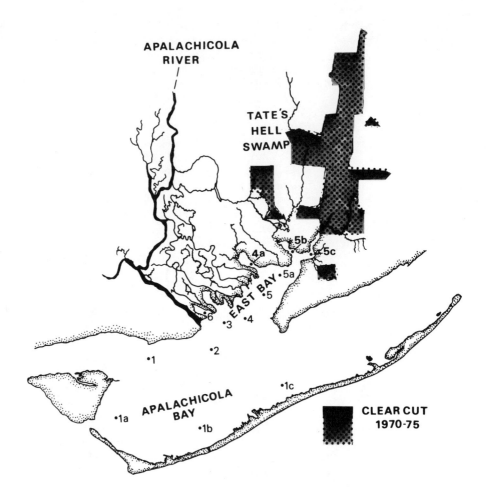

Fig. 1: Chart showing permanent sampling stations in the Apalachicola Bay System. Also included is the distribution of areas clearcut in the upper East Bay drainage system (Tate's Hell Swamp) from 1970-75. The relative positions of input from the Apalachicola River and overland runoff from upland creek systems into the East and West Bayous (5c, 5b) are also given.

operations in the East and West Bayou drainages together with the 1975 clearcutting in the West Bayou drainage (Fig. 2) qualified these stations as so-called experimental areas (5b, 5c). Originally the Round Bay drainage (4a) remained unaffected by upland forestry operations, and this area was chosen as a control. However, the immediate Round Bay drainage was clear-cut, ditched,

Fig. 2A: Composite of various forestry activities including roadbuilding, ditching (draining), clearing, plowing, and re-planting.

and drained during the 1976-77 period; this station was used as a control site as well as a test station for water quality impact during the final year of sampling. To test the hypothesis that stations 5b and 5c were stressed relative to station 4a during the 1975-76 period (within the context of temporal changes in forestry practices in associated upland drainage systems as described above), an intensive field program was established in this portion of East Bay. Monthly sampling of the 3 study sites was carried out from March, 1975, through February, 1978, while (additional) multiple samples were taken before, during, and after periods of heavy rainfall and runoff (May-November, 1976-77). Sampling frequency was determined by local rainfall patterns during the rainy season. Day collections (7 2-minute otter trawl tows) at stations 5c, 5b, and 4a were taken at 1400, 1530, and 1700 h. respectively while identical night collections were made at these stations at 1930, 2100, and 2230 h. Trammel net col-lections were made at stations 4a and 5b at night on a monthly basis. All other stations were sampled with multiple (7 2-minute) otter trawl tows during the day at monthly intervals. All organisms were preserved in 10% formalin, identified to species,

Fig. 2B: Road and canals draining areas of Tate's Hell Swamp which were clearcut in 1975 and drained into the upper West Bayou (5b) creek system.

measured, and counted. Dry weight conversions were carried out (*Livingston et al.*, 1977).

The permanent stations (Fig. 1) were established on the basis of habitat distribution and spatial relationships to terrestrial runoff patterns (*Duncan*, 1977; *Livingston*, 1978). Surface and bottom water samples were taken with a 1-ℓ Kemmerer bottle. Turbidity (J.T.U.) was determined with a Hach model 2100-A turbidimeter and water color (Pt-Co units) was measured by an A.P.H.A. platinum-cobalt test. Temperature (^oC) and dissolved oxygen (ppm) were measured with a Y.S.I. dissolved oxygen meter, while salinity was determined with a temperature-compensated re-fractometer calibrated periodically with standard sea water. All pH measurements were made with battery-operated field meters. Light penetration was estimated with a standard Secchi disk. Nutrient data were taken by Dr. R. L. Iverson (Department of Oceanography, Florida State University) according to standard procedures (*Livingston et al.*, 1974). Apalachicola River flow

data at Blountstown, Florida, were provided by the U. S. Army Corps
of Engineers (Mobile, Alabama) and the U. S. Geological Survey
(Tallahassee, Florida). Local rainfall information was provided
by the Environmental Data Service (N.O.A.A.) and the East Bay
forestry tower (Apalachicola, Florida).

Data analysis was carried out using an interactive computer
program (SPECS: *Livingston et al.*, 1977) under the KRONOS
operating system on a Cyber 74 computer (Florida State University
Computing Center). Numbers of individuals, dry weight biomass,
and numbers of species were used for all computations. Various
indices were calculated, including the Margalef Index (*Margalef*,
1958) and the Shannon Diversity index (*Shannon and Weaver*, 1963).
Relative dominance was determined by dividing the number of
individuals of the single most dominant species by the total
number of individuals. Statistical calculations were made using
the Statistical Package for the Social Sciences (*Nie et al.*,
1970). Pearson product-moment correlations (*Nie et al.*, 1970)
were calculated to determine the strength of the relationships
between pairs of variables taken during intensive sampling
periods. Significance levels were derived by use of Student's
t-test with N-2 degrees of freedom for the computed quantity:

$$r = [\frac{N-2}{1-r^2}]^{1/2}$$

PHYSICO-CHEMICAL FACTORS

River-flow and local rainfall patterns (Fig. 3) have been
analyzed by *Meeter et al.* (1979) and will not be discussed in
detail here. Over the 8-year period from 1970-78, peak river
flooding occurred during the winter-early spring months of 1973,
and local precipitation peaked during the summers of 1970, 1974,
and 1975. The heavy rainfall of 1974-75 coincided with major
clearcutting events in drainage areas leading to East and West
Bayous. Unusually high rainfall on consecutive days was noted
during the summer of 1975. From January, 1975, through February,
1978, there was a general decrease in rainfall (Fig. 4).
Sporadic occurrences of heavy precipitation occurred during the
winter and summer-fall periods of 1976. Drought conditions pre-
vailed during 1977 and early 1978. At this time, many of the
Tate's Hell Swamp drainage ditches were dry.

Visual observations of the distribution of highly colored
water from the Tate's Hell Swamp in East Bay are shown in Fig. 5;
runoff patterns were determined by tidal currents, local rainfall,
and wind direction and velocity. Such water movements have been
corroborated by LANDSAT projections of water quality (*Hill*, 1978).

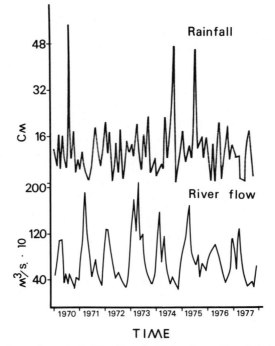

Fig. 3: Local rainfall (Apalachicola, Florida) expressed
as monthly totals (cm) and Apalachicola River flow expressed as
monthly means (m³/sec.) from January, 1970, through December, 1977.

The data indicate that during periods of heavy local rainfall,
eastern sections of East Bay (stations 5, 5a, 5b, 5c) were affected
by runoff from the Tate's Hell Swamp. The points of entry of
highly colored water appeared to be East and West Bayous, and
such water was characterized by low salinity and pH (*Livingston*,
1978).

The basic differences in water quality between East Bay
(station 5) and Apalachicola Bay (station 1) are shown in Fig. 6.
Salinity in East Bay was influenced by local rainfall patterns
while the outer bay salinity regime was dominated by Apalachicola
River flow as evidenced by the substantial decrease at station 1
during periods of river flooding. There were pronounced peaks of
turbidity at both stations during the heavy river flooding of the
winter of 1973. However, water color in East Bay reached 6-year
peaks during the summer rainfall of 1975 while Apalachicola Bay
color peaked during the winter river flooding of 1973. Secchi
readings tended to reflect both parameters, although, during the
period from 1975 to 1977, there were comparatively lower readings
in East Bay. These data indicate that, within the limits of high

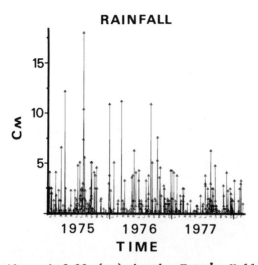

RAINFALL

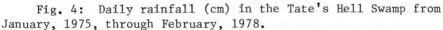

1975   1976   1977

TIME

Fig. 4: Daily rainfall (cm) in the Tate's Hell Swamp from
January, 1975, through February, 1978.

seasonal variability, the long-term trends of key physico-chemical
features of the habitat in the Apalachicola Estuary reflect
spatial differences in drainage. Apalachicola River flow and local
rainfall tend to control water quality in immediate receiving areas
of the bay; the long-term trends of river flow and local rainfall
determine spatial variability of key physico-chemical features of
the Apalachicola Estuary. While the river affects the entire
estuary, the Tate's Hell Swamp drainage controls key physico-
chemical functions in East Bay.

Upper East Bay stations showed considerable seasonal variation
in physico-chemical variables (Fig. 7). Peak temperatures occurred during late summer—early fall periods, and the winter of
1977 was particularly cold. Dissolved oxygen reflected seasonal
temperature fluctuations at all 3 stations and low dissolved
oxygen levels were often noted during summer months. There were
particularly low dissolved oxygen conditions during the summers
of 1974–75 at stations 5b and 5c while dissolved oxygen reached
its lowest level at station 4a during the summer of 1977. This
pattern tended to reflect the temporal pattern of clearcutting
activities in the respective upland drainages over the period of
study. The considerable seasonal changes of salinity were determined by river flow conditions while local (summer) rainfall was
associated with rapid reductions of salinity at the 3 stations in

the upper estuary. These data indicate short- and long-term
responses of the bay to combinations of controlling factors such
as temperature, river flow, and local rainfall. The physico-
chemical environment of upper portions of East Bay was influenced
to a considerable degree by conditions in contiguous upland areas.
Runoff from recently cleared uplands was associated with reductions
in dissolved oxygen in receiving portions of the bay.

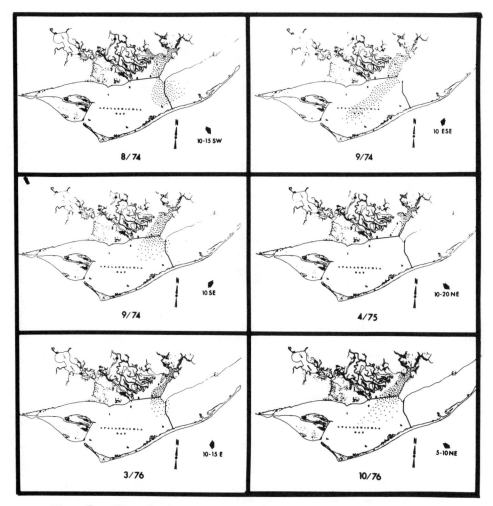

Fig. 5: Visual observations of the distribution of highly
colored water in the Apalachicola estuary following periods of
local rainfall. The influence of wind and tidal current fluctuations is evident, with primary runoff influence occurring in
eastern portions of East Bay. The entrance points of such
upland runoff appeared to be East and West Bayous.

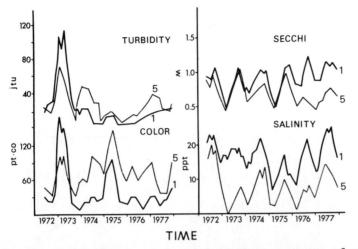

TIME

Fig. 6: A comparison of 6--month moving averages of surface
determinations of turbidity (Jackson Turbidity Units), color
(Platinum-Cobalt Units), Secchi readings and salinity ($^o/oo$)
between Apalachicola Bay (station 1) and East Bay (station 5)
from March, 1972, through February, 1978.

Analysis of long-term trends of pH in East Bay (Fig. 8)
showed little seasonal or long-term variation in river-dominated
(station 4) or lower East Bay areas (station 5).   There was
greater long-term variability in Round Bay (station 4a),
where there were periodic decreases in pH after the spring
of 1976. The lowest recorded pH in the control area (4a)
occurred during brief periods of rainfall following the 1977
drought. This reduction in pH coincided with clearing activities
in the Round Bay drainage as described above. At East and West
Bayous (stations 5c and 5b), after the relatively low pH levels
noted during the late summer rainfall of 1975, there was a subsequent increase in pH from 1975 to 1978. Despite heavy summer-fall
rainfall in 1970, no such decreases in pH occurred in these areas
prior to the clearing of Tate's Hell Swamp. Thus, the pH in the
upper East Bay system was influenced by local rainfall patterns
and clear--cutting activities in upland drainage areas. The
basic differences in the long--term patterns of pH in the various
bayous of East Bay indicate that, while patterns of rainfall are
an important determinant, the clearing of vegetation in a local
drainage is associated with short--term periods of low pH in
receiving aquatic systems. The differences in the long-term
patterns of pH in Round Bay (clear-cut in 1976-77) and East/West
Bayous (clear-cut prior to 1976) reflect the relative importance
of forestry operations on water quality in receiving water bodies.

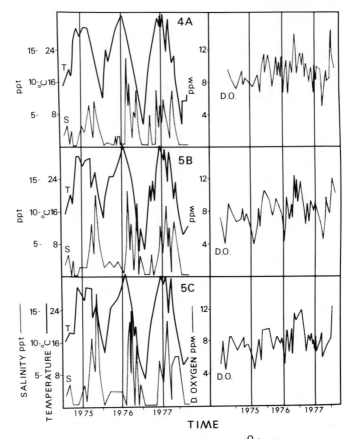

Fig. 7: Trends of surface salinity (°/oo), temperature (°C), and dissolved oxygen (ppm) in Round Bay (4a), West Bayou (5b), and East Bayou (5c) from 1974-75 through February, 1978.

A statistical analysis was made of the relationships of the physico-chemical features of upper East Bay (stations 4a, 5b, 5c) over the period of intensive sampling (Table 1). Overall, salinities tended to be reduced during 1975, with peaks in late fall; such peaks (more pronounced) occurred during late summer (August, September) over the next two years. Heavy local rainfall was generally associated with the extremely low pH levels at stations 5b and 5c during the late summer of 1975 and the winter of 1976. However, this relationship was not statistically significant ($p < 0.05$) in these areas, probably as a result of the lag time between successive inputs of rain in upper drainage areas and changes in water quality in the bay. This view is substantiated by the fact that rainfall was significantly associated with reduced pH

TABLE 1a: Pearson correlation coefficients for physico-chemical variables measured at the control station (4a) in East Bay. R^2 values are shown with the associated significance level. Significant correlations ($p<0.05$) are underlined.

STATION 4a

NIGHT \ DAY	pH	Salinity	Color	Temperature	Turbidity	Dissolved Oxygen	Secchi	River flow	Rainfall
pH		-.1832 p<.233	-.3942 p<.053	.5475 p<.009	.2565 p<.152	.3654 p<.068	.2894 p<.122	.4739 p<.023	-.4097 p<.046
Salinity	-.0721 p<.388		-.2138 p<.197	-.0814 p<.374	-.3693 p<.066	.0822 p<.373	-.2470 p<.162	-.4032 p<.049	-.1458 p<.282
Color	-.4933 p<.019	-.2145 p<.196		-.3557 p<.074	.2940 p<.118	-.3350 p<.087	-.5379 p<.011	.1173 p<.321	.4535 p<.029
Temperature	.4846 p<.026	.0287 p<.455	-.4441 p<.032		.3223 p<.096	-.1752 p<.243	.3258 p<.094	.3821 p<.059	-.2207 p<.189
Turbidity	.1383 p<.292	-.1884 p<.277	.2250 p<.185	-.2018 p<.211		-.1629 p<.259	-.0946 p<.354	.4358 p<.035	-.2250 p<.185
Dissolved Oxygen	.1971 p<.217	.0035 p<.494	-.2195 p<.191	-.4881 p<.020	-.2889 p<.122		.3794 p<.060	-.0880 p<.354	.0758 p<.382
Secchi								.0114 p<.482	-.2146 p<.196
River Flow	.2768 p<.133	-.4393 p<.034	-.0486 p<.424	.3153 p<.101	.0615 p<.404	-.1341 p<.298			-.1441 p<.284
Rainfall	-.4531 p<.029	.0528 p<.418	.4378 p<.035	-.2270 p<.183	-.5184 p<.014	.1384 p<.292		-.1441 p<.284	

TABLE 1b: Pearson correlation coefficients for physico-chemical variables measured at experimental station 5b in East Bay. R² values are shown with the associated significance level. Significant correlations (p<0.05) are underlined.

STATION 5b

DAY

NIGHT	pH	Salinity	Color	Temperature	Turbidity	Dissolved Oxygen	Secchi	River flow	Rainfall
pH		.3111 p<.704	-.9060 p<.001	.0373 p<.442	.3619 p<.076	.6082 p<.004	.4096 p<.046	.0762 p<.383	-.1123 p<.329
Salinity	.4785 p<.022		-.4172 p<.042	-.2138 p<.197	-.1227 p<.314	.3233 p<.095	.1818 p<.235	-.4351 p<.036	.0641 p<.400
Color	-.7549 p<.001	-.5051 p<.016		.0542 p<.415	-.2016 p<.211	-.6549 p<.002	-.3858 p<.057	.0320 p<.450	-.0100 p<.484
Temperature	-.3084 p<.107	-.2613 p<.147	.2562 p<.152		.2867 p<.124	-.5811 p<.006	-.1942 p<.220	.3941 p<.053	-.2394 p<.169
Turbidity	.2838 p<.127	-.2218 p<.188	.0046 p<.493	.0848 p<.369		-.2549 p<.154	.0091 p<.486	-.0123 p<.481	-.2110 p<.200
Dissolved Oxygen	.5336 p<.011	.3299 p<.091	-.4853 p<.021	-.8643 p<.001	.0007 p<.499		.5693 p<.007	-.0328 p<.438	-.0198 p<.469
Secchi								-.1844 p<.232	-.4032 p<.049
River Flow	-.3076 p<.107	-.5078 p<.016	.2379 p<.171	.3848 p<.057	-.0477 p<.425	-.3377 p<.085			-.1441 p<.284
Rainfall	-.3247 p<.094	.0207 p<.467	.0031 p<.495	-.2060 p<.206	-.5017 p<.006	.0371 p<.442		-.1441 p<.284	

TABLE 1c: Pearson correlation coefficients for physico-chemical variables measured at experimental station 5c in East Bay. R^2 values are shown with the associated significance level. Significant correlations ($p<0.05$) are underlined.

STATION 5c

DAY / NIGHT	pH	Salinity	Color	Temperature	Turbidity	Dissolved Oxygen	Secchi	River flow	Rainfall
pH		.1418 p<.287	-.7570 p<.001	.2342 p<.175	.1346 p<.297	.4073 p<.047	.3257 p<.094	.1664 p<.255	-.0633 p<.402
Salinity	.3365 p<.086		-.3420 p<.062	-.3045 p<.110	-.4421 p<.033	.2632 p<.146	-.1403 p<.289	-.4573 p<.028	.0537 p<.416
Color	-.6034 p<.004	-.5104 p<.015		-.2434 p<.165	-.0697 p<.392	-.3483 p<.078	-.4307 p<.037	-.0293 p<.454	.0899 p<.361
Temperature	.0442 p<.431	-.1803 p<.237	.0533 p<.417		.4855 p<.018	-.5700 p<.007	.1961 p<.218	.3910 p<.054	-.2424 p<.166
Turbidity	.1068 p<.333	-.2247 p<.185	.0859 p<.367	.0367 p<.443		-.1766 p<.242	.4404 p<.034	.5728 p<.006	-.5461 p<.010
Dissolved Oxygen	.3547 p<.074	-.0016 p<.498	-.3363 p<.086	-.7758 p<.001	-.0942 p<.355		.3186 p<.099	-.1538 p<.271	-.1255 p<.310
Secchi								-.0699 p<.391	-.3627 p<.070
River Flow	-.0143 p<.478	-.4666 p<.025	.1735 p<.245	.3988 p<.051	.1083 p<.334	-.2508 p<.158			-.1441 p<.284
Rainfall	-.2962 p<.116	.0708 p<.390	.1400 p<.289	-.1805 p<.237	-.2320 p<.177	-.1509 p<.275		-.1441 p<.284	

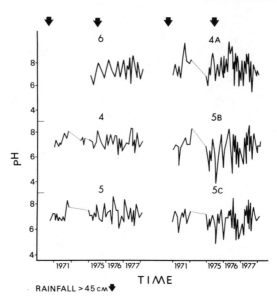

Fig. 8: Levels of surface pH in various areas of East
Bay prior to (October, 1970, through November, 1971) and after
(January, 1975, through February, 1978) the clearcutting of the
Tate's Hell Swamp. Data shown prior to 1975 were provided by
Mr. John Taylor, Jr., of the Florida Division of Health.

and increased color and turbidity in Round Bay which remained
in close proximity to upland clearing activities relative to
East and West Bayous. The most pronounced decreases in pH
in the bay occurred during prolonged periods of rainfall on
successive days; thus, the pattern of daily precipitation
appears to be an important factor in runoff considerations. At
stations 5b and 5c, there were statistically significant (direct)
relationships between dissolved oxygen and pH. Color was inversely
related to pH and salinity. Because of the importance of sequential periodicity and of local rainfall and the lagged response
of downstream water quality parameters in the bay, the use of
correlation analysis to test potential causal relationships was
not considered particularly useful. However, this approach did
prove valuable in the determination of system response to major
ecological determinants such as temperature and river flow (which
work over extended periods) and for association of groups of co-
occurring water quality functions in a given area of the bay at a
particular time. The short- and long-term trends of biologically
important variables such as pH and dissolved oxygen at the three
sampling sites indicate that clearcutting was associated with

sudden, temporary changes (reductions) in such features in receiving portions of the bay. This conclusion is consistent with
previous studies and our own background work in the drainage
ditches of the Tate's Hell Swamp (*Livingston,* 1978).

BIOLOGICAL RELATIONSHIPS: FISHES

    A list of fish and invertebrate species taken during the
study are given in Tables 2 and 3. Long-term changes in the numbers of individuals and species of fishes in the upland bayous
of East Bay are shown in Fig. 9. The highest numbers of

FISHES

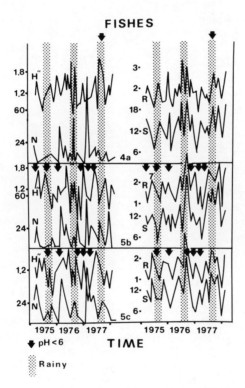

Fig. 9: Numbers of species (S) and individuals (N) of fishes
taken in Round Bay (4a), West Bayou (5b), and East Bayou (5c) from
March, 1975 through February, 1978. Also shown are monthly
Shannon diversity and Margalef richness indices. Periods when
the existing pH levels were less than 6.0 are as noted.

TABLE 2:   List of species of fishes and invertebrates taken in
           otter trawls in East Bay during the period of study.
           Organisms are listed in descending order of abundance.

FISHES

1. *Anchoa mitchilli*
2. *Cynoscion arenarius*
3. *Leiostomus xanthurus*
4. *Micropogonias undulatus*
5. *Bairdiella chrysura*
6. *Trinectes maculatus*
7. *Brevoortia patronus*
8. *Paralichthys lethostigma*
9. *Eucinostomus argenteus*
10. *Lagodon rhomboides*
11. *Prionotus tribulus*
12. *Porichthys porosissimus*
13. *Microgobius gulosus*
14. *Arius felis*
15. *Symphurus plagiusa*
16. *Gobionellus boleosoma*
17. *Chloroscombrus chrysurus*
18. *Cynoscion nebulosus*
19. *Dorosoma petenense*
20. *Menticirrhus americanus*
21. *Ictalurus catus*
22. *Lepisosteus osseus*
23. *Microgobius thalassinus*
24. *Dasyatis sabina*
25. *Caranx hippos*
26. *Citharichthys spilopterus*
27. *Gobiosoma bosci*
28. *Peprilus paru*
29. *Bagre marinus*
30. *Menidia beryllina*
31. *Gobionellus hastatus*
32. *Syngnathus scovelli*
33. *Lucania parva*
34. *Ictalurus punctatus*
35. *Micrognathus crinigerus*
36. *Sphoeroides nephelus*
37. *Archosargus probatocephalus*
38. *Enneacanthus gloriosus*

39. *Pogonias cromis*
40. *Gobiosoma robustum*
41. *Sciaenops ocellata*
42. *Strongylura marina*
43. *Elops saurus*
44. *Lutjanus griseus*
45. *Prionotus scitulus*
46. *Anchoa hepsetus*
47. *Opsanus beta*
48. *Micropterus salmoides*
49. *Morone chrysops*
50. *Myrophis punctatus*
51. *Paralichthys albigutta*
52. *Eucinostomus gula*
53. *Scomberomorus maculatus*

INVERTEBRATES

1. *Penaeus setiferus*
2. *Callinectes sapidus*
3. *Rangia cuneata*
4. *Penaeus duorarum*
5. *Rhithropanopeus harissii*
6. *Palaemonetes pugio*
7. *Neritina reclivata*
8. *Penaeus aztecus*
9. *Mactra fragilis*
10. *Palaemonetes vulgaris*
11. *Palaemonetes intermedius*
12. *Periclimenes longicaudatus*
13. *Hippolyte zostericola*
14. *Acetes americanus*
15. *Sesarma cinereum*
16. *Polymesoda caroliniana*

TABLE 3. Organisms taken by trammel nets at night at stations 4A (Round Bay) and 5B (West Bayou) in the Apalachicola estuary from March, 1975, through February, 1978. (* indicates sexes not separated)

ROUND BAY (TRAMMEL NETS: STATION 4A)

Date	3-75	4-75	5-75	6-75	7-75	8-75	9-75	10-75
Bagre marinus	7	1	3	7		22	8	7
Arius felis			1	3		11		2
Dorosoma cepedianum								
Dasyatis sabina	13	4	3			1	1	4
Mugil cephalus				3				
Lepisosteus osseus	1	2	2	1				1
Paralichthys lethostigma	3			1				
Micropogonias undulatus				2		3	1	
Ictalurus catus	2							
Brevoortia patronus						8		
Cynoscion nebulosus								
Morone saxatilis								
Carcharinus leucas			3	1				
Leiostomus xanthurus		1					1	
Sciaenops ocellata								
Ictalurus nebulosus								
Elops saurus				1				
Alosa chrysochloris								
Cyprinus carpio						1		
Dorosoma petenense								
Amia calva								
Lepomis microlophus								
Lepisosteus oculatus								
Micropterus salmoides								
Pogonias cromis								
Morone chrysops								
TOTALS	20	8	12	19		46	11	14
Callinectes sapidus (M)	9*	2*	20*	81*		12	5	2
Callinectes sapidus (F)							1	

TABLE 3 (con't).

ROUND BAY (TRAMMEL NETS: STATION 4A)

Date	11-75	12-75	1-76	2-76	3-76	4-76	5-76	6-76
Bagre marinus	11					3	9	8
Arius felis		5	1			2	8	16
Dorosoma cepedianum				1		1		2
Dasyatis sabina					8		4	1
Mugil cephalus		3						5
Lepisosteus osseus	1				2			2
Paralichthys lethostigma						3		1
Micropogonias undulatus	1			1	1	1	1	2
Ictalurus catus				5	1			
Brevoortia patronus								
Cynoscion nebulosus			1	1				
Morone saxatilis								
Carcharinus leucas	2							2
Leiostomus xanthurus		1						1
Sciaenops ocellata								1
Ictalurus nebulosus								
Elops saurus								1
Alosa chrysochloris								
Cyprinus carpio								
Dorosoma petenense								
Amia calva								
Lepomis microlophus								
Lepisosteus oculatus								
Micropterus salmoides								
Pogonias cromis								
Morone chrysops								
TOTALS	15	9	2	8	12	10	22	42
Callinectes sapidus (M)	7		2	6*	5	3	54	46
Callinectes sapidus (F)							4	9

TABLE 3 (con't).

ROUND BAY (TRAMMEL NETS: STATION 4A)

Date	7-76	8-76	9-76	10-76	11-76	12-76	1-77	2-77
Bagre marinus		3	5	3				
Arius felis	1	1	3	8	5			
Dorosoma cepedianum				39				
Dasyatis sabina			3	5	2	1		
Mugil cephalus		2		5				
Lepisosteus osseus			1	2				
Paralichthys lethostigma	1	2	1					
Micropogonias undulatus		1						
Ictalurus catus								1
Brevoortia patronus					5			
Cynoscion nebulosus					5			
Morone saxatilis								
Carcharinus leucas								
Leiostomus xanthurus					3			
Sciaenops ocellata					2			
Ictalurus nebulosus								
Elops saurus								
Alosa chrysochloris								1
Cyprinus carpio								
Dorosoma petenense					1			
Amia calva								
Lepomis microlophus								
Lepisosteus oculatus								
Micropterus salmoides								
Pogonias cromis								
Morone chrysops								
TOTAL	2	9	16	65	17	1		2
Callinectes sapidus (M)	12	20	8		5	4	1	4
Callinectes sapidus (F)	4	2	2		2	1	1	

TABLE 3 (con't).

ROUND BAY (TRAMMEL NETS: STATION 4A)

Date	3-77	4-77	5-77	6-77	7-77	8-77	9-77	10-77
Bagre marinus		6	30	10	15	7	4	
Arius felis	2	6	8	41	9	11	5	
Dorosoma cepedianum	1		6	8	4	1	1	
Dasyatis sabina		6			1			
Mugil cephalus	1	1			3	4	3	
Lepisosteus osseus	7	6			1	2	2	
Paralichthys lethostigma	4		2	3	1			
Micropogonias undulatus		2		2		2	2	
Ictalurus catus	3					1		
Brevoortia patronus								
Cynoscion nebulosus						1	2	
Morone saxatilis				5	5			
Carcharinus leucas					2			
Leiostomus xanthurus		1			1			
Sciaenops ocellata	2				1			
Ictalurus nebulosus								
Elops saurus								
Alosa chrysochloris								
Cyprinus carpio								
Dorosoma petenense								
Amia calva	1							
Lepomis microlophus	1							
Lepisosteus oculatus					1			
Micropterus salmoides					1			
Pogonias cromis								
Morone chrysops								
TOTAL	22	28	46	69	45	29	19	
Callinectes sapidus (M)		6	33	13	32	30	9	
Callinectes sapidus (F)		1	14	3	11	6	3	

TABLE 3 (con't).

ROUND BAY (TRAMMEL NETS: STATION 4A)

Date	11-77	12-77	1-78	2-78	TOTAL
Bagre marinus					169
Arius felis	4				153
Dorosoma cepedianum					64
Dasyatis sabina	1				58
Mugil cephalus	1		2	1	34
Lepisosteus osseus				1	34
Paralichthys lethostigma			3	1	26
Micropogonias undulatus					22
Ictalurus catus	2		1		16
Brevoortia patronus	2				15
Cynoscion nebulosus					10
Morone saxatilis					10
Carcharinus leucas					10
Leiostomus xanthurus					9
Sciaenops ocellata	2				8
Ictalurus nebulosus				4	4
Elops saurus					2
Alosa chrysochloris				1	2
Cyprinus carpio					1
Dorosoma petenense					1
Amia calva					1
Lepomis microlophus					1
Lepisosteus oculatus					1
Micropterus salmoides					1
Pogonias cromis	1				1
Morone chrysops				1	1
TOTAL	13		6	9	654
Callinectes sapidus (M)	2		3	3	339
Callinectes sapidus (F)	1		6		71

TABLE 3 (con't).

WEST BAYOU (TRAMMEL NETS: STATION 5B)

Date	3-75	4-75	5-75	6-75	7-75	8-75	9-75	10-75
Bagre marinus		2	1	5			7	2
Lepisosteus osseus	3	1				1	3	1
Arius felis			1	2		2	3	1
Dorosoma cepedianum							5	
Dasyatis sabina			1	1				1
Paralichthys lethostigma	9	1		2			1	
Mugil cephalus			1					1
Micropogonias undulatus							1	
Cynoscion nebulosus								
Ictalurus catus	5							
Carcharinus leucas							1	
Sciaenops ocellata			1					1
Amia calva	2							
Brevoortia patronus								
Micropterus salmoides								
Cyprinus carpio			1					
Alosa chrysochloris								
Ictalurus nebulosus								
Pogonias cromis								
TOTAL	19	4	7	8		3	21	7
Callinectes sapidus (M)	2*	1*	8*	10*			4	2
Callinectes sapidus (F)							1	1

TABLE 3 (con't).

WEST BAYOU (TRAMMEL NETS: STATION 5B)

Date	11-75	12-75	1-76	2-76	3-76	4-76	5-76	6-76
Bagre marinus	7				1	2		14
Lepisosteus osseus			1		2		1	1
Arius felis	3	2					1	2
Dorosoma cepedianum		2	1		3			2
Dasyatis sabina	1						1	3
Paralichthys lethostigma								
Mugil cephalus		8	7					
Micropogonias undulatus							1	
Cynoscion nebulosus	1	1			1			
Ictalurus catus					1			
Carcharinus leucas								2
Sciaenops ocellata								
Amia calva				1				
Brevoortia patronus		1						
Micropterus salmoides								
Cyprinus carpio								
Alosa chrysochloris								
Ictalurus nebulosus								
Pogonias cromis								
TOTAL	12	14	10		8	2	4	24
Callinectes sapidus (M)	2	1		1*	3	2	4	7
Callinectes sapidus (F)							1	3

TABLE 3 (con't).

WEST BAYOU (TRAMMEL NETS: STATION 5B)

Date	7-76	8-76	9-76	10-76	11-76	12-76	1-77	2-77
Bagre marinus	2	1	3	1				
Lepisosteus osseus	3		2	3				
Arius felis	1		2					
Dorosoma cepedianum			2	1	2			
Dasyatis sabina			4		2			1
Paralichthys lethostigma								
Mugil cephalus			2					
Micropogonias undulatus								
Cynoscion nebulosus								1
Ictalurus catus								2
Carcharinus leucas		1						
Sciaenops ocellata				2				
Amia calva								
Brevoortia patronus								
Micropterus salmoides								
Cyprinus carpio								
Alosa chrysochloris					1			
Ictalurus nebulosus								
Pogonias cromis								
TOTAL	6	2	15	7	5			4
Callinectes sapidus (M)	3	2	1	2	2			1
Callinectes sapidus (F)			1					

TABLE 3 (con't).

WEST BAYOU (TRAMMEL NETS: STATION 5B)

Date	3-77	4-77	5-77	6-77	7-77	8-77	9-77	10-77
Bagre marinus		1	4	5	11	9	1	
Lepisosteus osseus	1		4	19	1	21	2	
Arius felis			4	21	7	2	8	
Dorosoma cepedianum	1	11	10		1			
Dasyatis sabina			2	1	1	1	2	
Paralichthys lethostigma			1	1				
Mugil cephalus								
Micropogonias undulatus		2	1			1	3	
Cynoscion nebulosus								
Ictalurus catus								
Carcharinus leucas			1					
Sciaenops ocellata								
Amia calva								
Brevoortia patronus								
Micropterus salmoides			2					
Cyprinus carpio								
Alosa chrysochloris								
Ictalurus nebulosus	1							
Pogonias cromis		1						
TOTAL	3	14	30	47	21	34	16	
Callinectes sapidus (M)		3	21		5	16	10	
Callinectes sapidus (F)		2	8		1		1	

TABLE 3 (con't).

WEST BAYOU (TRAMMEL NETS: STATION 5B)

	11-77	12-77	1-78	2-78	TOTAL
Bagre marinus					79
Lepisosteus osseus					70
Arius felis	5				67
Dorosoma cepedianum					41
Dasyatis sabina					22
Paralichthys lethostigma		2			17
Mugil cephalus					17
Micropogonias undulatus					11
Cynoscion nebulosus	2	3	2		11
Ictalurus catus				1	9
Carcharinus leucas					5
Sciaenops ocellata					4
Amia calva					3
Brevoortia patronus		2			3
Micropterus salmoides					2
Cyprinus carpio					1
Alosa chrysochloris					1
Ictalurus nebulosus					1
Pogonias cromis					1
TOTAL	7	7	2	1	365
Callinectes sapidus (M)	2	1	8	1	125
Callinectes sapidus (F)		1	2		22

individuals were taken during winter months at all three stations. This pattern reflects the seasonal abundance of species such as spot (*Leiostomus xanthurus*) and Atlantic croaker (*Micropogonias undulatus*). Secondary peaks of numbers occurred during summer-early fall periods. There was a progressive increase in summer-fall totals of fish numbers at stations 5b and 5c over the 3-year period. This trend was not apparent at station 4a. Numbers were particularly low during the summer rainy season at stations 5b (1975) and 4a (1977). Relative numbers of fishes generally followed the progressive changes in water quality described above. In general, summer dominants such as the sand seatrout (*Cynoscion arenarius*) were more abundant in Round Bay than in either of the other areas of study.

Overall numbers of species of fishes tended to peak during summer-fall periods of high salinity and falling temperatures. During the heavy rainfall and associated water quality conditions in August, 1975, the numbers of species in East and (especially) West Bayous were low relative to Round Bay. The runoff from newly cleared uplands was associated with the lowest summer species richness values taken during 6 years of sampling. Significant reductions in numbers of fish species also occurred during winter periods. Thus, summer runoff was associated with a general evacuation of East Bay by numbers of fish species which tended to be controlled by seasonal and annual trends of local rainfall, upland clearing activities, temperature, and salinity.

During part of the intensive sampling effort (May–November, 1976), there were three primary episodic decreases in water quality in East and West Bayous (16 July; 10–14 September; 8–12 October). In Round Bay, there was one such incursion of low quality water on 7 September. Four species (*Anchoa mitchilli, Cynoscion arenarius, Leiostomus xanthurus,* and *Micropogonias undulatus*) dominated the collections, comprising 89.7% of the total dry weight biomass. The bay anchovy (*A. mitchilli*) was numerically the most abundant species (49.2%) and was associated primarily with higher salinity during fall months. The sand sea trout was second in abundance (17.0%) with a similar spatial and temporal (late summer–early fall) distribution. The spot (*L. xanthurus*) and Atlantic croaker (*M. undulatus*) were third (12.3%) and fourth (11.2%), respectively. These species were taken early in the study period (spring) during periods of low salinity as noted above.

Peak diurnal abundance of *A. mitchilli* occurred from early September through November. Decreases in total abundance took place during an episodic water quality decline at stations 5b and 5c (October 8). Such decreases were found in Round Bay while the pH was still relatively high (7.3). At this time, a significant decrease in salinity due to local rainfall occurred at all three

stations, which would explain the acidic conditions at stations
5b and 5c. This occurrence would also implicate rapid declines
of salinity as a stress factor. Significant decreases in
abundance of anchovies occurred at stations 5b and 5c following
reduced pH and increased color due to local runoff. A decrease
was also seen at station 4a with an immediate, rapid recovery
when water quality parameters improved.

Pronounced decreases of biomass and numbers of *C. arenarius*
occurred in East and West Bayous on 16 July, 7 September, and 8
October. All such instances were correlated with the input of
acidic runoff to the area. Declines in this species were also
observed at control station 4a on 7 September, when there were
decreases in pH and salinity. However, no such decrease was observed at station 4a on July 16, when there was no decrease in
pH. The reduction in numbers and biomass of this species at
stations 5b and 5c thus appears to be associated primarily with
the runoff conditions.

Large numbers of *L. xanthurus* were found primarily during
the early spring (May). Numbers and biomass declined during the
summer months at all three stations, except for an increase at
station 5c in late August. An increase of individuals and biomass occurred at station 4a in mid-July while water quality was
low at 5b and 5c. From August throughout the remainder of the
study period, considerable variability occurred within this
population. Biomass increased through the summer while numbers
remained relatively low, reflecting the maturing of spring
juveniles. In September and October, when there were periodic
intrusions of runoff, numbers and biomass of this species declined concurrently. The temporal distribution of *M. undulatus*
was similar to that of *Leiostomus*. After a decrease in abundance
in June, no recovery was noted until the recruitment of juveniles
in November. From these facts, it can be concluded that any
evaluation of the influence of summer/fall runoff on fish populations should be made within the context of seasonal species-
specific patterns of abundance. These relatively stable temporal
patterns of species successions have been analyzed by *Livingston
et al.* (1978).

A total of 53 species of fishes was collected during the
summer and fall of 1976. There were no significant differences
in the overall total number of species among the three study
sites. Transient decreases in numbers of species paralleled
trends of runoff in July, September, and October at stations 5b
and 5c. A decrease occurred at station 4a in early September,
although the number of species remained relatively high during
July and October. The September decrease coincided with the
appearance of dark water at station 4a. Fewer species were observed at all three stations in early August, possibly reflecting

high temperature. High variability was noted over the entire
study period with greater numbers of species generally found at
night.

Overall, there were high numbers and biomass of fishes at the
beginning of the intensive study (May, 1976). This result was due
primarily to the presence of adult *L. xanthurus* and *M. undulatus*.
Such peak numbers were followed by a decrease lasting until the
proliferation of *C. arenarius* and *A. mitchilli* in August. The
sizeable increase in numbers at stations 4a and 5b in September
and October, respectively, was caused by the presence of large
schools of anchovies. Reduced numbers and biomass at the end of
the intensive sampling effort was associated with declining water
temperature. Total numbers and biomass of fishes declined following peak runoff and decreases in pH. Abundance and biomass at
station 5b reflected water quality variations.in the form of
episodic changes in July, September, and October. Biological
recovery was noted after subsequent increases in pH and reduced
color levels.

The diversity of fish assemblages at all three stations was
relatively low because of the low numbers of species and high
relative dominance at any given time (Fig. 9). Day/night differences in diversity were noted as the community structure
changed. This index was highly variable because of rapid temporal
fluctuations of numbers and species. During the intensive sampling effort, the decline in diversity in July was due to the
establishment of a less even distribution of populations within
the community. This change occurred because the total number of
individuals decreased in this period of low water quality, while
A. mitchilli increased as a result of seasonal recruitment.
These facts corroborate previously observed relationships of
dominance and diversity (*Livingston et al.*, 1978). Reduced
diversity was observed in September at all stations. The opposite
was noted during periods of increased salinity, and diversity
tended to increase with relative decreases of anchovy numbers
because of the greater number of species and reduced dominance.

The above trends were noted over the three-year period of
study. Many times, diversity and species richness were inversely
related because of particular dominance relationships at the time
of sampling. Overall, the dependence of species diversity on
rapid fluctuations of dominant species seriously weakened its
value as an indicator of water quality. Margalef richness did
not follow the same pattern as the diversity indices; this index
increased significantly during periods of high salinity.
Considerable decreases were noted in July during periods of low
pH and increased color. The Margalef richness tended to follow
changes in species numbers and thus was determined by combinations
of forcing functions as described above.

BIOLOGICAL RELATIONSHIPS: INVERTEBRATES

Total numbers of individuals and species of invertebrates taken at upper East Bay stations are given in Fig. 10. Invertebrate abundance and species richness tended to peak during summer-fall periods of high local rainfall, declining temperature, and increased salinity. Reduced pH, especially during periods of 1975-76 at stations 5b and 5c and 1977 at station 4a, was directly related to sharp, short-term decreases in numbers of individuals. Overall, such numbers were higher in Round Bay than at either of the other study sites, and there were generally low numbers of

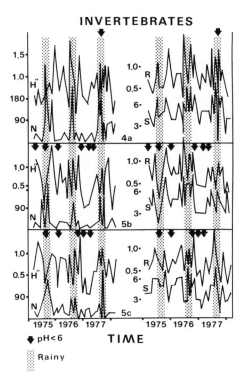

Fig. 10: Numbers of species (S) and individuals (N) of invertebrates taken in Round Bay (4a), West Bayou (5b), and East Bayou (5c) from March, 1975, through February, 1978. Also shown are monthly Shannon diversity and Margalef richness indices. Periods when the existing pH levels were less than 6.0 are also noted.

invertebrates at all three sites during the summer of 1976.
Species richness assumed different temporal patterns at the three
stations. In Round Bay (station 4a), the short-term impact of
local rainfall on various biological indices tended to increase
over the three-year period. This impact coincided with the incidence of clearcutting activities in the upland drainage in 1976-
77 and the changes in pH and dissolved oxygen with time. At
station 5b, adverse impact was particularly evident during the
heavy summer rainfall of 1975, with a general recovery of species
richness values during the succeeding year. At all three stations,
the cold winter of 1976-77 had a prolonged effect on numbers of
invertebrate species. The long-term invertebrate species richness
patterns thus tended to reflect trends of temperature, salinity,
rainfall, and local runoff conditions. The heavy rainfall of **1975**
(and associated declines in water quality at stations 5b and 5c) and
the cold winter of 1976-77 had a particularly pronounced effect on
the distribution of invertebrates in the upper estuary. Over the
three-year period of study, the lowest levels of invertebrate
species richness at all three study sites occurred during periods
of high local rainfall and recent upland clearing activities.
Thus, while numbers of individuals and species richness of nurserying invertebrates in East Bay followed seasonal cycles with peaks
during periods of high local rainfall, rapid reductions of water
quality parameters (pH, dissolved oxygen, salinity), often associated with recent forestry activities in upland drainages, had
pronounced short-term (adverse) effects on the estuarine invertebrate assemblages.

     During the intensive sampling effort of 1976, the white
shrimp (*Penaeus setiferus*) comprised 76% of total numbers and 53%
of the total invertebrate biomass. Periods of peak abundance of
this species occurred from August to October. Invertebrate
composition of the upland portions of East Bay was dominated by
this species during summer/fall periods. Reduced water quality
due to runoff in early September significantly reduced the numbers
and biomass of this species at all three stations. There were
considerable decreases in pH, salinity, and dissolved oxygen at
this time. Reduced numbers and biomass observed in early October
reflected such changes in the system. Low biomass (relative to
numbers of individuals) reflected the recruitment of juveniles into
the area in late July and August. Biomass increased rapidly until
late October. The mean individual size in August 1976, was 43 mm.
It increased to 100 mm by October. Reduced abundance in November
was attributed to low temperature, which apparently acts as a
stimulus for the mature penaeid shrimp to move out of upland
estuarine areas (*Barrett and Gillespie*, 1973; *Trent et al.*, 1976).
The white shrimp was taken most frequently at night, and was most
prevalent at the control station (4a).

The blue crab, *Callinectes sapidus*, was second in invertebrate
dominance (17% of total collection; 41.5% of total biomass) with
greatest numerical abundance at stations 5b and 5c. Larger individuals were found at station 4a throughout the study period
(mean carapace width = 58.4 mm). The highest number of individuals
occurred during the first two months of the intensive sampling
effort. A similar diurnal distribution was observed at stations 5b
and 5c; however, considerable nocturnal recruitment of juveniles
occurred at these stations beginning in August. In May and June,
the mean carapace width in East and West Bayous was 42.2 mm. It
decreased to 17.0 mm in August following summer recruitment in the
area. The spatial preference of juveniles for the East and West
Bayous was shown by the reduced average individual size in these
areas. Numbers of juvenile blue crabs thus tended to peak during
periods of high runoff (low pH and dissolved oxygen, high color).
This phenomenon was discussed by *Laughlin et al.* (1978), who
found that the highest numbers of young blue crabs were taken in
areas having the lowest recorded water quality. Over the entire
study period, the mean carapace widths at stations 5b and 5c were
31.1 and 32.6 mm respectively. At both stations, decreases in
numbers were observed in September and October. This decrease
occurred at station 4a in September.

Sixteen species of invertebrates were collected at station 5b,
including five relatively rare types. Mid-range and rare mollusks
(*Rangia cuneata* and *Mactra fragilis*) were also associated with East
and West Bayous. Species generally associated with grass beds
(*Palaemonetes* spp. and *Neritina reclivata*) were found primarily at
station 4a. The pink shrimp (*Penaeus duorarum*) also was more
abundant in Round Bay as noted above. The greatest numbers of
individuals were found at station 4a because of the dominance of
P. setiferus. Numbers of individuals and total biomass declined
during periods of high local rainfall and runoff. Total numbers
of species were relatively low and there was considerable variability in the richness component as a result of the appearance
of rare species. Numbers of individuals were generally low except
in August and early September, when juvenile *P. setiferus* and *C.
sapidus* moved into the area. Numbers of species increased during
months of peak productivity (August–October), when salinity was
relatively high. The numbers of species were generally higher at
night and there were considerable differences between day and night
diversity indices. Because *P. setiferus* and *C. sapidus* contributed
93% of the total invertebrate collection, the diversity indices
strongly reflected changes in the distribution of these two
species (Fig. 10). Although Shannon diversity tended to reflect
broad changes in the physico-chemical environment such as the
rainfall in 1975 and the cold winter of 1976–77, previously defined limitations of the applicability of such indices for impact
evaluation were operable here thus ruling out the use of such
indices for understanding short-term changes in the system.

Overall, the invertebrate assemblages in the upper East Bay system tended to reflect short-term fluctuations of water quality; the principal impact of low water quality (*i.e.* low dissolved oxygen and pH) was indicated by species richness indices and changes in individual populations such as penaeid shrimp. Such changes, however, were comprehensible only within the context of species-specific response to particular sets of controlling environmental factors and the seasonal progression of nurserying species.

TRAMMEL NET DATA

The results of the trammel net effort in Round Bay and West Bayou are given in Table 3 and Fig. 11. During the first year of

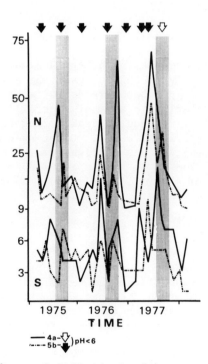

Fig. 11: Numbers of individuals (N) and species (S) taken monthly in trammel nets at night in Round Bay (4a) and West Bayou (5b) from March, 1975, through February, 1977. Periods when the pH at the respective sites was less than 6.0 are noted.

sampling, the heavy local rainfall was associated with relatively
low numbers of organisms in West Bayou relative to Round Bay. At
this time, numbers of species and individuals declined to low
levels at 5b while both functions remained high at 4a. In 1976,
increased summer rainfall was associated with declines in numbers
of individuals and species in both areas, with partial recovery
apparent in West Bayou and full recovery in Round Bay during the
following fall. During the early summer of 1977, there was a
pronounced increase in numbers of organisms in both areas; this
increase was associated with the low level of local rainfall at
this time. The numbers fell precipitously in Round Bay during the
intrusion of acid water. These changes thus tended to follow
general trends in salinity and specific water quality parameters
such as pH as indicated above. With minor variation, the long-
term changes in the trammel data showed relationships to the rainfall and water quality data similar to those described for the
trawl information. Low quality runoff and cold winter temperatures
tended to control the biotic indices. The trammel data indicated
recovery of numbers of individuals and species in West Bayou
relative to Round Bay over the three-year study period.

LONG-TERM TRENDS (1972-1978)

Various dominant species in the Apalachicola estuary showed
general trends of relative abundance over the six-year period of
study (*Livingston*, 1978). These trends reflected individual
species strategies relative to sets of controlling functions.
Anchovies (*A. mitchilli*) tended to peak during periods of high or
increasing salinity (1972, 1976-77). Other species such as *M.
undulatus* peaked during periods of high river flow (1973). Subsequent decreases (1976-77) of this species coincided with the
increased abundance of spot (*L. xanthurus*). Trout (*C. arenarius*)
peaked during periods of heavy rainfall (1974) and drought (1977).

Long-term trends of fish and invertebrate species richness in
eastern (station 5) and western (station 6) portions of East Bay
were compared with system-wide sampling efforts (Fig. 12).
Despite considerable seasonal variability, prolonged trends and
abrupt changes were evident over the 6-year period of study. The
numbers of fish and invertebrate species peaked during the first
year of high salinities at both East Bay locations while baywide
species richness peaked in 1975 (fishes) and 1976 (invertebrates).
The winter river flooding of 1973 influenced the long-term trends
of bay-wide species richness. However, the pattern of recovery
differed in the two East Bay areas. At station 6, fish species
richness gradually increased to a peak in 1976. Subsequent
declines were possibly related to the cold winter of 1976-77.
After a gradual increase in invertebrate numbers of species to a
1975 peak at this station, there were subsequent decreases which

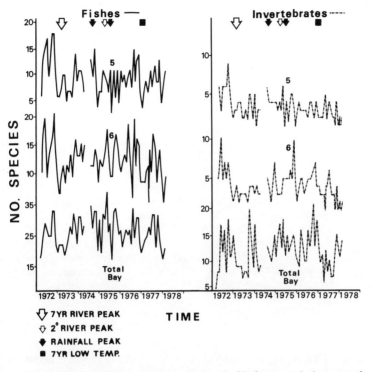

Fig. 12: The number of species of fishes and invertebrates taken monthly in otter trawls in eastern (5) and western (6) portions of East Bay and the Apalachicola estuary as a whole from March, 1972, through February, 1978. Also indicated are periods of major and minor river flooding and peaks of local rainfall.

could also have been associated with low winter temperatures. Station 6 thus followed, in a general way, the bay-wide trends. At station 5, however, substantial decreases in fish species numbers occurred during the 2-year period (1974-75) of heavy rainfall. Such differences were also noted in the numbers of invertebrate species from 1975 to 1976. The long-term trends of the biota in western (river-dominated) areas of East Bay thus differed from those in eastern sections, and such differences reflected local runoff patterns and water quality trends. These data indicate that key forcing functions such as river flow, rainfall, and temperature control the number of species of fishes and invertebrates in the estuary at any given time, and overall species richness is affected by the timing and quality of runoff from the upland drainage systems.

Estuaries are often affected by local runoff patterns; in this
case, the naturally variable physical environment was temporarily
destabilized by forestry operations in the Tate's Hell Swamp even
during periods of reduced overland runoff. Local rainfall patterns, together with recent clearcutting, contributed to short-
term reductions in salinity, pH, and dissolved oxygen which, in
turn, caused repeated shocks to sensitive species such as penaeid
shrimp. Since the regional productivity of such species remains
unknown, the absolute biological impact of the forestry management
practices in Tate's Hell Swamp on numbers of sensitive species remains undetermined. However, within the context of individual
reactions to repeated shocks, the upland forestry activities did
cause local disruptions in nursery habitat for key commerical
species during periods of usually maximal (habitat-specific)
activity. Such impact, qualified by local drainage characteristics and long-term fluctuations of temperature, river flow, and
rainfall, appeared to be temporary. In addition, the effects were
species-specific and were influenced by seasonal progressions of
species abundance and long-term trends of population variability.

SUMMARY OF CONCLUSIONS

1. Water quality and the biological associations of the
Apalachicola estuary are controlled by combinations of major
climatic features such as temperature, river flow, and local
rainfall. The spatial and temporal distribution of epibenthic
estuarine populations is determined by physico-chemical instability which increases with proximity to the land-water interface. Upland portions of bay systems such as East Bay serve as
nursery areas for certain key species which undergo regular
seasonal progressions of relative dominance. In the Apalachicola
estuary, this group of species includes various fishes (*A.
mitchilli, L. xanthurus, M. undulatus, C. arenarius*) and invertebrates (*Penaeus* spp., *C. sapidus*) which respond to environmental
perturbations according to individual (species-specific) strategies. Natural variability of the physical environment in such
areas is associated with high dominance at the species level and
high productivity at the systems level.

2. Forest management practices (ditching, draining, clearcutting) in upland areas contiguous with eastern portions of East
Bay have been associated with enhanced destabilization of the
physical environment. Such effects occur during periods of heavy
local rainfall and include periodic flashing of runoff into receiving aquatic areas. Water quality changes include increased
color, turbidity, and dissolved nutrients, and relatively rapid
decreases of salinity, pH, and dissolved oxygen. The extent
and duration of such impact on water quality depend on the magnitude and timing of the forestry operations relative to seasonal

and annual events of local precipitation. The reductions in pH
and dissolved oxygen tend to be relatively short-lived possibly
because of regrowth of upland vegetation. The impact of wetlands
as a modifying factor in flashing is indicated here.

3. Pulsed changes in the water quality of receiving estuarine systems have been associated with short-term reductions in
key populations and community complexity in affected sections of
the East Bay nursery. Such effects are directly attributable to
forest management practices such as clearing and ditching.
These changes, while inhibitory to certain populations, were
stimulatory to others, thus confirming the important role of
species strategy in impact evaluation. Community indices such
as Shannon diversity proved to be relatively unreliable as indicators of stress because of the influence of dominants on such
indices and the natural temporal fluctuations of key populations.

4. Impact due to modification of upland runoff should be
evaluated relative to both short-term and long-term fluctuations
of key climatic features. Repeated storm incidence together with
reduction of upland vegetation, flow channelization, and expansion
of the drainage area can lead to short-term shocks, which change
seasonal and supraanual progressions of biological associations.
Whereas forestry operations cause only temporary impact because of
revegetation, long-term impact is possible when more permanent
modification of drainage is made through urbanization and replacement of vegetation with impermeable surfaces.

ACKNOWLEDGMENTS

The authors acknowledge G. C. Woodsum, Jr., for his help
with the computer programs used for this project. Portions of
the data collection were funded by NOAA Office of Sea Grant,
Department of Commerce, under Grant No. 04-3-158-43. Data
analysis was supported by EPA Program Element No. 1 BA025 under
Grant No. R-803339. Matching funds came from the people of
Franklin County, Florida.

REFERENCES

Barrett, B. B. and M. C. Gillespie, 1973. Primary factors which
influence commercial shrimp production in coastal Louisiana,
Tech. Bull. 9: 3-11, La. Wildl. and Fish. Comm.

Bishai, H. M., 1960. The effect of hydrogen ion concentration on
the survival and distribution of larval and young fish, Z.
Wissenschftliche Zool. 164: 107-118.

Bormann, F. H., G. E. Likens, D. W. Fisher, and R. S. Pierce, 1968. Nutrient loss accelerated by clearcutting of a forest ecosystem, *Science 159*: 882-884.

Cherry, D. S., K. L. Dickson, and J. Cairns, Jr., 1975. Temperatures selected and avoided by fish at various acclimation temperatures, *J. Fish. Res. Bd Can. 32:* 485-491.

Collins, G. B., 1952. The lethal action of soluble metallic salts on fishes, *Brit. J. Exp. Biol. 4:* 378-390.

Copeland, B. J. and T. J. Bechtel, 1974. Some environmental limits of six Gulf coast estuarine organisms, *Contrib. Mar. Sci. 18:* 169-204.

Dickinson, W. T., M. E. Holland, and G. L. Smith, 1967. An experimental rainfall-runoff facility, Colorado State University Hydrology Paper No. 25, 78 pp.

Duncan, J. L., 1977. *Short-term Effects of Storm Water Runoff on the Epibenthic Community of a North Florida Estuary (Apalachicola, Florida)*, M.S. Thesis, Florida State University.

Harr, R. D., W. C. Harper, and J. T. Krygier, 1975. Changes in storm hydrographs after road building and clearcutting in the Oregon coast range, *Water Resour. Res. 11:* 436-444.

Hewlett, J. F. and J. D. Helvey, 1970. Effects of forest clearfelling on the storm hydrograph, *Water Resour. Res. 6:* 768-782.

Hewlett, J. F. and A. R. Hibbert, 1961. Increases in water yield after several types of forest cutting, *Quart. Bull. Internatl. Assoc. Sci. Hydrol. Louvain, Belgium:* 5-17.

Hill, J., 1968. *LANDSAT Assessment of Estuarine Water Quality with Specific Reference to Coastal Land-use*, Ph.D. Dissertation, Texas A & M University.

Hornbeck, J. W., 1973. Storm flow from hardwood forested and cleared watersheds in New Hampshire, *Water Resour. Res. 9:* 346-354.

Hornbeck, J. W., R. S. Pierce, and C. A. Federer, 1970. Streamflow changes after forest clearing in New England, *Water Resour. Res. 6:* 1124-1132.

Jones, J. R. E., 1964. *Fish and River Pollution*, 203 pp., Butterworth and Co., London.

Kwain, W., 1975. Effects of temperature on development and survival of rainbow trout, *Salmo gairdneri*, in acid waters,
J. Fish. Res. Bd Can. 32: 493-497.

Laughlin, R. A., C. R. Cripe, and R. J. Livingston, 1978. Field
and laboratory avoidance reactions by blue crabs (*Callinectes
sapidus*) to storm water runoff, *Trans. Amer. Fish. Soc. 107:*
78-86.

Likens, G. E., F. H. Bormann, and N. M. Johnson, 1969.
Nitrification: importance of nutrient losses from a cutover
forest watershed, *Science 163:* 1205-1206.

Likens, G. E., F. H. Bormann, N. M. Johnson, D. W. Fisher, and
R. S. Pierce, 1970. Effects of forest cutting and herbicide
treatment on nutrient budgets in the Hubbard Brook Watershed
Ecosystem, *Ecol. Monogr. 40:* 23-47.

Livingston, R. J., 1978. *Short- and Long-term Effects of Forestry
Operations on Water Quality and the Biota of the Apalachicola
Estuary (North Florida, U. S. A.)*, Florida Sea Grant Report
(unpublished).

Livingston, R. J., C. R. Cripe, R. A. Laughlin, and F. G. Lewis,
III, 1976. Avoidance response of estuarine organisms to
storm water runoff and pulp mill effluents, *Estuarine Proc.
1:* 313-331.

Livingston, R. J., R. L. Iverson, R. Estabrook, V. Keys, and J.
Taylor, 1974. Major features of the Apalachicola Bay system:
physiography, biota, and resource management, *Fla. Sci. 37:*
245-271.

Livingston, R. J., R. L. Iverson, and D. C. White, 1977. *Energy
Relationships and the Productivity of Apalachicola Bay*,
Florida Sea Grant Report (unpublished).

Livingston, R. J., N. P. Thompson, and D. A. Meeter, 1978. Long-
term variation of organochlorine residues and assemblages of
epibenthic organisms in a shallow north Florida (USA)
estuary, *Mar. Biol. 46:* 355-372.

Lloyd, R. and D. H. M. Jordan, 1964. Some factors affecting the
resistance of rainbow trout (*Salmo gairdneri* Richardson) to
acid waters, *Int. J. Air and Water Poll. 8:* 393-403.

Lull, H. W. and K. G. Reinhart, 1972. Forests and floods in the
Eastern United States, Res. Paper NE-226, 94 pp., Northeast
Forestry Exp. Sta.

Margalef, R., 1958. Information theory in ecology, *Gen. Systematics 3:* 36–71.

Nie, N. H., C. H. Hull, J. G. Jenkins, K. Steinbrenner, and D. H. Bent, 1970. *Statistical Package for the Social Sciences,* 663 pp., McGraw-Hill, Inc.

Packer, P. E., 1965. Forest treatment effects on water quality, In: *Forest Hydrology,* edited by W. E. Sopper and H. W. Lull, 687–698, Pergamon Press.

Pierce, R. S., J. W. Hornbeck, G. E. Likens, and F. H. Bormann, 1970. Effect of elimination of vegetation on stream water quantity and quality, Int. Assoc. Sci. Hydrol. Publ. 96: 311–328.

Powers, E. B., 1941. The variation of the condition of sea water, especially the hydrogen ion concentration and its relation to marine organisms, *Publ. Puget Sd Mar. Sta. 2:* 369–385.

Satterlund, D. R., 1967. Forest types and potential runoff, In: *Forest Hydrology,* edited by W. E. Sopper and H. W. Lull, 497–503, Pergamon Press.

Shannon, C. E. and W. Weaver, 1963. *The Mathematical Theory of Communication,* 125 pp., University of Illinois Press, Urbana.

Smith, W. H., F. H. Bormann, and G. E. Likens, 1968. Response of chemoautotrophic nitrifiers to forest cutting, *Soil Sci. 106:* 471–473.

Sokolovskii, D. W., 1968. River runoff, *Gid. Izdat., Leningrad,* Israel program for scientific translations, Jerusalem (1971).

Swank, W. T. and J. E. Douglass, 1974. Stream-flow greatly reduced by converting deciduous hardwood stands to pine, *Science 185:* 857–859.

Tebo, L. D., 1955. Effects of siltation, resulting from improper logging, on the infauna of a small trout stream in the southern Appalachians, *Prog. Fish-Cult. 12:* 64–70.

Trent, L., E. J. Pullen, and R. Procter, 1976. Abundance of macro-crustaceans in a natural marsh and a marsh altered by dredging, bulkheading, and filling, *Fish. Bull. 74:* 195–200.

Ziemer, R. R., 1964. Summer evapotranpiration trends as related to time after logging of forests in Sierra Nevada, *J. Geophys. Res. 69*: 615–620.

THE PASS AS A PHYSICALLY-DOMINATED, OPEN ECOLOGICAL SYSTEM

Rezneat M. Darnell

Texas A & M University

ABSTRACT

Despite their importance in the life histories of many coastal species, passes have received little ecological study. From scant literature sources the salient features of pass ecology are sketched out. It is suggested that, as physically-dominated nutritionally rich systems, passes are ecologically unique and that their detailed study should shed new light on a number of classical ecological paradigms.

INTRODUCTION

In a recent court case the author was called upon to testify concerning the ecological effects which might be anticipated if an engineering structure were placed across the mouth of an estuary. This structure would continuously restrict water exchange by reducing the depth and breadth of the opening. Gates could be closed in the event of a hurricane. Of particular interest were potential effects upon both the estuary and the shelf, as well as upon the intervening pass. Surprisingly, the American technical literature contains very few papers on the ecology of the pass, although some extrapolation can be made from studies on the ecology of estuaries and shelves and on biofouling and other communities of jetties and nearshore environments.

The present discussion, based on the literature of passes and related environments as well as on the author's experience, is put forth as a first step toward the understanding of the ecology of the pass. It is recognized that, because of geographic

location as well as local physical conditions, each pass exhibits
an ecological individuality, but attention here is focused on
the commonalities of passes. Hopefully, this presentation will
stimulate a series of local studies from which a real knowledge
of pass ecology can be developed.

GENERAL DESCRIPTION OF THE ECOLOGY OF PASSES

As the estuary is the expanded mouth of a river, so the
pass is the constricted mouth of an estuary. Actually, the
mouths of estuaries vary considerably in width, some approaching
or exceeding the width of the estuary itself. However, the pass
is typically a coastal constriction, and it is this narrowness
or channelization, along with its position (between estuary and
shelf), that endows the pass with its unique properties.

Passes, as such, have not been the subject of much ecological
study. Hydrographers and sedimentary geologists have given them
fair attention, but most biological work has been limited to
occasional trawl or dredge collections of benthic organisms and
to study of the species attached to lateral walls and to channel
buoys. Only a few workers have attempted to correlate biological
information with physical, chemical, or geological data. The
summary paper of *Kitching and Ebling* (1967) in Scotland is an
important exception. Fortunately, much of the literature
treating estuaries is of relevance (see especially *Lauff*, 1967,
and *Cronin*, 1975), as is some of the work on species and
communities which attach to firm substrates (see *Redfield and
Deevey*, 1952; *Costlow*, 1977; and *Sutherland and Karlson*, 1978).
Of special interest is information derived from experimental
study of fouling and other communities (*Paine*, 1974; *Dayton*,
1975; *Sutherland*, 1977, 1978).

Much of our present knowledge of the ecology of passes has
been reviewed and interpreted by *Odum et al.* (1974), and the
following statement taken from their work provides an excellent
introduction to the subject.

> *"In channels where sea waters flow at high velo-
> cities, 3-20 miles per hour or more, bottoms are swept
> clean of fine sediments and reef-like accumulations,
> and specialized encrusting organisms develop growths,
> taking advantage of the foods available in the rapidly
> passing waters . . . . In nature the system occurs
> on hard coarse bottoms where tidal flow passes through
> narrow passages and inlets.*

*"The very strong current dominates the system and
allows dense patterns of attached organisms but also is
a source of stress requiring energies to be expended
by the organisms in adaptation. If the surface is
within range of light, heavy algal growths develop
facilitated by the rapid renewal of nutrients for
photosynthesis. High velocity channels are favorite
collecting locations for dredging of marine organisms
in quantity. Because the high currents often occur
in high salinities at the entrance of estuaries where
conditions, other than current, are uniform, species
diversities tend to be moderately high, but diminishing
with current. This system is found in every state of
the U.S. . . . . "*

THE DOWNSTREAM SERIES

Actually, the pass is not just the connection between the
estuary and the shelf; it is the conduit through which all the
upstream influences of the basin enter the sea and through which
the sea, in turn, influences various portions of the basin.
All the rain which falls in the drainage basin and which is not
otherwise lost through seepage and evaporation must enter the
sea through the pass, and with it go soluble organic and inorganic
materials derived from various areas of the basin. Much fine
particulate matter also passes to the sea in this manner. The
downstream transport phenomenon is a well-known aspect of
biogeochemical cycles and need not be elaborated upon here, but
it is important to note that, because of the large number of
factors involved and their complex interactions, the discharge
of each pass is unique in terms of volume, timing, and rates of
flow, as well as in composition. Studies of *Hasler* (1966) on
the olfactory response of homing salmon dramatically underscore
the unique quality of the discharge water of each individual
pass. Marine waters, which traverse the pass and enter the
estuary, also undoubtedly have complex histories and may possess
locally unique properties, but this phenomenon has not been well
documented.

THE CURRENTS

The physics of water movement through the passes represents
a dynamic and shifting balance between the forces from the two
directions. Involved are relative water levels, temperatures,
and salinities. These act within the context of confined
cross-sectional area, wind stress, and roughness characteristics
of bottom and sidewalls. Wind-driven surface waves pound the
exposed edges and, to some extent, the sidewalls throughout the

length of the pass. For example, *Whitten, Rosene, and Hedgpeth*
(1950) note that, even when passes are protected by jetties, wave
action is effective far up the channel between the jetties
during storms and periods of strong onshore winds acting against
an outgoing tide. Large quantities of suspended organic and
inorganic particles are transported by the flowing water. At
slack water of the intertidal period this material may settle
out only to be raised again at the next tidal flow. Although
the pass is basically an erosional environment, deposition
does take place in protected situations where the current is
slower (as, for example, in the cracks between blocks of the
jetty wall). Since the physical forces which interact within
the pass are subject to both short and long-term variation,
equilibrium conditions within the pass constantly change, and
the local water is more or less constantly in a state of readjust-
ment to a new set of imposed conditions. The frequent and severe
shifts in water quality, flow rate, and direction create a
demanding and stressful environment to which the living systems
must either adjust or perish.

MORPHOMETRY AND GEOLOGY

From the standpoint of morphometry and geology there are
two primary types of passes, the fjord-type and the soft substrate type. The fjord-type pass has been scoured, generally by
glacial ice, in granitic or other hard rock. Hence, the shape
and substrate have largely been determined by physical forces
acting in the past and are not related to present water and
substrate conditions. Most such passes are deep and have steep
sides, and they often possess shallow sills at the mouths.
Fjord-type passes are limited to high latitudes.

The soft-substrate pass predominates throughout most of the
warmer regions of the world. Here, the substrate is generally
composed of sediments brought down by rivers and reworked by
winds, waves, and water currents. The dimensions (depth, width,
and cross-sectional area) are determined by interactions of the
normal current flow with the substrates. Owing, in part, to
the Bernoulli effect, currents travel faster through the pass
than they do beyond the pass. Hence, the pass tends to be
deeper than the waters beyond either end, and the bottom is
composed of coarser materials, often coarse sand and much
loose shell debris. Sides may be steep if the current is
swift. Shallow bars generally develop on both ends where
sedimentary material is dropped by the slower current and where
longshore currents sweep perpendicular to the mouth of the pass.

A third type of pass is sometimes encountered, in which the
substrate is hard, stratified sedimentary rock or in which there

are major rock outcrops and boulders. To the extent that these
have been worn down and covered with sedimentary material, some
measure of current/sediment equilibrium has been reached. To
the extent that the erosion process is still young, the substrate dominates the morphometry and surface sediment types.

From the above conditions it is clear that the currents
and the nature of the substrate together determine the morphometry and surface sediments of most passes. In no significant
way are these factors under biological control or influence.

BIOLOGY

Table 1 presents an ecological classification of living
organisms which may be found in passes. The categories overlap,
and the species which represent the different categories vary
on a geographic, often latitudinal, basis (*Hedgpeth*, 1953;
Redfield and Deevey, 1952). Within a given geographic area,
however, the presence of a particular species or an entire
category clearly depends on local circumstances. Phyto- and
zooplankton tend to be universally present. The species are
derived from both the estuarine and littoral marine plankton
groups. Most individual species are seasonal in occurrence.
Zooplankton are dominated by larval forms of benthic and
nektonic species, and the appearance of each larval species is
also highly season-specific. Since the current of the pass is
often swift, many small benthic organisms are temporarily
suspended, and they traverse the pass in the water column as
pseudoplankters.

The attached species are also derived largely from neighboring estuaries and littoral marine environments, but some may
come from salt marshes and even from stranded sargassum blown
in from the open ocean (*Whitten, Rosene, and Hedgpeth*, 1950).
Some of the species are imported in biofouling communities
attached to the hulls of commercial ships. The presence and
distribution of attached algal species depends upon the availability and distribution of suitable hard substrate for attachment
within the euphotic zone. The presence of attached animal species
also depends upon the availability of hard substrate, but the
species composition of any given spot relates in some measure
to the time at which suitable substrate is available. In
experimental studies, *Sutherland* (1977, 1978) and *Sutherland and
Karlson* (1978) demonstrated that for a given species to gain a
foothold, firm substrate of suitable quality (*i.e.*, not inhabited
by inhibitory species) must be present at the time the larvae
are seeking places of attachment. Once a given species becomes
established, it may then inhibit settlement by larvae of other
species. The presence of infauna depends upon the availability

TABLE 1.   Subsystems and biological components of the pass
           environment (compiled from many sources).

Plankton

     -Phytoplankton (diatoms, dinoflagellates, etc.)

     -Zooplankton

          Holoplankton (copepods, chaetognaths, ctenophores,
             medusae, etc.)

          Meroplankton (larvae of fishes, tunicates, echinoderms,
             crustaceans, mollusks, annelids, etc.)

          Pseudoplankton (small benthic crustaceans, annelids,
             protozoans, etc.)

Nekton   (dolphins, fishes, adult crabs and shrimp, cephalopods,
             etc.)

Benthos

     -Attached (algae, bryozoans, barnacles, some mollusks, etc.)

     -Infauna (some mollusks and annelids, small crustaceans,
        nematodes, protozoans, etc.)

     -Relatively non-mobile fauna (some echinoderms, crustaceans,
        mollusks, annelids, etc.)

     -Relatively mobile fauna (benthic fishes, some crabs and
        shrimp, etc.)

Lateral wall species

     -Attached (algae, bryozoans, mollusks, sponges,
        tunicates, etc.)

     -Infauna (some crustaceans, mollusks, annelids, etc.)

     -Crypt fauna (small crabs and other crustaceans, some
        mollusks, annelids, etc.)

     -Relatively non-mobile fauna (some small crustaceans,
        echinoderms, etc.)

     -Relatively mobile fauna (some fishes, crabs, shrimp, etc.)

of protected substrate soft enough for burrowing, yet stable
enough so that there is not constant burial or erosion. The
presence of species in the intertidal and supratidal zones
relates to the force of wave action and the length of the
desiccation period. The presence of a crypt fauna depends on
the availability of cracks, crevices, and other sheltered nooks
(as between the stones of jetties).

Nekton and mobile benthic species are somewhat independent
of the current, and, in any event, they can delay movement
through the pass until the current is favorable. The most
detailed information on the movement of mobile species through
the pass has been provided by *Copeland* (1965). For a period of
a year he studied the passage of animals into and out of the
Aransas Pass, Texas. Samples were taken on maximum flood and
ebb tides every week by means of a tide trap. More than 24
species of invertebrates and 55 species of fishes were collected.
Of this number, 17 species were considered to be "common" and the
remainder "occasional." The great majority of individuals
captured were taken on ebb tide. Only a few individuals appeared
in flood-tide collections even though the flood was fished for
the same length of time as the ebb. The appearance and abundance of most species was highly seasonal, and the only species
collected with some consistency throughout the year was the
anchovy (*Anchoa* spp.). Exceptionally high capture rates were
noted in October immediately following the passage of a severe
cold front. Calculations suggest that, at the time, 352.2 kg
sec^{-1} biomass of fishes and invertebrates were leaving the bays
and estuaries (as opposed to an annual average of 22.15 kg sec^{-1}).
Interestingly, the peak abundance of migrating brown shrimp
(*Penaeus aztecus*) during the summer months of May through August
occurred during the time of full moon (*i.e.*, when the tides were
highest and currents fastest). The same general pattern
appeared in the pink shrimp (*Penaeus duorarum*), although they
were much less abundant than the brown shrimp. Comparable
studies are needed for other passes to place this information
in perspective.

Thus, the flora and fauna which enter the pass are derived
chiefly from the estuary and the littoral marine environments
supplemented by a few vagrants from other areas. Some are
carried passively by the water masses, while others move
partially or largely under their own power. Some use the pass
only as an avenue from estuary to shelf (or *vice versa*), while
others take up temporary or permanent residence there. The
fates of all which enter are determined ultimately by their
mobilities, attachment capabilities, and tolerance to the
variable and often extreme physical factors.

COMMUNITY ORGANIZATION AND DYNAMICS

There is ample evidence that the physical environment is the
primary determinant of which species may survive in a given pass
or in a given area of a pass (*Whitten, Rosene, and Hedgpeth*, 1950;
Hedgpeth, 1953; *Redfield and Deevey*, 1952). However, among those
species which can tolerate the physical environment, biological
interaction has a great deal to do with which species actually
populate and thrive in an area. In some passes there is evidence
for biological organization among the attached species involving
succession and leading to relatively stable local climax communities (*Redfield and Deevey*, 1952). The stable association may be
dominated by *Mytilus* (Fig. 1) or a *Balanus-Mytilus* complex, or by
attached algae, depending upon location. However, experimental
studies by *Sutherland* (1977, 1978) and by *Sutherland and Karlson*
(1978) clearly demonstrate that, for the North Carolina situation,
there may be more than one stable association. In some cases
the dominant species was a bryozoan *(Schizoporella)*, and in
others it was a tunicate *(Styela)*, depending upon when the bare
experimental plates were set out. In both cases stability was
temporary because of death or sloughing off of the dominants
and because the dominants inhibited development of their own
young. Such interactions by short-lived animal species which
produce seasonal larvae should lead to mosaics of species
assemblages varying from place to place and changing from
season to season and from year to year. This, in turn, would
be reflected in local species monotony but great species diversity
in the larger perspective of the total pass environment.

As pointed out by a number of workers, the pass is rich in
nutrients and organically-bound energy. For plant life the
currents bring abundance of dissolved nutrients, and they wash
away wastes and other metabolic products. Filter-feeding animals
are bathed in a stream rich in plankton and particulate organic
detritus. Benthic animals on loose shell bottoms and crypt
fauna are supplied with an abundance of precipitated particulate
organic material. Hence, passes offer exceptional trophic
opportunities, and it is likely that the plants and animals are
in no way nutrient- or energy-limited.

CONCLUSION

As physically-dominated nutritionally rich ecological systems,
passes are unique, and their study should throw new light on
classical ecological paradigms such as succession, diversity,
stability, and trophic balance. They provide exceptional opportunities for investigating tolerance and growth under variable and
extreme physical conditions, species interactions, and trophic
interdependence of different general system components (plankton,

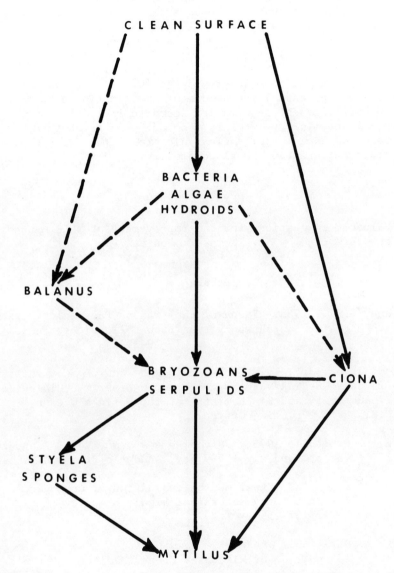

Fig. 1. Sequences of dominant organisms on surfaces exposed in Newport Harbor, California. (From *Redfield and Deevy*, 1952, after *Scheer*, 1945).

benthos, attached species, etc.). Of especial importance is the opportunity for examining life history strategies for species in which the time and place of success can be predicted only in a loose probabilistic sense. These matters are of some practical importance as well. Most commercially and recreationally important marine species must utilize or traverse the pass at some stage of their life histories, and the significance of the pass as a potential bottleneck merits analysis. Furthermore, we must be in a position to provide an intelligent evaluation of the actual or potential intrusions of civilization upon the pass environment. Present information seems to indicate that ecological details vary greatly from one pass to another, but nowhere do we have a really clear picture of the pass in all its structural and functional complexity.

ACKNOWLEDGMENTS

Valuable suggestions for the improvement of this manuscript have been provided by Mr. Thomas M. Soniat.

REFERENCES

Copeland, B. J., 1965. Fauna of the Aransas Pass Inlet, Texas. I. Emigration as shown by tide trap collections, *Publ. Inst. Mar. Sci., Univ. Texas, 10:* 9-21.

Costlow, J. D., Jr., 1977. *The Ecology of Fouling Communities,* 425 pp., US-USSR Workshop within the program, "Biological Productivity and Biochemistry of the World's Oceans," U.S. Gov't. Printing Office.

Cronin, L. E. (ed.), 1975. *Estuarine Research,* Vols. I and II, xiv + 738 pp. and xiv + 587 pp., Academic Press, N. Y.

Dayton, P. K., 1975. Experimental evaluation of ecological dominance in a rocky intertidal algal community, *Ecol. Monogr. 45:* 137-159.

Hasler, A. D., 1966. *Underwater Guideposts, Homing of Salmon,* xii + 155 pp., U. of Wisconsin Press, Madison.

Hedgpeth, J. W., 1953. An introduction to the zoogeography of the northwestern Gulf of Mexico with reference to the invertebrate fauna, *Publ. Inst. Mar. Sci., Univ. Texas, 3:* 107-224.

Kitching, J. A. and F. J. Ebling, 1967. Ecological studies at Lough Ine, In: *Advances in Ecological Research, Vol. 4,* edited by J. B. Cragg, 197-291, Academic Press, N. Y.

Lauff, G. H. (ed.), 1967. *Estuaries,* xv + 757 pp., Publ. No. 83. Amer. Assoc. Adv. Sci., Washington, D. C.

Odum, H. T., B. J. Copeland, and E. A. McMahan (eds.), 1974. *Coastal Ecosystems of the United States,* The Conservation Foundation, Washington, D. C. Vols. I-IV.

Paine, R. T., 1974. Intertidal community structure. Experimental studies on the relationship between a dominant competitor and its principal predator, *Oecologia 15:* 93-120.

Redfield, A. C. and E. S. Deevey, Jr., 1952. Chapter 3: The Fouling Community, Chapter 4: Temporal Sequences and Biotic Successions, Chapter 5: The seasonal sequence, and Chapter 6: Quantitative Aspects of Fouling, In: *Marine Fouling and Its Prevention,* prepared by Woods Hole Oceanographic Institution for the U. S. Naval Institute, Annapolis, Md., 37-90.

Scheer, B. T., 1945. The development of marine fouling communities, *Biol. Bull. 89:* 103-121.

Sutherland, J. P., 1977. Effect of *Schizoporella* removal on the fouling community at Beaufort, North Carolina, In: *Ecology of Marine Benthos,* edited by B. C. Coull, 155-176, U. of S. Carolina Press, Columbia, S. C.

Sutherland, J. P., 1978. Functional roles of *Schizoporella* and *Styela* in the fouling community at Beaufort, North Carolina, *Ecol. 59:* 257-264.

Sutherland, J. P. and R. H. Karlson, 1978. Development and stability of the fouling community at Beaufort, North Carolina, *Ecol. Monogr. 47:* 425-446.

Whitten, H. L., H. F. Rosene, and J. W. Hedgpeth, 1950. The invertebrate fauna of Texas coast jetties; a preliminary survey, *Publ. Inst. Mar. Sci., Univ. Texas, 1:* 53-87.

PERTURBATION ANALYSIS OF THE NEW YORK BIGHT

Arthur G. Tingle, Dwight A. Dieterle, and John J. Walsh

Oceanographic Sciences Division, Brookhaven National

Laboratory

ABSTRACT

The physical transport of pollutants, their modification by the coastal food web, and their transfer to man are problems of increasing complexity on the continental shelf. In an attempt to separate cause and effect, a computer modeling technique is applied to problems involving the transport of pollutants as one tool in assessment of real or potential coastal perturbations. Approaches for further development of models of the biological response within the coastal marine ecosystem are discussed. Our present perturbation analyses consist of 1) a circulation sub-model, 2) a simulated trajectory of a pollutant particle within the flow field, and 3) a time-dependent wind input for each case of the model. The circulation model is a depth-integrated, free surface formulation that responds to wind stress, bottom friction, the geostrophic pressure gradient, the Coriolis force, and bottom topography. The transport diffusion model is based on Lagrangian mass points, or "particles," moving through an Eulerian grid. The trajectories of material moving on the surface and in the water column are computed. It has the advantage that the history of each particle is known. With these models, we have been able to successfully 1) reproduce drift card data for determining the probabilities of a winter oil spill beaching within the New York Bight, 2) analyze the source of floatables encountered on the south shore of Long Island in June, 1976, and 3) predict the trajectory of oil spilled in the Hudson River after it had entered the New York Bight Apex. For future analyses, the shallow water model can be modified or replaced with a numerical model that contains a more sophisticated parameterization of the physical circulation. Second, the particle-in-cell model can be modified to include explicitly chemical reactions and interactions with the biota. Any model, however, should be used in the context of the level of resolution or aggregation to which the ecosystem is known and the management decision required as an aid in selecting situations that merit further analysis with more comprehensive ecological reasoning.

INTRODUCTION

". . . the high piled scow of garbage, bright-colored, white-flecked, ill smelling, now tilted on its sides, spills off its load into the blue water, turning it a pale green to a depth of four or five fathoms as the load spreads across the surface, the sinkable part going down and the flotsam of palm fronds, corks, bottles, and used electric light globes, seasoned with an occasional condom or a deep floating corset, the torn leaves of a student's exercise book, a well-inflated dog, the occasional rat, the no-longer distinguished cat; all this well shepherded by the boats of the garbage pickers who pluck their prizes with long poles, as interested, as intelligent, and as accurate as historians; they have the viewpoint; the stream with no visible flow, takes five loads of this a day when things are going well in La Habana and in ten miles along the coast it is clear and blue and unimpressed as it was ever before the tug hauled out the scow; and the palm fronds of our victories, the worn light bulbs of our discoveries and the empty condoms of our great loves float with no significance against one single, lasting thing - the stream."

Hemingway (1935)

Over the last 40 years, we have come to realize that the above graphic description of dilution can no longer be considered a simple or permanent removal process within either the open ocean or nearshore waters. The increasing utilization of the continental shelf for oil drilling and transport, siting of nuclear power plants, and various types of planned and inadvertent waste disposal, as well as for food and recreation, requires a careful analysis of the impact of these activities on the coastal marine ecosystem. The continental shelves comprise 10% of the area of the world's ocean and yield 99% of the global fish catch. This ecosystem is presently subject to both atmospheric and coastal input of pollutants in the form of heavy metals, synthetic chemicals, petroleum hydrocarbons, radionuclides, and other urban wastes. Overfishing is an additional man-induced stress. For example, the present finfish stocks on the U. S. northeast continental shelf are 25% of their virgin biomass while the Peru anchovy fishery has now been reduced

to less than 10% of its peak harvest in the late sixty's.
Elucidation of cause and effect within a perturbation response
of the food web of this highly variable and complex continental
shelf ecosystem is thus a difficult matter.

Physical transport of pollutants, their modification by the
coastal food web, and their transfer to man are problems of
increasing complexity on the continental shelf. After 30 years
of discharge of mercury into the sea, the Minimata neurological
disease of Japan was finally traced to consumption of fish and
shell-fish containing methyl mercuric chloride, while the *Itai
itai* disease is now attributed to high cadmium ingestion. Discharges of chlorinated hydrocarbons, such as DDT off California,
PCB in the Hudson River, New York, and within Escambia Bay, Florida,
mirex in the Gulf of Mexico, and vinyl chloride in the North Sea,
have led to inhibition of photosynthesis, large mortality of
shrimp, and reproductive failure of birds and fish. Oil spills,
such as in the Santa Barbara and Ekofisk oil fields and from the
tankers *Torrey Canyon, Argo Merchant, Amoco Cadiz, Metula, Florida,*
and *Arrow,* constitute an estimated annual input of 2 million metric
tons of petroleum to the continental shelves with an unresolved
ecological impact; another 2 million tons of petrochemicals is
added each year from river and sewer runoff (*Goldberg,* 1976).
Fission and neutron-activation products of coastal reactors such
as San Onofre, Hanford, and Windscale are concentrated in marine
food chains with, for example ^{137}Cesium found in muscle tissue
of fish, ^{106}Ruthenium in seaweed, ^{65}Zinc in oysters, and
^{144}Cerium in phytoplankton; their somatic and genetic effects
on man are unknown. Disposal of dissolved and floatable material
from New York City has been implicated as a possible factor in
both a $60 million shellfish loss off New Jersey and the closure
of Long Island beaches during the summer of 1976 (*Swanson et al.,*
1978). One approach to quantitative assessment of the above
pollutant impacts is construction of simulation models of the
coastal food web in a systems analysis of the continental shelf
(*Walsh,* 1972).

A research goal of the Atlantic Coastal Ecosystem (ACE)
program at Brookhaven National Laboratory (BNL) is to link
models of varying complexity in 1) a heuristic analysis for
design criteria of future experiments on the continental shelf
and 2) a management analysis as a credible tool for planning
societal responses to perturbation events. The management
computer models require at least two criteria:

A) They should have three operational modes:

    1) A climatological or statistical mode for use in
       ecological assessments similar to those of present
       environmental impact statements,

2) A dynamic mode for analysis of individual past events,
    and

3) A forecast mode to guide human response to real time
    events.

B) They should be validated against available data:

1) To establish confidence limits on the computations,
    and

2) To evaluate requirements for inclusion of additional
    sophistication drawn from the heuristic models.

Both the heuristic ecosystem research models and the management
models require a transport field to separate the conservative and
non-conservative properties of the spatially and temporally inhomo-
geneous continental shelf (*Walsh*, 1976; *Walsh et al.*, 1978). It
is our present purpose to describe the use of the three modes of
a management model in actual case studies of the transport and
dispersal of two types of pollutants, oil and urban waste, within
the New York Bight (Fig. 1). As a natural extension of this work,
biological and chemical interactions can be included within these
models (*Walsh*, 1975; *Walsh*, 1977; *Walsh et al.*, 1979a).

METHODS

Our present perturbation analyses consist of 1) a circulation
submodel, 2) a simulated trajectory of a pollutant particle within
the flow field, and 3) a time dependent wind input for each case
of the model. We describe the first two methods in this section
and the third wind input with respect to separate climatological,
dynamic, and forecast uses of the model in the results section.

The Shallow Water Circulation Model

The currents on the New York continental shelf were computed
using a linearized, single-vertical-layer, barotropic, free surface model on a two-dimensional horizontal grid. This model
simulates a time-dependent, spatially homogeneous ocean, driven
by the wind stress, the bottom friction and topography, the
Coriolis force, and the geostrophic pressure gradient. The
vertically integrated equations of motion and continuity are:

$$\frac{\partial M}{\partial t} - fN = -gH\frac{\partial \zeta}{\partial x} + \tau^x - \tau^{bx} \tag{1}$$

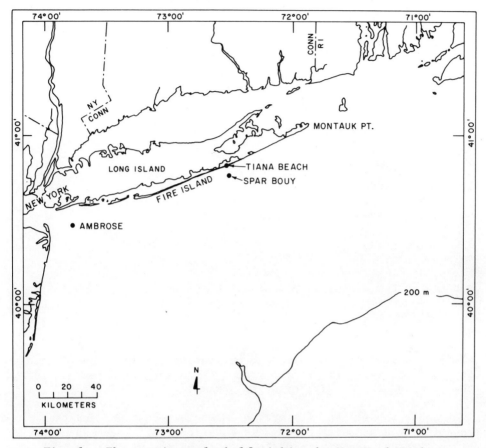

Fig. 1.   The continental shelf within the New York Bight.

$$\frac{\partial N}{\partial t} + fM = -gH\frac{\partial \zeta}{\partial y} + \tau^y - \tau^{by} \tag{2}$$

$$\frac{\partial \zeta}{\partial t} + \frac{\partial M}{\partial x} + \frac{\partial N}{\partial y} = 0 \tag{3}$$

where

$$M = \int_{-H}^{\zeta} u dz, \quad N = \int_{-H}^{\zeta} v dz$$

$+u$ = eastward current speed

$+v$ = northward current speed

H = depth (cm)

g = gravity = 986 cm sec^{-2}

ζ = free surface elevation in cm

$\tau^{x,y}$ = wind stress in E-W, N-S direction in dynes cm^{-2}

$\tau^{bx,by}$ = bottom stress in dynes cm^{-2}

f = Coriolis parameter = 10^{-4} sec^{-1}

The time-dependent wind stress is computed by taking the square of the wind speed at 10 m above sea level and multiplying by the air density and a drag coefficient, namely

$$\tau_{air} = (CD) \; (\rho_{air}) \; (V_a^2) \simeq 1.5 \; x \; 10^{-6} \; V_a^2 \tag{4}$$

and the bottom stress by a similar procedure

$$\tau_{sea} = (CD) \; (\rho_{water}) \; (V_s^2) \simeq 2 \; x \; 10^{-3} \; V_s^2 \tag{5}$$

where V_a and V_s are the wind and current speeds in cm sec^{-1}, and the τ^x in Equation (1) is the vector component of the stress in the x, or eastward, direction, for example.

Equations (1) and (2) are the momentum equations and the equation of continuity of mass is Equation (3). These equations are solved using a finite difference technique similar to that of *Platzman* (1972). The above state variables are discretized in space on a grid known as a Richardson lattice with a spacing ($\Delta x = \Delta y$) of 5.67 km over the New York Bight (Cape May, New Jersey, to Montauk Point, New York) in a staggered manner. Although the velocities $M_{i,j}$, $N_{i,j}$ and the free surface elevation and depth $(\zeta,H)_{i,j}$ are thus placed at different physical locations on the grid, they have the same subscripts for compactness of computer storage within our CDC 6600-7600 system at BNL.

The time integration of equations (1)-(3) is carried out by the following forward difference scheme:

$$M_{i,j}^{n+1} = M_{i,j}^n + 0.25 \; (\Delta t) \; (N_{i,j}^n + N_{i,j+1}^n + N_{i-1,j}^n + N_{i-1,j+1}^n)$$

$$-\frac{g(\Delta t)}{2\Delta X} \; (H_{i,j} + H_{i-1,j}) \; (\zeta_{i,j}^n - \zeta_{i-1,j}^n)$$

$$+ \Delta t (\tau_{i,j}^x - \tau_{i,j}^{-bx}) \quad , \tag{4}$$

$$N_{i,j}^{n+1} = N_{i,j}^{n} - 0.25 \quad (\Delta t) \quad (M_{i,j}^{n+1} + M_{i+1,j}^{n+1} + M_{i,j-1}^{n+1} + M_{i+1,j-1}^{n+1})$$

$$- \frac{g(\Delta t)}{2 \Delta X} (H_{i,j} + H_{i,j-1}) \quad (\zeta_{i,j}^{n} - \zeta_{i,j-1}^{n})$$

$$+ \Delta t(\tau_{i,j}^{y} - \bar{\tau}_{i,j}^{by}) \quad \text{and} \tag{5}$$

$$\zeta_{i,j}^{n+1} = \zeta_{i,j}^{n} - \frac{\Delta t}{\Delta X} (M_{i+1,j}^{n+1} - M_{i,j}^{n+1}) - \frac{\Delta t}{\Delta y} (N_{i,j+1}^{n+1} - N_{i,j}^{n+1}) \tag{6}$$

where

$$\bar{\tau}_{i,j}^{bx} = \frac{2\alpha C D \rho}{(H_{i,j} + H_{i-1,j})} M_{i,j}^{n} \tag{7}$$

with a similar expression for $\bar{\tau}_{i,j}^{by}$ in which $\alpha = 10$ cm sec^{-1} is an assumed velocity used to linearize the bottom stress in order to simplify the computation. In the numerical solution, (4) is computed for the entire spatial grid, then (5) and then (6); that is only one time level is stored for any variable at each grid point. The time step for the difference scheme must satisfy the requirement that $\Delta t < \Delta X / \sqrt{2gH_{max}}$ where H_{max} was set to 200 meters.

The boundaries of the model contain both coastline and connections to the adjacent ocean. The land boundary is drawn in a north-south and east-west stair-cased manner so that the coastline passes through either an M or an N grid point where the normal component of the velocity is set to zero. At the seaward boundaries, the free surface elevation was also set to zero. The physical location of the seaward boundaries on the continental shelf varies according to the scale of the problem, but the edge of the shelf is set at the 200 m depth contour and no computations are made beyond the shelf break.

The complexity of the physical processes that govern the circulation in the New York Bight is discussed by *Beardsley et al.* (1976). They include tides, estuarine exchanges, local wind forcing, topographic steering, input of low salinity water from the upstream Gulf of Maine, an alongshore pressure gradient, and shelf-slope water exchange. It is doubtful that a single physical model can be credibly used to predict all of these circulation details. However, the present simple model does include the effects of local wind and bottom topography (Fig. 2). Within the model, for example, the bend in the coastline at New York frequently causes a clockwise gyre in the inner Bight and the

LONGITUDE

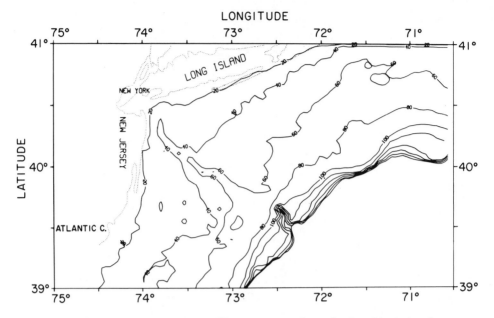

Fig. 2.   Contour map at 20 m intervals of the digitized
New York Bight bathymetry used in the shallow water model.

Hudson Canyon typically forces the model flow to be up or down
the Canyon; these features do correspond to the observed circulation (*Hansen*, 1977).

The present circulation model is most applicable during
winter in the coastal region when the currents tend to be barotropic, *i.e.*, to have the same speed and direction at depth.
The performance of the model in the near shore zone is shown in
Fig. 3 where the computed alongshore current at one of the grid
points is compared against observations at a spar buoy (Fig. 1)
south of Long Island in February and March, 1976, (see *Scott
and Csanady* (1976) for details of the current measurements).
Data from only one wind station (Tiana Beach; see Fig. 1) was
used to drive the model and this input is not valid during fast
moving storms as shown for example for March 6 in Fig. 4.   In
this case, the model was computing 5 cm sec^{-1} eastward flow
rather than the observed 10 cm sec^{-1} westward flow. We believe
that this disagreement is caused by the complex geostrophic forces
set up by a cold front. (Note that the storm on March 10 was
modeled quite well.)   Furthermore, the small high-frequency peaks
in Fig. 3 are the diurnal tides which were not completely filtered
out of the observations. Such difficiencies mean that, on one hand,
a larger-scale wind field must be used in computing all details of
the general shelf circulation but that, of course, on the other,

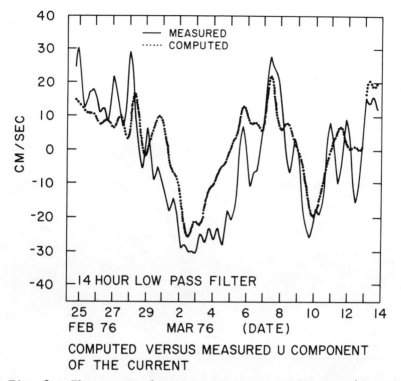

COMPUTED VERSUS MEASURED U COMPONENT
OF THE CURRENT

Fig. 3.   The computed currents at one grid point (dotted
line) versus the observed current at the BNL spar buoy south of
Long Island.

the tides must be considered close to the beach. The importance
of correctly predicting the wind forcing is discussed in more
detail with respect to the third problem involving the forecast
mode of the model. The comparison of computed and observed
currents (Fig. 3) does indicate, however, that there is fidelity
in the computed barotropic currents, usually within 10 cm sec^{-1},
except in certain complex meteorological situations.

The Particle Trajectory Model

The particle trajectory model is derived from the state
equation for the distribution of a non-conservative variable
in the sea, with turbulent diffusion of material parameterized
using "K theory" or eddy coefficients. In a cartesian coordinate
system, the three-dimensional form of this equation for the mean
concentration of material C can be writen as:

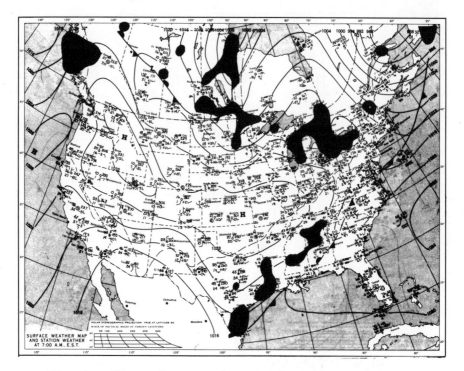

Fig. 4. The surface weather map for the continental U. S.
at 0700 on March 6, 1976, with a complex wind field over the New
York Bight.

$$\frac{\partial C}{\partial t} + u\frac{\partial C}{\partial x} + v\frac{\partial C}{\partial y} + w\frac{\partial C}{\partial z} = \frac{\partial}{\partial x}(K_x\,\frac{\partial C}{\partial x}) + \frac{\partial}{\partial y}(K_y\,\frac{\partial C}{\partial y})$$

$$+ \frac{\partial}{\partial z}(K_z\,\frac{\partial C}{\partial z}) + Q + S\ ,\qquad\qquad (8)$$

where t is time, u, v, and w are the velocity components along
the respective x, y, z coordinate axes, and K_x, K_y, and K_z are
the turbulent eddy diffusivities. The terms Q and S represent
sources and sinks which may include chemical and biological processes (see *Walsh* (1977) for a review of this type of biological
model). The vertically integrated current velocities computed
from the above circulation model are entered as u and v in the
two-dimensional form of Equation (8), while, if desired, the
weathering of oil, uptake of radionuclide, or dosage response
to synthetic chemicals can be included in the Q and S terms.

   Assuming an incompressible fluid, the two dimensional form
of Equation (8) can then be written as:

$$\frac{\partial C}{\partial t} + \frac{\partial}{\partial x} \ (u+u_D) \ C + \frac{\partial}{\partial y} \ (v+v_D) \ C = Q + S \ , \tag{9}$$

with $u_D = - \frac{K_x}{C} \frac{\partial C}{\partial x}$, etc., representing turbulent flux velocity and
$u+u_D$, etc., the total equivalent transport velocity. The eddy
diffusivities in Equation (9) are usually of the order of 10^6
$cm^2 \ sec^{-1}$.

   For the solution of Equation (9), a particle-in-cell (PIC)
numerical technique has been used previously (*Dieterle and Tingle*,
1976) in a study of the dispersal of conservative atmospheric
pollutants within the coastal sea-breeze circulation off Long
Island. The PIC method was originally developed by *Harlow* (1963)
for the solution of the hydrodynamical equations and first adapted
by *Sklarew et al.* (1971) for environmental problems. The method
is an accurate general technique for the numerical solution of
partial differential equations such as Equation (9). It overcomes
the problem of phase and amplitude errors within standard definite
difference techniques for numerical approximation of the advective
process. These errors may be large if the spatial distribution of
the advected material is highly peaked (*Molenkamp*, 1968). In
addition, an integral part of the PIC method is the computation
of trajectories with the history of any parcel of material available during the simulation.

   From our point of view, the PIC method is preferable to other
techniques which are of comparable accuracy, such as the moment
method suggested by *Eagan and Mahoney* (1972) or the pseudospectral
method of *Christensen and Prahm* (1976), because it is a hybrid
of Lagrangian and Eulerian techniques. Recall that, in a Lagrangian
formulation, the concentration of material refers to a coordinate
system which is fixed in relation to the mean flow of the fluid,
whereas in an Eulerian formulation the concentration of material
refers to a coordinate system which is fixed in space through
which the fluid is moving. Most previous marine ecosystem models
were of the Eulerian type (*Walsh*, 1977), where patchiness and
trajectories of pollutants or organisms could not be explicitly
considered.

   In the present simulations, the physical space of the continental shelf is first divided into a number of cells to make a
fixed Eulerian grid on which the velocity fields and external
sources of pollutant are represented, and on which the turbulent
fluxes can be evaluated. The Eulerian grid for this purpose is
the same grid as that of the circulation model described above.
The spatial distribution of the pollutant is then represented as

a discrete number of Lagrangian particles which undergo transport
from cell to cell as they are moved by the velocity field.

The numerical intergration of Equation (9) is performed in
two steps. In the Lagrangian step, the new particle position at
time $t+\Delta t$ are calculated using explicit forward differencing:

$$x(t+\Delta t) = x(t) + u_p \Delta t,$$

$$y(t+\Delta t) = y(t) + v_p \Delta t, \tag{10}$$

where u_p and v_p are the velocities of one particle as determined
from the Eulerian grid calculation of the circulation model and
interpolated to the particle position. The interpolation procedure consists of a linear area (A_i) weighting scheme in which
each particle is arbitrarily assigned the same area as the
Eulerian grid cells. The depth-averaged velocity at the position
of the particle is then obtained as the weighted sum of the
velocities in the neighboring Eulerian cells, where the weights
correspond to the amount of overlap of the particle within the
respective cells, $i.e.$ in the two dimensions of the depth-
integrated model,

$$u_p = \frac{u_1 A_1 + u_2 A_2 + u_3 A_3 + u_4 A_4}{\Delta x \Delta y} \tag{11}$$

In (11), u_1 through u_4 represent the u velocity components which
have been averaged to the ζ, H grid of the above circulation model.
Also, although there is no computational stability requirement for
the time step Δt in (10), it would be undesirable for a particle
to be moved too far from the grid location of the velocities doing
the advection. Thus a time increment which restricts particle
movement to no more than one half of a grid cell per time step
is imposed.

Within the Eulerian part of the computation, a concentration
field of C on the Eulerian grid can be determined each time step
from the particle positions, using the interpolation procedure
outlined above. The turbulent velocities are then evaluated
with an eddy coefficient and the local gradient of particles
following $Lange$ (1973). For present purposes, however, the
following simplifying assumptions have been made:

(1) All pollutants are treated as conservative quantities,
     $i.e.$ there are no biological or chemical sources and
     sinks to modify the state of the pollutant.

(2) The turbulent diffusion of the pollutants is not
     included, and thus K_x and K_y in Equation (9) are
     set to zero.

With these assumptions, the present computation is reduced to a
trajectory calculation of each pollutant particle, in contrast to
tha additional sophistication of the biological models (*Walsh
et al.*, 1979a).

A pollutant discharge to the shelf ecosystem can be represented as either an initial condition of a single, puff-type
release, or as a continuous source. Since the PIC method uses
Lagrangian particles to represent pollutant mass, the discharges
are simulated through the generation of particles at the respective source locations over the grid. A continuous source is
modeled as a succession of individual puffs, and the number of
particles generated per time step is simply

$$M = R\Delta t/C_m$$

where R is the steady release rate and C_m the particle mass.

The boundaries for the particle trajectory calculations
are the same as those of the circulation model. Along the open
ocean, a transmittive boundary condition is used and any particle
which leaves the spatial domain is deleted from the model. Along
the coastline, the option to use either an absorptive or a reflective boundary has been incorporated. For the absorptive boundary,
any pollutant particle which strikes the boundary is frozen and
treated as a hit, whereas for the reflective boundary, the
particles are allowed either to slip along the beach or to
move offshore.

RESULTS

Climatological Use of the Model

Our initial interest in applying the above computer model
to a systems analysis of the coastal zone was stimulated by the
announcement of proposed federal leasing of oil drilling rights
on the northeast Atlantic continental shelf. Our main concern
was the potential impact either of site-specific oil spills in
the shelf ecosystem resulting from blowouts, leaks and accidents
at offshore platforms, or of larger oil spills from pipelines
and tankers, depending upon which mode of transport is to be
used. Spills from some tanker accidents are of considerable
volume, as demonstrated by the recent wreck of the *Argo Merchant*
in the fishing area near Nantucket shoals. Presumably only the
fact that this accident occurred during offshore winds in winter,

when food web productivity is low, prevented a major ecological impact on the open shelf or within an enclosed area, such as the Barnstable marsh.

The complexity and variability of the degradation mechanisms involving the fate of oil in the marine environment are not included in the present model, however. Basically, when oil is introduced into the sea, three general processes occur. The oil reacts chemically and physically with the air, it spreads on the water, and it reacts chemically, physically and biologically beneath the surface (*Ahearn*, 1974; *Nelson-Smith*, 1973). Before the biological response and a fine resolution of the above oil degradation processes can be added to an ecosystem model, confidence must first be established in the transport and diffusion submodels. The surface dispersion data for testing the trajectory of hypothetical oil spills on the continental shelf south of Long Island consist of a series of drift card releases in 1973-74 by aircraft in the shipping lanes and at proposed drilling sites (*Hardy et al.*, 1976). When washed ashore, the cards were retrieved by beach walkers who filled in the required information and returned them for data processing as discussed by *Hardy et al.* (1975). Therefore, the climatological computations were based on a similar scenario of a series of hypothetical winter oil spills that either washed ashore at some point along the coast or drifted into the open ocean.

The initial spread of an individual oil spill and the details involved in the beaching process are not specifically considered in this analysis. Rather, simulated trajectories of the oil particles are computed with the assumption that a spill had beached if a particle came within 5 km of shore. Even these simplifications require further analysis as there are several factors involved in the transport of oil on the surface of the sea and in the water column. The travel of an oil spill appears to be directly related to the local wind (*James*, 1966; *Schwartzburg*, 1971). The wind stress on the upper ocean causes a thin layer at the sea surface to move almost twice as fast as the bulk surface water below, in a direction closely parallel to that of the wind (*Sverdrup et al.*, 1942). The ratio of the sea surface drift to the wind is called the wind factor, defined as $u = kW$, in which u is the wind-induced surface current, W is the wind speed 10 m above sea level, and k is the wind factor. Numerous calculations have been made for the wind factor (see Table 1), but some were not related to just the upper surface layers of the sea. More recent estimates range from 3.3% to 4.3% of the wind speed for the upper centimeter of the sea surface.

Such a wind factor does not include either transport of oil due to wind-driven currents, which are highly responsive to changing wind stress over the continental shelf, or transport of oil

TABLE 1. Estimates of the wind factor in oil transport (after *Hardy et al.*, 1976).

Date	Wind Factor (%)	Method of Determination	Water Column Depth Sampled
1914	1.44	ship's drift	ship's draft
1905	1.85	current measurement	surface to 5 m depth
1935	2.53	theory of hydrodynamics	"surface layer"
1954	2.9	drifting buoys	surface to 1 m depth
1956	3.3	drift cards	thin surface layer
1967	3.3–3.4	floating oil	thin surface layer
1953	3.6	paraffin flakes	thin surface layer
1971	3.7	floating oil in wind tunnel tank	thin surface layer
1974	3.4–3.8	drift cards	thin surface layer
1964	4.3	oil spill	thin surface layer
1975	4.4	drift cards	thin surface layer

due to residual currents when the wind is light. In the latter
case, wind factors can be anomalously large. Since our circulation model includes time-dependent, wind-driven currents which
can be modified to include net background drift, we assumed that
the surface oil moves as the vector sum of the vertically integrated current speed and 3% of the wind speed, *i.e.*

$$\vec{V}_{oil} = \vec{V}_{current} + 0.03 \; \vec{V}_{wind}, \tag{12}$$

and that the path of oil in the water column follows that of the
currents.

Wind input for the climatological study was the twice daily
(0000 and 1200 GMT) upper air observations for the stations
(Wallop Island, Va.; JFK airport, N. Y.; Philadelphia, Pa.; and
Chatham, Mass.) of the U. S. radiosonde network around the New
York Bight from October, 1973, to April, 1974. These winds were
treated as a first estimate of balanced (geostrophic) atmospheric
flow at the 1000 m level. To obtain an estimate of low level
winds and then the stress at the sea surface, the wind profile
between the 1 km and 10 m levels was first assumed to vary in
accordance with the Ekman equations,

$$f(v-v_g) + \frac{\partial}{\partial z} \left(K \frac{\partial u}{\partial z} \right) = 0 \tag{13}$$

$$-f(u-u_g) + \frac{\partial}{\partial z} \left(K \frac{\partial v}{\partial z} \right) = 0 \tag{14}$$

where u and v are the horizontal wind components, and u_g and v_g
are the geostrophic, 1 km wind components. The eddy viscosity,
K, was calculated as a function of height, z, by (*Shir*, 1973)

$$K = k \, U_* Z e^{-4 \, z/H}, \; H = .45 \, U_*/f;$$

where $k = .35$ is von Karman's constant, and $U_* = (\tau_{air}/\rho_{air})^{\frac{1}{2}}$.
Within the layer below $h = 10$ m, a logarithmic profile of the
wind velocity was assumed,

$$U_h = (u^2 + v^2)^{\frac{1}{2}} = \frac{U_*}{k} \left[ln \left(\frac{h+z_0}{z_0} \right) \right] \tag{15}$$

where the roughness height $z_0 = .01$ cm, and $U = 0$ at z_0. To
couple the Ekman layer above the sea surface with the velocity
at the 10 m level the condition was imposed that the gradient
of U be continuous at this interface with no change in wind
direction, *i.e.*

$$\frac{U(h+\Delta z) - U(h)}{\Delta z} = (\frac{\partial U}{\partial z})_h = \frac{U_*}{kh} . \tag{16}$$

Using the above set of equations, each value for the wind at 10 m and the surface stress, recall Equation (4), is then obtained with the Ekman equations rewritten in complex notation as

$$ifV + \frac{\partial}{\partial z} K \frac{\partial V}{\partial z} = ifV_g \tag{17}$$

where $V = u + iv$, $V_g = u_g + iv_g$, $i = \sqrt{-1}$. The finite difference form of this equation, using centered differences, represents a tridiagonal matrix which was inverted using a Gaussian elimination process. Obtaining U at $z = h+\Delta z$ from this procedure, we use (15) and (16) to obtain a value for U_*, and then (15) to obtain a value for the 10 m wind. Initially we let $V = V_g$, and iterate until the solution converges. The 10 m wind produced from this solution is generally 50-60% of the 1000 m wind, with a "cross-isobaric" angle of 10-15°.

Using the above approach, the wind input to the model was updated every 12 hours of the 180-day period (October, 1973- April, 1974). Between updates, the wind forcing function was linearly interpolated each time step, Δt, of the model. Two types of oil spills were released each 12 hours of the six month period at all 14 stations in the Bight, corresponding to several of the release locations used by *Hardy et al.* (1975) (the tri-angles in Fig. 9). One of the spills represented a surface slick which was tracked as the sum of the current vector and 3% of the wind vector as discussed above. The second spill represented oil in the water column which was tracked following just the current vector. Each set of oil spills was followed for 60 days until the oil particles either came within 5 km of the shoreline or went out to sea. Two trajectories are shown in Fig. 5 for surface slicks released on October 10, 1973. In this case, one spill had washed ashore on eastern Long Island by October 16 whereas the other, released nearby, narrowly skirted the coastline and then drifted out to sea by October 26.

When a spill hit the shore, information was retained on its release position, landing position, travel time, and wind and current history preceding the landing. For example, the number of shore hits of oil from all release locations as a function of the wind history is shown in Fig. 6 for December, 1973, and Fig. 7 for February, 1974. The largest number of hits was recorded following a south through east wind history. Even prolonged southwest winds did not result in as many hits as shown on December 26 and February 19 and 23. Similar results

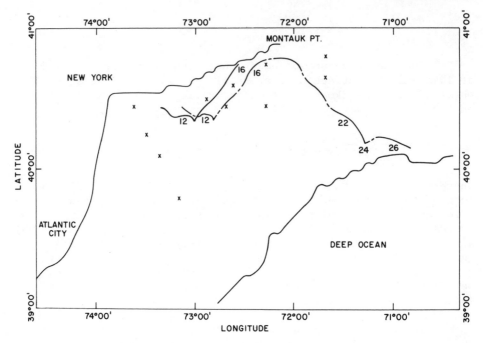

Fig. 5. Surface trajectories of oil particles released from two stations on October 10, 1973, with one impacting the shore on the 16th and another drifting out to sea by the 26th.

were found by *Hardy et al.* (1975), with the majority of drift card strandings on Long Island caused by east and southeast winds. Southwest winds led to stranding of drift cards within the Bight apex.

A similar analysis was also made for each of the release positions. There was a trend for the oil particles to beach eastward of their release positions, as a result of westerly winds followed by a period of strong south or southeast winds. The time-to-shore analysis indicated that, from those release positions which lead to more than 10% beach hits, 50% of the particles hit within 2 days and 95% within 6 days. The surface slick particles from all release positions hit either within 10 days or not at all. In contrast, only two of all the oil particles released in the water column came within 5 km of the shore after 15 days.

The surface slick results for the entire 6-month study are summarized in Fig. 8, with the 14 release stations marked by a "+". In this analysis, the New York-New Jersey coast was divided into nine sections, labeled A through I. The number under a letter (Fig. 8) is the probability that that section of the

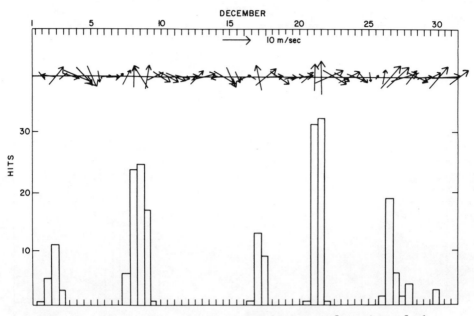

Fig. 6. Shore hits of oil particles as a function of the wind input during December, 1973, with a wind from the south indicated by an arrow pointing to the top of the figure.

shore will be hit by an oil spill. The coast east of the Bight Apex (Sections C, D, and E) had high probabilities of being hit, whereas the probability of a spill reaching the New Jersey shoreline was very small. The recovered drift cards (*Hardy et al.*, 1975) also tended to wash ashore east of their release position, with most found on the north side of the Apex and east of Fire Island to Montauk Point.

The numbers under the release positions of the model (Fig. 8) represent the probability that an oil spill from that location would reach any part of the shore. The model results replicate the drift card observations for the winter of 1974 (Fig. 9), where a beach "hit" was scored independent of the number of drift cards reaching the shore from any release station. As in the model, the data indicated that drift cards released 18 km south of Long Island had a zero probability of recovery within ten days. In contrast, cards released within 13 km of Long Island had a 33% probability of recovery within 10 days. The drift card probabilities also showed no significant change after sixty days, *i.e.* an oil spill within 17 km of the coast would strand either within ten days or not at all.

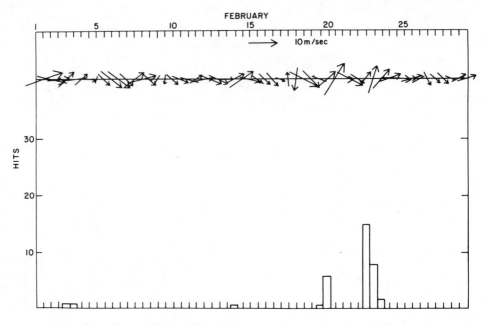

Fig. 7. Shore hits of oil particles as a function of wind during February, 1974.

There are many questions that can be addressed with a model of this type. For example, we computed the probability that a given section of the shore would be hit from a particular release position and the stranding limits for each release position. We could also have computed the probable tracks of a spill from each position in relation to the seasonal impact on marine productivity (*Walsh et al.*, 1978). For example, we used this model to predict successfully the offshore track of the *Argo Merchant* spill. The changing composition of the oil as a function of age and weathering could also be computed if these factors were known. The climatological use of the model can thus be improved readily with further investigation and understanding of the coastal circulation, the oil, the biology, and the weather.

Dynamic Use of the Model

During June, 1976, large quantities of garbage and other diverse debris were washed ashore on the beaches of Fire Island (Fig. 1). The material consisted of tar balls, oil, charred wood, plastic wastes, bits of fingernails, fresh chicken heads, fruit crates, and sewage. The filth forced the closing of more than 100 km of Long Island beaches, threatening the summer tourist industry and raising fears that the clambeds in Great

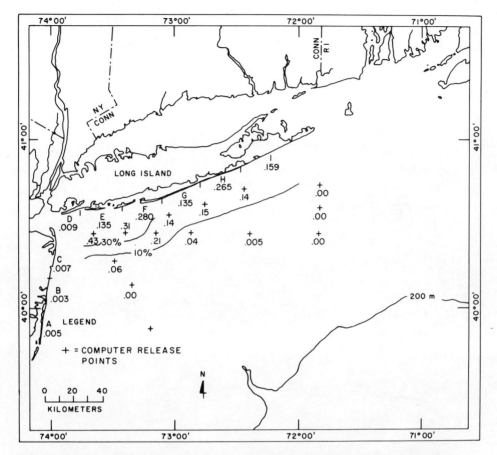

Fig. 8.   Summary of winter oil spill model results with the probability that oil will strand on the coast from several hypo-thetical release positions (the numbers under the +'s).   The positions correspond to the triangles in Figure 9.

South Bay, north of Fire Island, would be infected.   Both of these activities generate multi-million dollar revenues for Nassau and Suffolk Counties of Long Island, and ∿ $20 million was lost during this event to the recreational industry (*Swanson et al.*, 1978).

The waste material, which floated on the ocean surface, could have come from a variety of sources, including sewage outfall pipes, wastes dumped from commerical and pleasure boats, floatables illegally released at the sludge dump site southeast of Ambrose Light Tower, a sewage tank explosion at Bay Park (between New York City and Fire Island) and pier fires on the

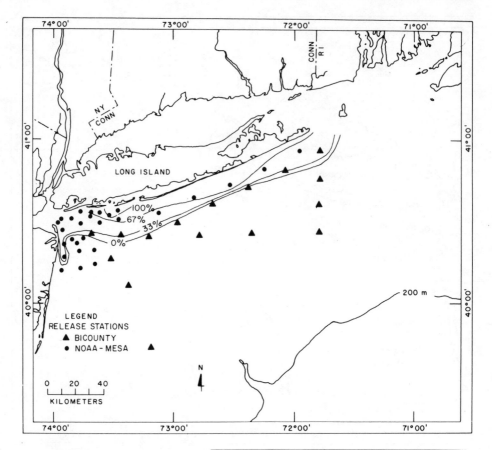

Fig. 9. Per cent probability contours that some fraction of
oil spills will strand within 10 days on Long Island in winter
(January–March), 1974, based on drift card return frequencies
per station (after *Hardy et al.*, 1975).

Hudson River. It is somewhat surprising that a beaching event of
such magnitude had not occurred previously, since much of this
waste dumping is done routinely in the New York Bight and fears
of the "creeping sludge monster" had been raised a few years
previously.

As part of an effort to trace the source, or sources, of
this debris the above model was adapted to address the following
questions:

  Assuming an initial source at sea, what were the landing
  limits (dispersion cones), times, and probabilities of

pollutant beaching associated with a wind switching
from northerly to southerly before the 1976 event?

• Assuming sources from the Hudson River estuary, what
  distribution of material would be expected in the New
  York Bight at the time of the wind reversal and what
  was the subsequent fate of such material?

• Assuming that a pulse of material from the Bay Park
  sewage tank explosion on June 2 was released into the
  ocean, what was the fate of such material?

• Assuming that the spoils from the clean-up operations
  (June 3) of the Bay Park explosion were instead deposited
  at the sewage sludge dump site near Ambrose Light Tower
  (Fig. 1), what was the fate of this material?

The complexity of mechanisms involved in the transport and
fate of this debris could not all be simulated in detail and the
effects of waves, tides, and estuarine discharges were not
explicitly included. Furthermore, when the ocean is stratified
during the summer the flow pattern is different above and below
the thermocline as a result of vertical shear. Our single layer
model could thus not be used to simulate either the lower layer
transport or the movement of material deposited on the ocean
floor. We assumed that the computed currents were representative
only of the upper layer flow, since most of the barotropic response
of the flow field to summer winds is above the thermocline. With
these caveats, the dynamic use of the model is presented in terms
of the spatial and temporal resolution required to distinguish
the sequence of events leading to the beach closures (*Swanson
et al.*, 1978).

The 10 m hourly winds from the BNL tower at Tiana Beach
(Fig. 1) were used both to drive the circulation model and to
compute the surface transport of floatables at 3% of the wind
speed. The time history of these wind data is plotted as
daily transport (km day^{-1}) in the progressive vector diagram
of Fig. 10 which shows a persistent southwesterly wind after
June 6. Within the time dependent flow field, 45 sources of
debris were continuously released south of Long Island, down
the New Jersey shore, and in the New York Bight Apex. The
time histories, after release, of the debris trajectories,
moving as the vector sum of the current speed and 3% of the
wind speed, were recorded every four hours from 0000 EST June 1
to 2400 June 28. These numerical experiments resulted in a data
base of the trajectories of 7560 debris particles. A hit was
counted if a particle came within 2 km of the shore because we
assumed that the material would not spread as much as oil. Once
a debris particle washed ashore, it was frozen at that location,

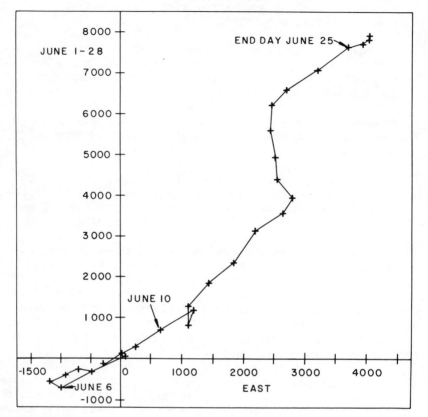

Fig. 10. Total June, 1976, wind transport for each day as observed at 10 m on Tiana Beach.

i.e., the particles were not allowed to float along the coast. Finally, particles were not allowed to hit the New Jersey shore as far as 40 km south of Sandy Hook, because it was assumed that the Hudson River plume would prevent offshore floatables from reaching this part of the coast.

On June 13, the computed currents (Fig. 11) were still flowing to the southwest except for a clockwise gyre in the Bight Apex, which is a typical summer circulation pattern in the New York Bight. However, by June 15, the flow pattern had reversed (Fig. 12), with the model currents along New Jersey flowing north and those along the south shore of Long Island flowing to the east. The observed currents at 5 locations in the New York Bight had also reversed by this time (*Mayer et al.*, 1979). This flow reversal is thought to be an essential part of

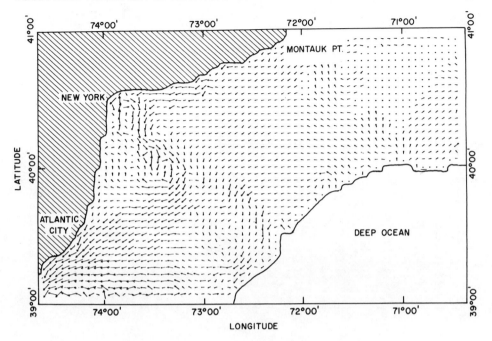

Fig. 11.   Computed barotropic currents on June 13, 1976,
showing a southwest flow with a clockwise gyre near New York.

a sequence of events also leading to anoxia along the New Jersey
coast in July, 1976, (*Walsh et al.*, 1979b).   Examples of the
different debris trajectories within these flow fields are shown
in Figs. 13 and 14.   Particles floating within the Apex at 0000
EST June 5 circulated around to reach Fire Island by about June
11, although some material missed Long Island.   The picture is
quite different by June 21, when particles close to Long Island
reached the shore within a day and particles as far south as 80
km hit Fire Island within 5 days.

     Sea Source.   The results of the dynamic use of the model are
summarized in Fig. 15 with the times shown as hours of travel time
from 0000 EST June 1.   Anything floating in the shaded area from
June 1 to June 24 would presumably have hit Long Island.   The travel
times ranged from less than one to more than ten days, but 90% of
the released debris beached within three days.   The release times
of those particles that hit Long Island also varied over a wide
range.   All particles floating in a large region southeast of
New York City on June 3 and 4 reached Fire Island within a week.
In contrast, no debris released southwest of Montauk Point before
June 12 hit Long Island, but 71% did after that date.

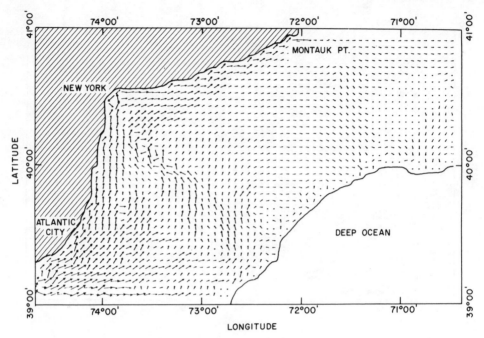

Fig. 12.   Computed barotropic currents for June 15, 1976, with the flow changed to an easterly direction.

        All beach hits for each 12-hour period of the model are
plotted against the 12-hour resultant winds in Fig. 16. Stranded
debris prior to June 6 was from particles released near the New
Jersey coast and which hit the southern New Jersey beaches. The
first Long Island beaching began in the model around June 8 and
was from those particles released after June 5. The beach hits
in the model after June 11 were from these particles released
earlier, from June 1 to June 4. A third wave of material washed
ashore after June 19 (Fig. 16) and by June 26 little debris was
reaching the beaches, as was true of observations (*Swanson et al.*,
1977). The predicted increase of hits after June 12 is very
important for model validation, because the debris was first
noticed on June 12, none was found on June 13, and large amounts
actually beached on June 14 (*Swanson et al.*, 1977). Since no
debris was observed on the beaches prior to June 12, these results
suggest that the initial closure of the beaches was probably due
to material floating in the Apex during the first few days of June.

        Sludge Dump Site. All debris particles released in the
vicinity of the sludge dump site from June 2 through June 10
landed on the western half of the south shore of Fire Island.
Material released during June 3 landed on June 10 or 11, however,

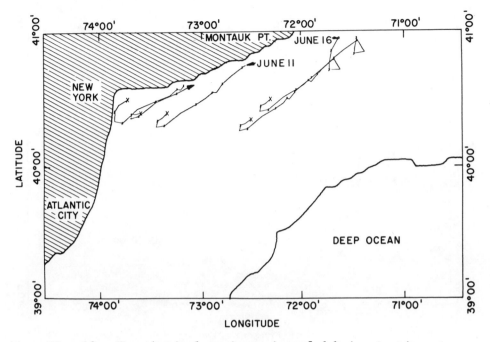

Fig. 13. Hypothetical trajectories of debris starting at
0000 EST on June 5, 1976, for selected release points.

while debris released from June 4 to June 7 arrived earlier,
mostly before June 9. Particles released after June 11 landed
on beaches nearer New York City, *i.e.* to the west of Fire Island,
within 1 to 3 days. The model results thus suggest that the
initial beach closing on Long Island was not caused by debris
from the dump site.

The Bay Park Explosion. Possible estuarine paths of debris
from East Rockaway Inlet, near the exploded sewage tank, into
the ocean could not be explicitly tracked on the grid scale of
this model. Instead, simulated releases were made at 20 points
in a 300 km^2 area of the ocean south of the Inlet. All particles
released in this area on June 3 and 4 ended up scattered along
Fire Island in a week or so. These particles were first computed
to be floating down the New Jersey coast for a couple of days
and then, after the southerly wind shift, the debris took 5 or
6 days to reach Fire Island. The model results imply that, if
floatables from the Bay Park explosion had entered the ocean,
they could have contributed to the beaching problem farther
east on Long Island.

The Hudson Estuary. This part of the analysis suggests
that the estuarine debris material released near Ambrose Light

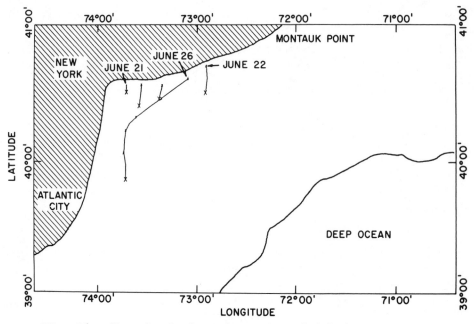

Fig. 14. Hypothetical trajectories of debris starting at
0000 EST, June 21, 1976.

Tower from June 1 through June 6 was scattered down the New
Jersey coast and dispersed within a large area of the Bight
Apex by the time of the wind reversal on June 6. In particular,
debris particles, released on June 3 and 4, also reached Fire
Island in a week or so. Estuarine particles released after this
time hit the beaches closer to New York City, however. Since
these beaches were mostly unaffected by the pollution episode,
the continuous Hudson River debris sources in June were presumably not an important part of the beaching episode. High river
runoff in May, 1976, (*Walsh et al.*, 1979b), however, could have
flushed urban wastes into the area of the sea sources (Fig. 15).
Our model results thus indicate that either the historical (May)
sea sources of waste and/or the sewage tank explosion rather than
the recent (June) dump site and the Hudson River sources (except
for pier fires on June 2), might be possible causes of the fouling
of Long island beaches during the wind reversal of June, 1976.

Fourteen years of hourly wind data from the 108 m Experimental Tower at BNL were analyzed (Table 2) for persistence to
see if there was anything particularly rare about the wind
forcing during June, 1976. Briefly, the wind situation prior
to June 14, 1976, was not an infrequent one. However, during
the period June 13-26, 1976, the winds had a steadiness that had

TABLE 2. Southwest wind events with 12-day steadiness of the stress greater than 0.7.

Year	Events	Month	Direction
1960	0		
1961	0		
1964	0		
1965	0		
1966	0		
1967	4	June	$209^\circ - 213^\circ$
1968	0		
1970	0		
1971	1	July	212°
1972	4	July	$209^\circ - 211^\circ$
1973	0		
1974	0		
1975	4	July	$195^\circ - 203^\circ$
1976	2	June	$211^\circ - 215^\circ$

occurred only four times since 1960. Frequency distributions of

1) direction vs. the steadiness of the wind speed,

2) direction vs. the wind speed,

3) direction vs. the steadiness of the stress,

4) direction vs. stress

were calculated for 3, 6, 12, 18, and 24-day averages. The steadiness of the wind speed (recall Equation (4) for the stress computation) is the ratio of the vector wind and the wind speed

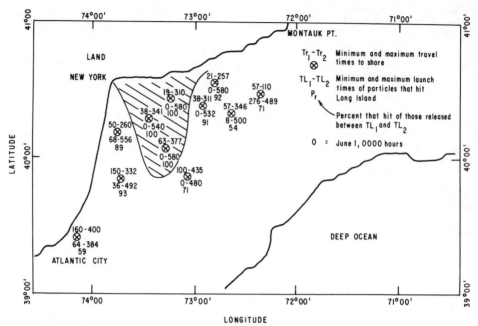

Fig. 15. A statistical summary of all trajectories for 11 of the 45 release points during June, 1976.

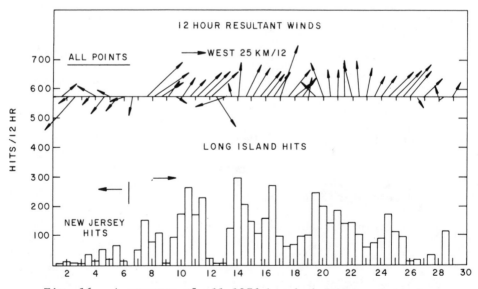

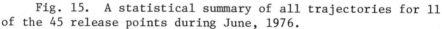

Fig. 16. A summary of all 1976 beached debris plotted against the wind vector with a flow from the south indicated by a vector pointing to the top of the figure.

summed for each hour of the period as $R = \overrightarrow{(V)}/\overline{|V|}$ and then
transformed by the equation $S = (2 \ arsin \ (R))/\pi$ in order to
give enhanced resolution near unity ($Singer$, 1967). If the
vectors are random, the steadiness value is zero, and if there
is no change its value is one.

The winds in the New York Bight are generally south-westerly
over 60% of the time in the summer and a six-day steadiness of up
to 0.7 is not unusual. The results for June, 1976, show that the
steadiness did not reach a high value until June 18 and then it
remained high through the 25th. The 12-day steadiness is shown
in Table 2 for similar events in the past. It is interesting
that July, 1975, was "worse" than June, 1976. However, we are
not aware of any pollutant stranding on the beaches then, or for
the other three cases during 1967, 1971, and 1972.

During July, 1976, anoxia was observed along most of the
New Jersey coast and a $60 million dollar shellfish loss ensued
($Walsh \ et \ al.$, 1979b). One of the contributing factors is
thought to have been the June wind and subsequent current
reversal, which could have retained 1976 aphotic populations
of the dinoflagellate $Ceratium \ tripos$ within the New York Bight
Apex. Their unusual abundance and respiratory oxygen demand
below the pycnocline that year was sufficient to generate the
anoxia off New Jersey when these algae were no longer normally
flushed southward. These dinoflagellates were also abundant
before the summer of 1974, a year of no summer southwest wind
events (Table 2), while they were scarce in 1975, a year of
4 summer events. Anoxia of much less spatial extent occurred
in the Bight Apex during 1974, moreover, and none in 1975. We
conclude that, given an initial spring condition of an abundant
dinoflagellate population, the anoxia off New Jersey and the
beach fouling off Long Island in 1976 are both linked to the
unusual June, 1976, wind forcing (Table 2).

Interannual changes in spring abundance of $C. \ tripos$, preceding the different sequence of summer wind events each year
(Table 2), can be linked to changes in timing of the normal
phytoplankton species succession in relation to the varying
depth of the spring thermocline ($Walsh \ et \ al.$, 1979b). Interannual variations of summer abundance of floatables in the New
York Bight is more perplexing, however, unless a large pulsed
input can be detected in any one year above the background
pollution. Early and high river runoff in 1976 could have
placed an unusual waste input within the New York Bight ($Swanson$
$et \ al.$, 1978). It is also possible that the Bay Park sewage tank
explosion was such an input; a model of greater spatial resolution and physical sophistication might resolve these questions.
The uncertainties of the present calculations prohibit exclusion

of the Hudson River pier fires, however, as another event source
which could also have contributed to the closing of Long Island
beaches. Additional circulation studies are required to confirm
that the source of contaminants was from the general area outlined
in Fig. 15, but the present simulation results are sufficiently
encouraging to suggest that such a dynamic perturbation analysis
can be used with increasing credibility in the future.

Forecast Use of the Model

On 11 February 1977, we were informed by the NOAA-MESA New
York Bight Project that oil from a grounded barge in the Hudson
River (carrying 400,000 gallons) might impact the New York Bight
Apex. Oil had been observed on the Rockaway side of the New
York Harbor, but it was not known where or at what rate it
might be entering the Apex. The winds were from the southwest,
and, from our climatological use of the model, we knew there was
a high probability that any oil within the Apex would wash ashore
on the Long Island beaches. To provide daily forecasts of this
potential oil hazard, the computer model was initiated by
immediate release of nine simulated oil spills in a 150 km^2
area to the east and south of Ambrose Light Tower. It was
hoped that the "correct" oil spill track would then be updated
for the forecast mode of the model after a helicopter located
the actual position of the oil spill in the apex. Unfortunately,
a helicopter was not available for validation observations and
the "nine spill" procedure was followed throughout this study
with releases every 6 hours. Because we were unable to reduce
the uncertainty of the initial condition of the oil spill, the
forecast product was of the format: "If the oil is at Ambrose
Light Tower it will reach Long Beach in twelve hours; if it is
15 km SE of Ambrose, then it should remain at sea." Despite
this initilization problem, however, we were able to predict
correctly the time and location of the beaching of this oil
spill.

The other inputs required for this use of the model were
the observed and forecast winds from Ambrose Light tower in the
New York Bight Apex. These data were obtained by telephone from
the NOAA Weather Service Forecast Office in New York City. The
observed winds are recorded every three hours and kept for about
three days. The forecast winds are for the next 42 hours at
6-hour intervals and are updated each 12 hours (using 0000Z
and 1200Z meteorological data). However, the forecasts come
over the teletype and are not available until about 9 hours
after the observed data. Therefore, the model's forecast
procedure was:

1)   Call WSFO in New York City for the observed winds over
     the last 24 hours and for the forecast winds (about
     1430 EST each day);

2)   Run the model and interpret the results; and

3)   Call MESA and discuss the updated oil hazard (about
     1600 EST);

The currents of the model were "spun-up" for two days prior
to a forecast using the observed winds. The following day's forecast would then be made by restarting the current model using both
updated and forecast winds in order to keep the model running in
as real time as possible. The effects of tides and the Hudson
estuary on the currents were not included. The surface oil was
again assumed to move as the vector sum of the current speed and
3% of the wind speed. The transport of oil within the water
column was also computed, but the results indicated no stranding
of oil from these sources. Oil was assumed to beach if a surface
particle came within 3 km of the shore, and the slicks were then
allowed to move either along shore or offshore depending upon
the winds and currents.

The most important part of each predicted oil trajectory
was the wind forecast, since it is the major forcing function of
this case of the model. Examples of two successive 24-hour wind
forecasts for February 13 and 14 are shown as progressive vector
diagrams in Fig. 17B and 17C, compared to the actual wind observations for the entire week of February 11-18 (Fig. 17A). The
wind has been multiplied by 3% to show its effect on the surface
transport of oil. Forecast #1 (Fig. 17B) projected an oil
transport of 20 km east and 10 km north from 0100 on 13 February
to 0100 on the 14th (the part of the PVD between the second and
third squares). The observed wind (Fig. 17A) instead indicated
a transport of 10 km east and 5 km north (this is the part of
the PVD between the second and third squares of Fig. 17A).
Similar errors are shown in Forecast #2 of the second day
from 0100 on February 14 to 0100 on the 15th. The effect of
the wind forecast error on the computed currents is shown in
Fig. 18. Although the computed current vectors show the same
pattern under either wind, the maximum difference in speed
between the forecast (Fig. 18A) and observed (Fig. 18B)
wind forcing is about 13 cm sec^{-1}.

These calculations suggest that the average current speed
could be in error by 6 cm sec^{-1} because of the wind forecast
error. If we add this to the wind transport error, then the
predicted position of the oil from Forecast #2 is about 15 km
east and 5 km south of where the same computation with observed
wind data would have placed the spill. This discrepancy in

distance is about the length of Long Beach (Fig. 19) and demonstrates the fundamental importance of temporal and spatial
resolution in this type of modeling. For example, to make a
48 hr wind forecast for the northeast continental shelf, the
weather patterns of the whole United States (Fig. 4) must be
considered. Because of computer storage and time constraints,
the weather forecast models must thus utilize a coarse spatial
grid pattern in order to cover the North American continent.
Such a coarse grid sacrifices more intense spatial resolution
within the New York Bight region. Recall that wind input at
one grid point, under certain meteorological conditions, was
insufficient to predict currents south of Long Island accurately
(Fig. 3). Thus the uncertainty in the initial location of the
oil spill and in the wind forecast both placed constraints on
the reliability of our own forecast of the oil spill trajectory.

The first model forecast for MESA was made on 12 February,
1977, using observed winds from 1300 EST on 9 February (for model
spin up) to 1600 on 12 February, with forecast winds extending
to 0400 on 14 February. Particles were released from all nine
positions each 6 hours starting on 11 February. Oil floating
at the northern position (only 3 release points are shown) on
February 11 at 1600 (Fig. 19A) was expected to reach the eastern
half of Jones Beach about 0400 on the 13th (the trajectories are
marked with a "+" each six hours), whereas there was less threat
from the other two positions. In general, the series of updated
forecasts also indicated that oil floating in the Apex on the
11th and 12th could wash ashore during the early hours of the
13th. By the evening of the 12th, oil at Ambrose Light Tower
(the north "x" in Fig. 19B) could reach Rockaway in 6 hours and
oil at the middle position could reach Long Beach in 12 hours.
Oil was found on Rockaway Beach on Sunday, the 13th of February,
1977 (C. *Parker*, pers. comm.).

Later releases during this forecast period indicated that
there was little oil hazard after about noon on the 13th from
any of the assumed positions within the Bight Apex. Four more
36-hour forecasts were done for MESA, the last being on Wednesday,
the 16th (Fig. 20). By this time, the surface trajectories from
Ambrose Light Tower and the two southern release points for
February 15 and 16 (Fig. 20 A,B) were to the southeast. Since no
more oil was observed entering the Apex, no further simulation runs
were warranted, and the forecast use of the model was then terminated.

Fig. 17. The observed Ambrose Light Tower winds (A), the
projected winds used in the first forecast for MESA (B), and the
projected winds used in the second forecast (C) during February
1977. ⟶

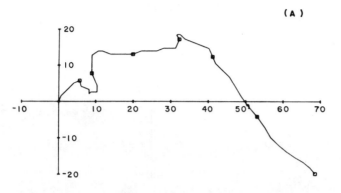

(A)

PVD (KM) OF TRANSPORT WIND II FEB 1600 - 18 FEB 0400

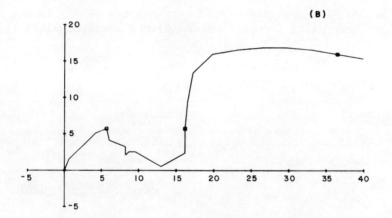

(B)

PVD (KM) OF TRANSPORT WIND II FEB 1600 - 14 FEB 0400

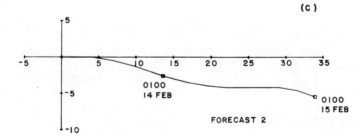

(C)

0100
14 FEB

FORECAST 2

0100
15 FEB

PVD (KM) OF TRANSPORT WIND 13 FEB 1300 - 15 FEB 0100

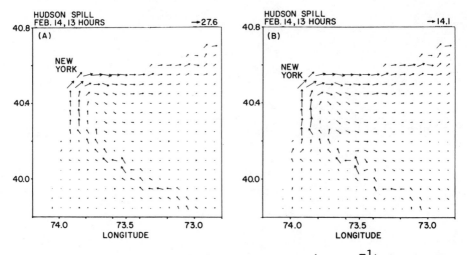

Fig. 18.   The computed winter currents (cm sec^{-1}) in the New York Bight Apex using forecast winds (A) and observed winds (B).

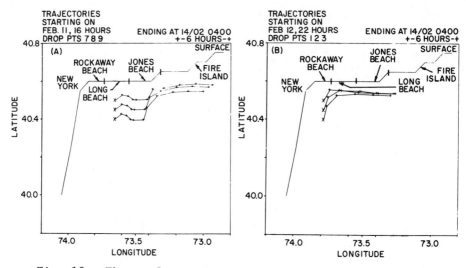

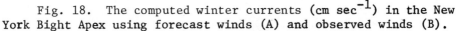

Fig. 19.   The surface oil trajectory forecast for February 11 (A) and 12 (B), 1977.

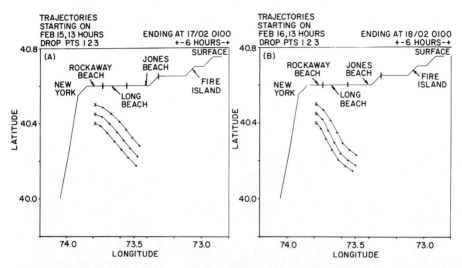

Fig. 20.   The surface oil trajectory forecast for February 15 (A), and 16 (B), 1977.

CONCLUSION

We have been able to successfully 1) reproduce drift card data for determining the probabilities of a winter oil spill beaching within the New York Bight, 2) analyze the source of floatables encountered on the south shore of Long Island, and 3) predict the trajectory of oil spilled in the Hudson River. The physical dynamics of our present circulation model are admittedly simple; we are in the process of updating this component of the calculations. Recall that the present model is depth integrated, has a free surface wind stress, and contains physical dynamics of a barotropic pressure gradient, Coriolis force, and a bottom friction term. The baroclinic field is being added with a diagnostic circulation model that interpolates current meter and density data (*Han et al.*, 1979), to give this component of the flow at specified depths. This additional sophistication will allow us to deal with stratified water column conditions. Also recall that the open ocean boundary condition excludes whatever physics occur at the artificial boundaries of the model's domain. A sea surface slope is now being generated at this boundary (*Hopkins*, 1972) and matched to the interior solution, allowing us to incorporate far-field forcing of the shelf system.

Incorporation of biological and chemical terms (*Walsh*, 1975, 1977) within the previous analyses requires 1) dosage-response functions of the organisms to each class of pollutants, and 2) a quantitative description of the "normal" food web interactions of the continental shelf. Toxicity levels in terms of median lethal concentrations (LC50) of metals, pesticides, biofouling agents (*e.g.* chlorine), PCB's, and petroleum fractions have been determined for a number of hardy organisms, (*i.e.* those that can be cultured in the laboratory). The actual form of the pollutant (*e.g.* methyl mecuric chloride or chloramine) and its concentration in the marine environment, however, are not always known. Furthermore, enclosed, large-scale environmental studies of pollutant responses (*e.g.* the IDOE-CEPEX (Controlled Environmental Pollution Experiments) program) suggest that species replacement occurs more often than total loss of a trophic level. The realized contribution of a pollutant to differential mortality of parts of a coastal food web is additionally confounded by our lack of understanding of natural mortality. Death on the continental shelf is a poorly known process, and despite a number of ongoing food chain studies, we can specify neither the fate of phytoplankton nor the three-dimensional trajectory of a particle in the coastal zone (*Walsh et al.*, 1978).

The present results on the transport of hypothetically inert particles within the New York Bight illustrate, nevertheless, the utility of computer modeling as one tool in the future assessment of real or potential coastal perturbation problems. First, the shallow water model must be modified or replaced with a numerical model that contains a more sophisticated parameterization of the physical circulation. Second, the particle-in-cell model can be modified to include explicitly chemical reactions and interactions with the biota. Of course, any model should be used in the context of the level of resolution or aggregation to which the ecosystem is known and the management decision required as an aid in selecting situations that merit further analysis with more comprehensive ecological reasoning. Recall the successful prediction of oil beaching on Rockaway Beach at a gross level of resolution despite the uncertainties in February, 1977, of the initial oil spill location, lack of ground truth, and inadequate wind forecast. We do believe, moreover, that our ability to track the physical dispersal of quasi-conservative variables within the New York Bight in a number of validation studies demonstrates that the additional chemical and biological complexity of perturbation responses of the coastal zone can now be addressed with a more firm foundation.

ACKNOWLEDGMENTS

This research was supported mainly by the Department of Energy (DOE) with Contract No. EY-76-C-02-0016 as part of our Atlantic Coastal Ecosystem (ACE) program. Additional support was made available by the NOAA-MESA New York Bight Project and the NSF-IDOE Coastal Upwelling Ecosystem Analysis (CUEA) program.

REFERENCES

Ahearn, D. G., 1974. The sources, fates, and effects of oil in the seas, In: *Pollution and Physiology of Marine Organisms,* edited by F. J. Vernberg and W. B. Vernberg, 247-251, Academic Press, N. Y.

Beardsley, R. C., W. C. Boicourt, and D. V. Hansen, 1976. Physical oceanography of the middle Atlantic Bight, *Am. Soc. of Limnol. Oceanogr. Spec. Symp. 2:* 20-34.

Christensen, O. and L. P. Prahm, 1976. A pseudospectral model for dispersion of atmospheric pollutants, *J. Appl. Meteor. 15:* 1284-1294.

Dieterle, D. A. and A. G. Tingle, 1976. A numerical study of mesoscale transport of air pollutants in sea-breeze circulations, *Proceeding of Third Symposium on Atmospheric Turbulence, Diffusion, and Air Quality,* Raleigh, North Carolina, Am. Meteor. Soc.

Egan, B. A. and J. R. Mahoney, 1972. Numerical modeling of advection and diffusion of area source pollutants, *J. Appl. Meteor. 11:* 312-322.

Goldberg, E. D., 1976. *The Health of the Oceans,* 172 pp., The UNESCO Press, Paris.

Han, G., D. V. Hansen, and A. Cantillo, 1979. Diagnostic model of water and oxygen transport in the New York Bight, In: *Anoxia in the New York Bight, 1976,* edited by C. J. Sinderman and R. L. Swanson, NOAA Prof. Pap. (in press).

Hansen, D. V. 1977. *Circulation,* MESA N. Y. Bight atlas monograph 3. N. Y. Sea Grant Inst. Albany, N. Y.

Hardy, C. D., E. R. Baylor, and P. Moskowitz, 1976. Sea surface circulation in the northwest apex of the New York Bight with appendix: bottom drift over the continental shelf, NOAA Technical Memo. ERL MESA 13.

Hardy, D. C., E. R. Baylor, P. Moskowitz, and A. Robbins, 1975. The prediction of oil spill movements in the ocean south of Nassau and Suffolk Counties, New York, Technical Report Series, No. 21, Marine Sciences Research Center at SUNY, Stony Brook, New York.

Harlow, F. H., 1963. The particle-in-cell method for numerical solution of problems in fluid dynamics, *Proceedings of Symposia in Applied Mathematics 15:* 269.

Hemingway, E., 1935. *Green Hills of Africa,* 149-150, Charles Scribner and Sons, New York.

Hopkins, T. S., 1972. On time dependent wind induced motions, *Rapp. P.-V. Cons. Int. Explor. Mer 167:* 21-36.

James, R. W., 1966. Ocean thermal structure forecasting, SP-105, *ASWEPS Manual Series Vol. 5.* U. S. Naval Oceanographic Off., Wash., D. C.

Lange, R., 1973. ADPIC, a three-dimensional computer code for the study of pollutant dispersal and deposition under complex conditions, Lawrence Livermore Laboratory, Report TID-4500, UC-32.

Mayer, D. A., D. V. Hansen, and S. M. Minton, 1979. A comparison of water movements on the New Jersey shelf during 1975, and 1976, with a view toward understanding the development of anoxic conditions during 1976, In: *Anoxia in the New York Bight, 1976,* edited by C. J. Sindermann and R. L. Swanson, NOAA Prof. Pap. (in press).

Molenkamp, C. R., 1968. Accuracy of finite-difference methods applied to the advection equation, *J. Appl. Meteor. 7:* 160-167.

Nelson-Smith, A., 1973. *Oil Pollution and Marine Ecology,* 260 pp., Plenum Press, New York.

Platzman, G. W., 1972. Two dimensional free oscillations in natural-basins, *J. Phys. Oceanogr. 2:* 117-138.

Schwartzberg, H. G., 1971. The movement of oil spills, In: *Proceedings of Joint Conf. on Prevention and Control of Oil Spills,* 489-494, June 15-17, 1971, Sheraton Park Hotel, Wash., D. C., sponsored by A.P.I., E.P.A., and U.S.C.G.

Scott, J. T. and G. T. Csanady, 1976. Nearshore currents off Long Island, *J. Geophys. Res. 81:* 5401-5409.

Shir, C. C., 1973. A preliminary numerical study of atmospheric turbulent flows in the idealized planetary boundary layer, *J. Atmos. Sci. 30:* 1327-1339.

Singer, I. A., 1967. Steadiness of the wind, *J. Appl. Meteor. 6:* 1033-1038.

Sklarew, R. C., A. J. Fabrich, and S. E. Pruger, 1971. A particle-in-cell method for numerical solution of the atmospheric diffusion equation and applications to air pollution problems, Division of Meteorology, National Environmental Research Center, Report 3SR-844.

Sverdrup, H. U., M. W. Johnson, and R. H. Fleming, 1942. *The Oceans, Their Physics, Chemistry and General Biology*, 1087 pp., Prentice-Hall, Englewood Cliffs, N. J.

Swanson, R. L., G. M. Hansler, and J. Marotta, 1977. Long Island beach pollution: June 1976, MESA Spec. Rept.

Swanson, R. L., H. M. Stanford, J. S. O'Connor, S. Chanesman, C. A. Parker, P. A. Eisen, and G. F. Mayer, 1978. June 1976 pollution of Long Island ocean beaches, *J. Environ. Eng. Div. EE6:* 1067-1085.

Walsh, J. J., 1972. Implications of a systems approach to oceanography, *Science 176:* 969-975.

Walsh, J. J., 1975. A spatial simulation model of the Peru upwelling ecosystem, *Deep-Sea Res. 22:* 201-236.

Walsh, J. J., 1976. Herbivory as a factor in patterns of nutrient utilization in the sea, *Limnol. Oceanogr. 21:* 1-13.

Walsh, J. J. 1977. A biological sketchbook for an eastern boundary current, In: *The Sea, Vol. VI,* edited by J. H. Steele, J. J. O'Brien, E. D. Goldberg and I. N. McCave, 923-968, Wiley Interscience, New York.

Walsh, J. J., T. E. Whitledge, F. W. Barvenik, C. D. Wirick, S. O. Howe, W. E. Esaias, and J. T. Scott, 1978. Wind events and food chain dynamics within the New York Bight, *Limnol. Oceanogr. 23:* 659-683.

Walsh, J. J., P. G. Falkowski, and T. S. Hopkins, 1979b. Climatology, phytoplankton species succession and oxygen depletion within the New York Bight, submitted to *J. Mar. Res.*

Walsh, J. J., C. D. Wirick, and A. G. Tingle, 1979a. Environmental constraints on larval fish survival within low and high latitude ecosystems, *Proc. Symp. Early Life History of Fish,* 2-5 April 1979, Woods Hole, Mass (in press).

ECOLOGICAL SIGNIFICANCE OF FRONTS IN THE SOUTHEASTERN BERING SEA

R. L. Iverson,[1] L. K. Coachman,[2] R. T. Cooney,[3] T. S. English,[2]

J. J. Goering,[3] G. L. Hunt, Jr.,[4] M. C. Macauley,[2] C. P.

McRoy,[3] W. S. Reeburg,[3] and T. E. Whitledge[5]

[1]*Florida State University;* [2]*University of Washington;*

[3]*University of Alaska;* [4]*University of California, Irvine;*

[5]*Brookhaven National Laboratory*

ABSTRACT

A series of three fronts divides the continental shelf of the southeastern Bering Sea into two interfrontal zones which contain different food webs. Large stocks of birds, mammals, and pelagic fish, primarily walleye pollock, occur in the outer shelf zone between the 200 meter isobath and the middle front near the 100 meter isobath. Large stocks of benthic infauna, demersal fish, and crabs occur in the middle shelf zone between the middle front and the inner front at the 50 meter isobath. Very low cross-shelf advection and the presence of the middle front which acts as a diffusion barrier restrict large oceanic herbivores to the outer shelf zone. Large diatoms are not grazed by the small coastal herbivores which inhabit the middle shelf zone, resulting in an accumulation of phytoplankton biomass which settles to the benthos.

INTRODUCTION

The Bering Sea has been recognized as a biologically productive region for several hundred years. Large populations of birds, seals and whales inhabit the continental shelf of the southeastern Bering Sea which contains well developed pelagic fisheries (*Wilimovsky,* 1974), demersal fisheries (*Alton,* 1974; *Bakkala and Smith,* 1978), and shellfish fisheries. The primary productivity

of the southeastern Bering Sea is not disproportionately high in
comparison with other north temperate marine systems (*McRoy and
Goering*, 1976), suggesting that other factors are responsible
for the productivity of higher trophic levels. PROBES (Processes
and Resources of the Bering Sea) was designed to investigate
mechanisms which lead to transfer of energy from primary to higher
trophic levels in an attempt to elucidate causes for the productivity of higher trophic levels of the Bering Sea food web.

PROBES field work has established the existence of several
fronts in the southeastern Bering Sea (Fig. 1). A front observed
along the continental slope near the shelf break at the 200 m
isobath is persistent for periods on the order of years and marks
a transition between oceanic and shelf water of the Bering Sea
(*Kinder and Coachman*, 1978). Shoreward of this shelf break front
lies a middle shelf front near the 100 m isobath (*Coachman and
Charnell*, 1979). A broad transition region separates the two
fronts (*Coachman and Charnell*, 1977). An inner front is located
near the 50 m isobath (*Schumacher et al.*, 1979).

We report observations which support a hypothesis that major
food webs leading to large stocks of pelagic fauna and benthic
fauna are separated in space and are organized in relation to the
fronts which exist in the southeastern Bering Sea.

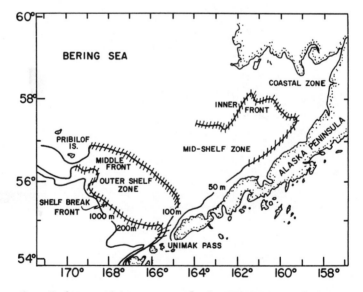

Fig. 1. Hydrographic zones of the PROBES study area in the
southeastern Bering Sea.

METHODS

Hydrography and Chemistry

A Bissett Berman CTD equipped with a rosette sampler was used to collect physical data and water samples from various depths at each hydrocast station. Physical data were processed on board with a PRIME computer system using methods given in *Coachman and Charnell* (1979).

Nutrient samples were collected from the rosette sampler using polypropylene bottles. The samples were processed with minimal storage time using a Technicon Autoanalyzer II. Chemical methods were modified for small-volume manifold glassware to optimize stability and sensitivity (*Whitledge et al.*, in press). The basic techniques and reagents for phosphate were described by *Murphy and Riley* (1962); for silicate, nitrate and nitrite by *Armstrong et al.* (1967) and *Bendschneider and Robinson* (1952); and for ammonium by *Slawyk and MacIsaac* (1972) and *Patton and Crouch* (1977).

Plankton

Chlorophyll was measured with the extracted fluorescence method (*Strickland and Parsons*, 1972) for water samples taken from the rosette sampler, filtered through Gelman GF/AE glass fiber filters, and immediately processed. Samples preserved with Lugol's solution were processed by the Utermohl method to obtain phytoplankton identifications and estimates of cell numbers. Particle counts were made with a Coulter Counter equipped with a 200 μm aperture.

Zooplankton biomass samples were obtained with Bongo nets (0.505 and 0.333 mm mesh size) equipped with flow meters. The nets were fished obliquely to 50 meters. Samples were preserved with buffered formalin after which they were transferred to shore for processing. Samples for characterization of the cross-shelf distribution of copepod species were obtained with a one-meter net equipped with 0.333 mm Nitex netting. The net was fished from the bottom or from 200 m when the bottom was deeper than 200 m. Samples were preserved with formalin and were transferred to shore for processing.

Acoustic Assessment of Zooplankton and Fish

A high frequency sonar system is necessary for acoustic assessment of zooplankton. Frequencies from 50 to 400 kHz produce Reyleigh scattering from most zooplankton and from individual fish

and can be used in acoustic stock assessment. The acoustic system
used in PROBES research consisted of two Ross model 200 echosounders,
one operating at 105 kHz and the other at 205 kHz. The output from
each echosounder was converted to 5 kHz and digitized in real time
with a Hewlett-Packard model 1000 computer system. Fish abundance
was estimated from the digitized data using the following acoustic
equation:

$$n = antilog\ [0.1\ (TS - c - c'log\ L)]$$

where n is the number of targets, TS is the target strength in dB,
c and c' are constants for any given species and L is target length.

   Beamish (1969, 1971) developed a method for approximating
target volume for non-spherical targets which considered the effective volume as composed of the summation of smaller spheres chosen
based on theoretical calculation of their contribution to total
scattering cross section. Macauley (1978) showed that quantitative
acoustic assessment of euphausiids could be achieved in the field
using the equation of Urick (1967) with the approximations of
Beamish (1969, 1971). The equation used for estimating zooplankton
abundance was:

$$n = \frac{1}{s}\ antilog\ [0.1\ (RL-SL + 20\ log\ r - ar - 10\ log\ (\frac{ct}{2})$$

$$+\ DI\ +\ 3.85)]$$

where n is the number of scatterers returning a signal of level
RL, s is the scattering cross section, SL is the acoustic source
level, r is the range to scatters in meters, a is the attenuation
with range coefficient, c is the speed of sound, t is the pulse
length in seconds, and $DI + 3.85$ is the directivity index corrected
for ideal beam pattern.

   Either the target strength-size relationship or the target
strength-biomass relationship must be known to obtain quantitative
estimates of scattering organisms. A five-meter opening-closing
net (Frost and McCrone, 1974) was used to sample depths where
characteristic acoustic signals were observed so that the signals
could be attributed to organisms of particular size and species.
A 3-meter beam trawl was fished on the bottom in areas where juvenile pollock were anticipated near the seabed.

                              Birds

   Bird observations were made during a series of ten minute
time periods from dawn to dusk while the ship was moving between
stations. Counts of birds were made to 300 meters from directly

ahead to 90⁰ abeam. Garbage and refuse were dumped only at the
completion of a day's observations and an effort was made to count
ship-circling and ship-following birds only once to minimize bias
introduced by the presence of the ship. Bird densities were plotted
by 10' x 10' blocks of latitude and longitude. When more than one
ten minute time period of observations fell within a block, data
for those time periods were averaged.

RESULTS

Phytoplankton, Physical, and Chemical Factors

Cross-shelf physical, chemical and phytoplankton data from
three transects illustrate the temporal sequence of cross-front
distribution of variables in the southeastern Bering Sea. Station
locations for the transects are shown in Fig. 2. Although there

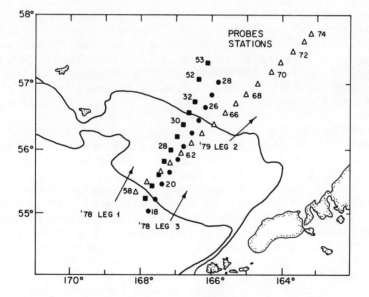

Fig. 2. Station locations for PROBES cross-shelf transects
in the southeastern Bering Sea. Leg 1, 1978, extended from 8 April
through 1 May; Leg 3, 1978, extended from 26 May through 12 June;
Leg 2, 1979, extended from 2 May through 22 May.

had been some phytoplankton productivity by 19 April, 1978, as
indicated by a particle maximum at station 33 (Fig. 3), nutrients
were present in high concentration at all depths. Density
structure was vertically homogeneous in the photic zone (44 m $\pm$
8 m, n = 11). Ammonium concentrations were less than 1 μg-at N ℓ^{-1},
suggesting minimal nitrogen regeneration had occurred. Chlorophyll
values were uniformly less than 1 μg ℓ^{-1} except in the particle
maximum region where values reached 2 μg chlorophyll ℓ^{-1}.

Transect data were not taken during late April and early May
of 1978, although the spring phytoplankton bloom occurred during
that period. Data taken along transects just after the development of the spring phytoplankton bloom in May, 1979, indicate
that salinity (Fig. 4) was the leading variable in control of
density (Fig. 5) below the region of influence of insolation
(Fig. 6). The shelf break front was evident between stations
59 and 60 in the salinity field. The middle front was marked by
the 25.60 isopycnal and by isohaline packing between stations 62
and 63, while vertical isohalines typical of the inner front
were evident between stations 72 and 73. Nitrate was depleted
in the upper 10 meters in the middle shelf zone (Fig. 7). The
nitrate and chlorophyll data (Fig. 8) suggest that the bloom
in the middle shelf zone occurred prior to the bloom in the shelf
break zone. Photic zone depth in the middle shelf zone was
13 m $\pm$ 3 (n = 11). The phytoplankton of the spring bloom were
dominated by *Thalassiosira aestivalis, T. nordenskioldii,* and
Chaetoceros debilis, all members of the high latitude stage I
successional group identified by *Guillard and Kilham* (1977).
Chlorophyll maxima were observed near the bottom at stations 65 and
69. Near-bottom maxima of up to 65 μg chlorophyll ℓ^{-1} were observed
at several stations in the middle shelf zone throughout the late
spring and early summer. Near-bottom maxima were not observed in
the outer shelf zone.

By the end of May, 1978, the distribution of particles (Fig.
9) and chlorophyll (Fig. 10) exhibited a pattern in relation to
the various hydrographic zones which was repeated in 1979. A chlorophyll maximum at station 18 in the shelf break front was not observed
in the particle data. This chlorophyll maximum persisted in the
shelf break front for about one month, slowly sinking at a rate
of about 1 meter per day (*Iverson et al.,* in manuscript). The dominant phytoplankton species in near-surface waters of the shelf
break front, *Pheocystis poucheti*, was a colonial chrysophyte composed of hundreds of 5-micron-diameter cells embedded in a gelatinous
matrix. *Pheocystis* was not readily counted with the Coulter Counter
geometry available. The outer shelf zone (stations 19 to 22)
contained surface particulate material but very low chlorophyll
values throughout the photic zone. Biological and chemical data
suggest the middle shelf front was located between stations 22 and
24 at about the 100 m isobath, consistent with the description of

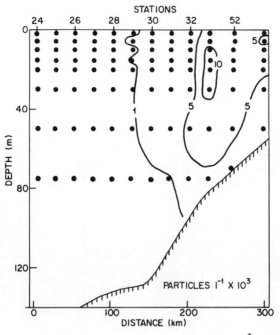

Fig. 3. Particle distribution (Particles ℓ^{-1} x 10^3), Leg 1, 1978.

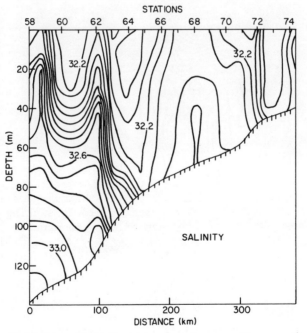

Fig. 4. Salinity distribution (parts per thousand), Leg 2, 1979.

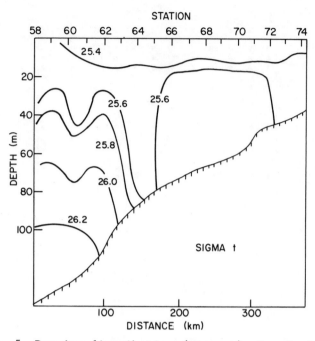

Fig. 5. Density distribution (sigma t), Leg 2, 1979.

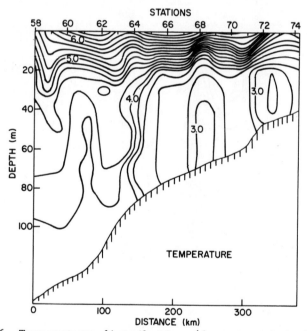

Fig. 6. Temperature distribution (degrees Celsius), Leg 2, 1979.

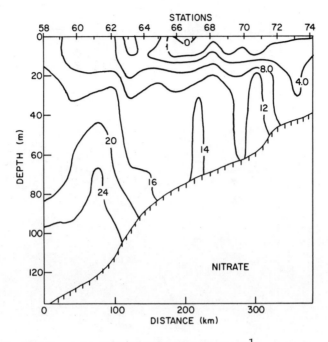

Fig. 7. Nitrate distribution (μg-at N ℓ$^{-1}$), Leg 2, 1979.

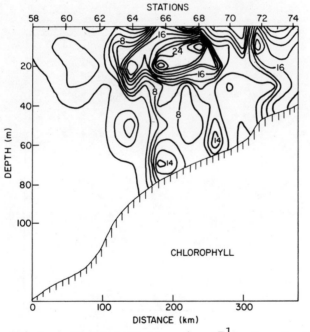

Fig. 8. Chlorophyll distribution (μg ℓ$^{-1}$), Leg 2, 1979.

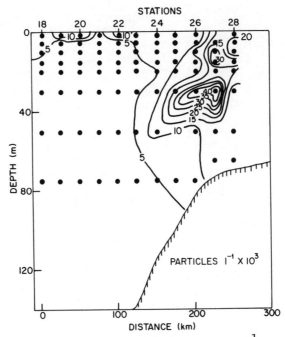

Fig. 9. Particle distribution (Particles ℓ^{-1} x 10^3), Leg 3, 1978.

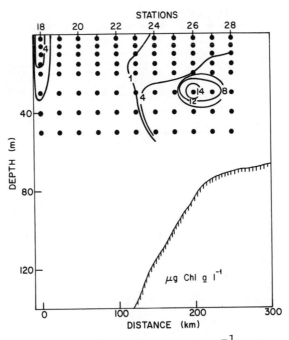

Fig. 10. Chlorophyll distribution (μg ℓ^{-1}), Leg 3, 1978.

Coachman and Charnell (1979). *Pheocystis poucheti* was present in
large numbers in near-surface waters of the middle shelf front.
The middle shelf zone contained maxima of chlorophyll and particles
at about 30 meters which extended to the inner front (as they did
in May of 1979), as evidenced in data collected during July, 1978.
The maxima persisted through September, 1978, when another cross-
shelf transect was made. The mean photic zone depth during May,
1978, in the middle shelf zone was 27 m $\pm$ 4 m (n = 6), a reflection
of the sinking of the spring bloom. An ammonium maximum (Fig. 11)
was observed at the depth of the chlorophyll maximum but was out
of phase seaward of the chlorophyll maximum, a phenomenon observed
along most transects during Leg 3, 1978. *Chaetoceros convolutus,*
C. decipiens, Corethron hystrix, and *Rhizosolenia alata* formed the
Stage II successional group which dominated the middle shelf zone
phytoplankton during this period. *Rhizosolenia* rapidly became
numerically dominant, reaching over 90 percent of the phytoplankton
numbers at some stations. Apparently *R. alata* can tolerate low
silicate concentrations (*Guillard and Kilham,* 1977). *Rhizosolenia*
alata was the stage III successional dominant in the middle shelf
zone during both 1978 and 1979. Nutrients were depleted in the
photic zone of the middle shelf zone with nitrate concentration
below detection in the upper 20 meters (Fig. 12) and with silicate
concentration below detection at stations 24 and 25 (Fig. 13).
Phosphate (Fig. 14) and nitrate concentrations were both reduced
in the photic zone of the outer shelf in contrast to silicate, which
suggests that diatoms were not primarily responsible for nitrate
and phosphate uptake in the outer shelf zone as they were in the
middle shelf zone.

Vertically integrated chlorophyll values for several cross-
shelf transects taken during May, 1978, (Fig. 15) exhibit the
shelf break front (100 mg m^{-2} chlorophyll maximum), the outer shelf
zone (chlorophyll less than 50 mg m^{-2}), and the middle shelf zone
(400 mg m^{-2} chlorophyll maximum). The middle front was located
along the 100 mg m^{-2} chlorophyll isopleth near the 100 m isobath.
The inner front was not sampled during May, 1978.

Zooplankton

Several kinds of acoustic targets were identified by correlation of different echogram or digitized pattern displays with
species present in net catches taken from depths where characteristic echos were observed. Euphausiid layers composed of *Thysanoessa*
spp. and *Euphausia pacifica* were observed with both the 105 and
205 kHz frequencies. Large clusters of medusae migrated from about
40 meters during the day to the surface at night.

The zooplankton distributions reported here are based on 105
kHz acoustic data. The 205 kHz frequency was used to check for

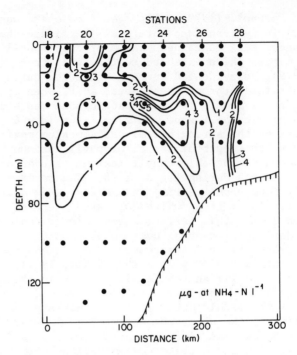

Fig. 11. Ammonium distribution (μg–at N ℓ$^{-1}$), Leg 3, 1978.

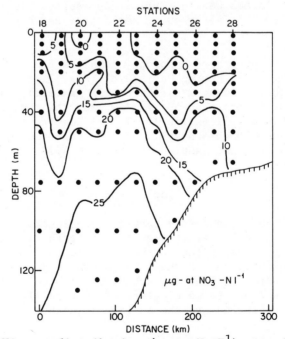

Fig. 12. Nitrate distribution (μg–at N ℓ$^{-1}$), Leg 3, 1978.

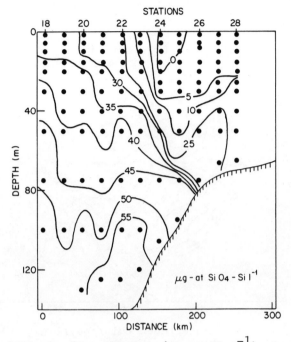

Fig. 13. Silicate distribution (μg–at Si ℓ^{-1}), Leg 3, 1978.

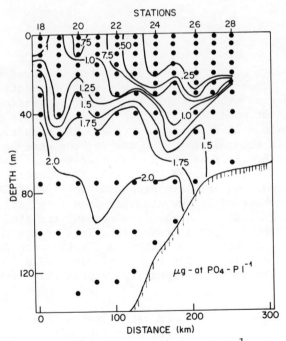

Fig. 14. Phosphate distribution (μg–at P ℓ^{-1}), Leg 3, 1978.

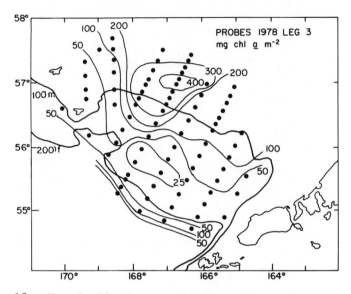

Fig. 15. Vertically integrated chlorophyll (mg Chl a m^{-2}), Leg 3, 1978.

presence or absence of zooplankton layers not observed with the 105 kHz system. There were no cases where 205 kHz data revealed signals not observed with the 105 kHz system, suggesting that zooplankton were never more abundant than 150 individuals per cubic meter, the threshold of detection for a 205 kHz acoustic detection system. Acoustic data suggest that zooplankton were concentrated around the shelf break front and in the middle front and middle shelf region during April, 1978 (Fig. 16). Zooplankton biomass estimates obtained from net samples during April exhibited a pattern similar to that observed in the acoustic data (Fig. 17). Zooplankton biomass had generally increased throughout the study area by the end of May with a major increase localized around the shelf break (Fig. 18). Acoustic zooplankton estimates were largest in the outer shelf zone and in the region of the middle front during May (Fig. 19). The acoustic signals were primarily attributed to euphausiids.

Copepod species in the southeastern Bering Sea were clustered into two general groups which were distributed in oceanic-outer shelf and middle shelf-inner shelf habitats (*Cooney and Geist*, 1978).

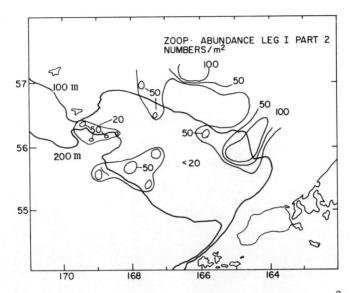

Fig. 16. Acoustic zooplankton estimates (numbers m^{-2}), Leg 1, 10-20 April, 1978.

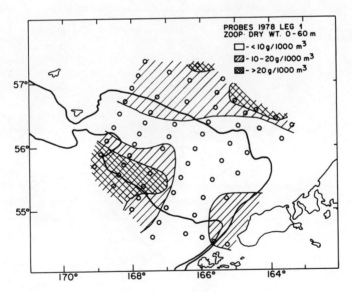

Fig. 17. Zooplankton dry weight for samples taken with a 0.333 mm net, Leg 1, 1978.

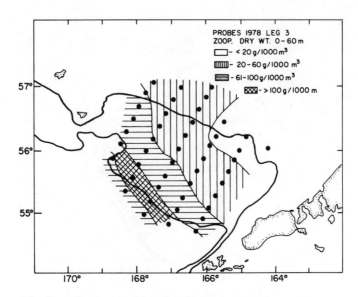

Fig. 18. Zooplankton dry weight for samples taken with a 0.505 mm net, Leg 3, 1978.

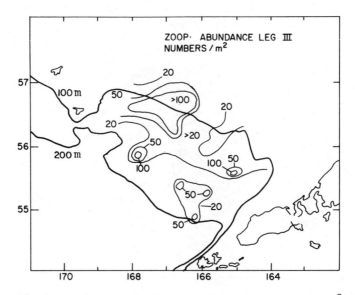

Fig. 19. Acoustic zooplankton estimates (numbers m^{-2}), Leg 3, 1978.

An oceanic community dominated by *Metridia lucens, Calanus plumchrus, C. cristatus,* and *Eucalanus bungii bungii* remained seaward of the middle front. *Metridia lucens* and *Calanus plumchrus* were major constituents of the oceanic community (27% and 14% of total numbers, respectively). The cross-shelf distribution pattern of *Metridia lucens* (Fig. 20) was exhibited by all of the large oceanic herbivores. *Pseudocalanus* spp. was a cosmopolite with a similar pattern of distribution among the different hydrographic zones throughout the season. *Pseudocalanus* spp. comprised 22 percent of the numbers of the oceanic community and 57 percent of the numbers of the shelf community. *Acartia longerimus* comprised 20 percent of total shelf copepod numbers while *Calanus marshallae* numbers were about 10 percent of the total.

Fish and Benthic Infauna

Concentrations of fish targets were observed near the bottom and at various depths in the water column. Demersal targets clustered close to the bottom were predominantly yellow fin sole, *Limanda aspera.* Schools of midwater fish were occasionally observed and a net haul made through one of the schools contained a capelin, *Mallotus villosus.* When acoustic targets exhibited considerable vertical and horizontal extent, walleye pollock, *Theragra chalcogramma,* were a major portion of net catches. Pollock appeared to be distributed as widely-spaced individuals as well as in patches ranging in size from a few kilometers to over 50 kilometers.

There appeared to be a seasonal change in the pattern of distribution of fish, which were mostly pollock, in the outer shelf zone. A few patches containing large numbers of fish observed in mid-April in the southern end of the middle shelf zone were associated with near-surface aggregations of pollock eggs and larvae. Fish were distributed in the center of the outer shelf zone, in the shelf break front at the northern end of the study area, and in a large school in the middle front zone during late April (Fig. 21). During May, most of the fish were located in the vicinity of the middle front (Fig. 22). Foreign vessels known to be fishing pollock were distributed in the region of acoustic signal maxima, indicating that most of the acoustic fish signals could be attributed to pollock. Juvenile pollock (length less than 20 cm) were only caught in nets in near bottom hauls east of the Pribilof Islands. Juveniles were not observed in net hauls taken in Bristol Bay or along the Alaskan Peninsula. Individuals collected from larger fish schools were predominantly adults and sub-adults of mixed sizes and sex.

Benthic collections were not part of PROBES research, which was directed toward the pelagic food web of the Bering Sea. However, when patterns in phytoplankton were observed which suggested that

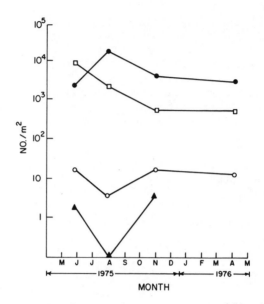

Fig. 20. Cross-shelf distribution of *Metridia lucens*, 1975 and 1976. Closed circles are numbers in the oceanic zone; open squares are numbers in the outer shelf zone; open circles are numbers in the middle shelf zone; closed triangles are numbers in the coastal zone. Refer to Fig. 1 for location of the different hydrographic zones.

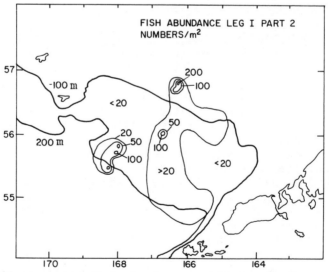

Fig. 21. Acoustic fish abundance estimates (numbers m^{-2}), Leg 1, 10–20 April, 1978.

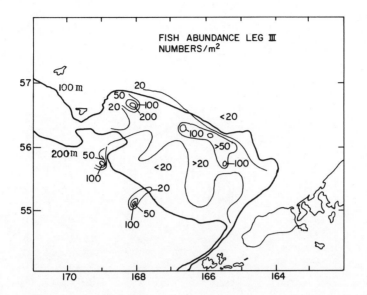

Fig. 22. Acoustic fish abundance estimates (numbers m^{-2}),
Leg 3, 27 May-4 June, 1978.

a major portion of the middle shelf chlorophyll reached the bottom,
data obtained from other investigations were examined to see if
there were patterns in benthic biomass which might be of interest.
Benthic infauna biomass was significantly larger in the middle
shelf zone than in other zones of the southeastern Bering Sea
(Fig. 23).

Birds

Northern fulmars, *Fulmarus glacialis,* were distributed widely
in the study area with concentrations near the shelf break and the
outer shelf zone throughout late May and early June, 1978 (Fig.
24). These birds eat a variety of pelagic foods including squid,
euphausiids, and fish, all of which they take near the surface by
gathering directly from the surface or by making shallow plunge
dives. Fulmars were almost always associated with fishing boats;
however, they were also observed over shelf break waters removed
from fishing boats. The density of fulmars was low shoreward of
the 100 meter isobath in the middle shelf zone except for an area
northeast of Unimak Island where they were seen feeding on natural
foods in association with fork-tailed storm petrels. Distribution
of fulmars in June and July was similar to that observed in May.

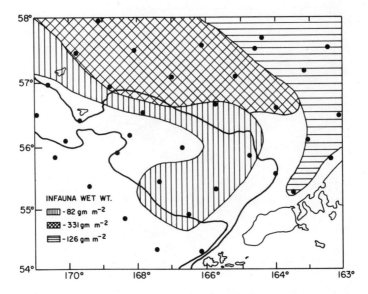

Fig. 23. Benthic infauna biomass for samples obtained with a van Veen grab, 1976.

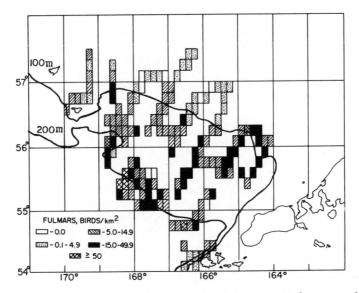

Fig. 24. Distribution of northern fulmars, *Fulmarus glacialis*, 25 May–12 June, 1978.

Fork-tailed storm petrels, *Oceanodroma furcata*, were relatively rare in the southeastern Bering Sea until mid June. Storm petrels were observed in greatest numbers in the vicinity of the shelf break front and in the outer shelf zone (Fig. 25). Few storm petrels were observed in water shallower than 100 meters. These birds feed exclusively on neuston, taking small items by hovering above the water or by sitting on the surface and picking at small objects within the top several centimeters.

The distributions of black-legged kittiwakes, *Risa tridactyla*, and red-legged kittiwakes, *R. brevirostris*, were significantly different throughout the study period. Both species were present in low numbers in the outer shelf zone; however, black-legged kittiwake numbers were not strongly localized. Red-legged kittiwakes were most consistently present at the shelf break front with a few individuals observed near the middle front or in the middle shelf region (Fig. 26). Black-legged kittiwakes feed widely on amphipods, euphausiids, and several kinds of fish, including *Mallotus*, myctophids, and gadids, particularly *Theragra chalcogramma* (*Hunt*, 1979). Red-legged kittiwakes are more specialized and feed on mesopelagic fish, primarily myctophids, with some *T. chalcogramma* in their diet.

Murres, primarily *Uria lomvia*, were present in the study area, but were withdrawing to their breeding colonies, located on the Pribilof Islands, during late May and early June.

DISCUSSION

The oceanic and outer shelf zones of the southeastern Bering Sea contain a mixture of Bering Sea and Alaska Stream water, while the continental shelf water mass is strongly affected by winter cooling, sea ice formation, and seasonal variations in river input. These water masses interact through a relatively broad transition zone bounded at the surface by well-defined salinity fronts which occur at the shelf break and near the 100 meter isobath. The entire region is hydrographically sluggish. Warmer oceanic waters rarely intrude onto the shelf to depths shallower than 100 m.

PROBES data has revealed three fronts and two interfront zones that persist throughout the biologically active portion of the year in the southeastern Bering Sea (Fig. 27). The shelf break front persists for periods on the order of years (*Kinder and Coachman*, 1978) while the other two fronts are rendered inactive in a biological sense by the presence of ice cover during some winters (*McRoy and Goering*, 1974). The fronts occur where there is a change in lateral flux rates due to changes in mixing energy and topographic features of the shelf. The shelf break front occurs in the upper 50 m near the 200 m isobath and is separated

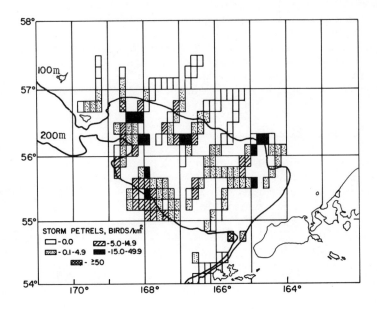

Fig. 25. Distribution of fork-tailed storm petrels, *Oceanodroma furcata*, 25 May–12 June, 1978.

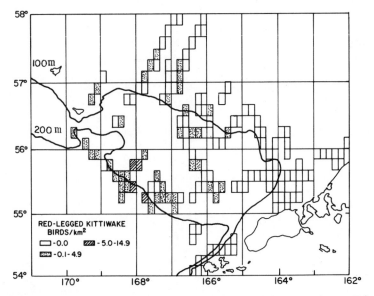

Fig. 26. Distribution of red-legged kittiwakes, *Rissa brevirostris*, 17 June–10 July, 1978.

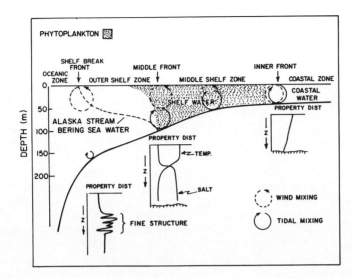

Fig. 27.   Scenario of cross-shelf hydrographic features and relative phytoplankton biomass for the southeastern Bering Sea shelf.

from the middle shelf front by the outer shelf zone which is about 100 km wide. Waters of the open Bering Sea extend onto the shelf in a 30 m thick bottom layer while shelf waters extend seaward in the surface and mid-layers in the outer shelf zone, which contains enhanced lateral and vertical fluxes. Vertical mixing results from wind mixing in the upper 50 m and tidal mixing on the bottom. The surface and bottom mixed layers are separated as a consequence of the depth of the outer shelf zone by a mid-depth region of inter-leaving water layers that results in vertical finestructure of water column properties (*Coachman and Charnell*, 1979).

The middle shelf front occurs near the 100 m isobath. Shore-ward of this front is the middle shelf zone which extends from the 100 m isobath to the 50 m isobath, a distance of 200 to 300 km. A strong seasonal thermocline forms a two-layered water column in this zone. There is minimal net advection in this zone, but the waters are shallow enough that wind mixing from the surface can interact with tidal mixing on the bottom during storms to cause some exchange between the two layers. The bottom layer in the middle shelf region is cold and relatively nutrient-rich, a remnant of the winter ocean. The shoreward edge of the middle shelf zone

is bordered by an inner front which is located at the 50 m isobath
where the water column is vertically mixed by tidal energy
(*Schumacher et al.*, 1979).

The physical system of the southeastern Bering Sea shelf is
structured into two interfront regions separated by three fronts.
This physical system is the context within which biological interactions lead to separate, highly productive pelagic and benthic
faunal assemblages. The source of nutrients for the shelf biological system is limited to deeper waters of the outer shelf zone
or to bottom water of the middle shelf zone. High nutrient concentrations occur across the shelf during winter. With the onset of
water column stability caused by surface heating, phytoplankton
blooms occur. Spring storms mix the water column terminating some
of the blooms, a phenomenon which explains near-bottom chlorophyll
maxima observed early in the season (Fig. 8). When the critical
depth criterion is met (*Sverdrup*, 1953), the major spring phytoplankton bloom occurs first in the shallow waters of the middle
shelf zone. Some of the more intense spring and summer storms mix
the water column to depths sufficient to carry new nutrients into
the photic zone (*Iverson et al.*, 1974). Following the spring
bloom, phytoplankton sink to about 30 m where a chlorophyll maximum
layer persists in the middle shelf zone into the fall.

A chlorophyll maximum was observed in the shelf break front in
July and September, 1978 cross-shelf transect data. The role of
fronts in transport of nutrient-rich water into the Bering Sea
photic zone is not yet clear. Vertical transport of nutrients may
be associated with vertical diffusion which is enhanced in areas of
interleaving of temperature and salinity (*Joyce*, 1977) such as the
Bering Sea outer shelf zone (*Coachman and Charnell*, 1979). Nutrient-
rich water is transported from lower layers into the photic zone in
fronts in the English Channel (*Pingree et al.*, 1977) and in the
Scotian shelf break front (*Fournier et al.*, 1977). *Herman and
Denman* (in manuscript) reported vertical intrusion of nutrient-rich
water into the Scotian shelf break photic zone coherent with the
M_2 tidal cycle.

The patterns of phytoplankton productivity and standing crop
levels are directly influenced by the distribution and abundance
of herbivores. There appear to be two distinct copepod communities in the southeastern Bering Sea. A shelf group is confined to
the region shoreward of the middle front while an oceanic group
exists seaward of the front. The shelf group consists of small
animals such as *Pseudocalanus* and *Acartia* which are year-round
residents of the shelf and which reproduce and develop large populations after the major spring phytoplankton bloom. Experimental
evidence suggests these animals are ineffective in grazing the
large chain-forming diatoms which dominate the phytoplankton of the
middle shelf zone (*Vidal and Dagg*, unpublished). The oceanic group

is composed of euphausiids and large calanoid copepods which winter
in deep water beyond the shelf and which move into the surface
layers of the outer shelf zone during the spring. These animals
are effective grazers of large phytoplankton, including those
species which dominate the middle shelf zone. Large calanoid
copepods were observed to graze on *Rhizosolenia alata*, the mid-
summer dominant species of the middle shelf phytoplankton.
Euphausiids grazed on *Pheocystis poucheti* as evidenced by cell
remnants in gut contents and in fecal pellets. *Pheocystis poucheti*
was the dominant phytoplankton species in near surface waters of the
shelf break front and the middle front. Physical and biological
factors interact to confine the major distribution of the large
oceanic herbivores seaward of the middle front. Flow over the
continental shelf of the southeastern Bering Sea is very sluggish,
with long-term mean speeds of the order of 1 cm s^{-1} from SE to NW,
parallel to the bathymetry. Mean flow laterally across the shelf is
negligible. Transport of mass laterally across the shelf is accomplished by diffusion defined to include tidal scales (*Coachman and
Charnell*, 1979). The fronts arise in water depths where there are
significant changes in the role of tidally-generated mixing in the
total mixing energy balance of the water column. Lateral mixing is
significantly reduced in the fronts, leading to enhanced horizontal
property gradients. Biological utilization acts on the chemical
concentration fields leading to a significant reduction in the
amount of these properties fluxed across the fronts. In the absence
of cross-shelf advection, the large oceanic herbivores, which
exhibit primarily vertical swimming behavior in diel migrations,
are generally confined to the region seaward of the middle front.

It is difficult to infer feeding relations from apparent
similarities between general distributions of zooplankton and fish.
Fish may have recently moved into a region of zooplankton abundance
without having fed, may not feed as a consequence of preference, or
may not feed while spawning. *Takahashi and Yamaguchi* (1972)
reported that the food of adult pollock is primarily of pelagic
origin. Guts of large adult pollock caught on PROBES cruises contained euphausiids, amphipods, and shrimp. Juvenile pollock guts
contained large oceanic copepods, while larval pollock guts contained eggs and nauplii of shelf group copepods (*Clarke*, 1978).
The pattern in PROBES data is that both biomass and numbers of large
zooplankton and fish appear to be concentrated offshore of the
middle shelf zone. This observation is corroborated in National
Marine Fisheries Service finfish censuses where populations of
walleye pollock, pacific cod, and sable fish, all of which are
pelagic fishes, were observed primarily in waters deeper than 100 m
in the southeastern Bering Sea (*Bakkala and Smith*, 1978). Bird
densities were greater seaward of the middle shelf zone. Fin whales
were observed in greatest numbers between the 100 m and 200 m
isobaths around the Pribilof Islands (*Nasu*, 1974).

In contrast to the pelagic food web, the benthic food web
which supports the largest benthic biomass in the southeastern
Bering Sea is localized in the middle shelf zone. Benthic infauna
biomass is maximal in the region of the middle shelf zone where
vertically integrated chlorophyll was maximal. This region is
located away from the Alaska Peninsula in an area where frontal
structure is particularly well developed (*Coachman and Charnell*,
1979). The benthic food web location is also corroborated in
finfish census data. Yellowfin sole, a demersal fish, is seldom
found deeper than 100 m in the southeastern Bering Sea (*Bakkala
and Smith*, 1978). King crabs and tanner crabs are caught exclusively in the middle shelf zone of the southeastern Bering Sea.

Although equatorial and coastal upwelling fronts have long
been known to be biologically productive, supporting large stocks
of fish (*Cushing*, 1971) and birds (*King and Pyle*, 1957; *Ashmole*,
1971), the significance of fronts for the food web dynamics on
continental shelves has only recently been recognized. Phytoplankton blooms were associated with fronts in the English Channel
(*Pingree et al.*, 1975) and in the Celtic and Irish Seas (*Savidge*,
1976). Phytoplankton productivity and chlorophyll values were
enhanced in the Nova Scotian shelf break front (*Fournier et al.*,
1977) which is a region of maximum fishing effort on the Scotian
shelf (*Fournier*, 1978). Not only are biological processes enhanced
in the fronts of the southeastern Bering Sea but the fronts play
a role in the control of biological processes which leads to a
spatial separation of pelagic and benthic food webs on the continental shelf of the southeastern Bering Sea.

ACKNOWLEDGMENTS

PROBES is funded by the Division of Polar Programs under grant
no. DPP7623340 to the University of Alaska. Bird data were
collected under BLM-NOAA (OCSEAP) contract no. 03-5-022-72. We
appreciate the superb performance of the captain and crew of the
R/V Thomas G. Thompson as well as that of students and technicians
involved in PROBES.

REFERENCES

Alton, M. S., 1974. Bering Sea benthos as a food resource for
    demersal fish populations, In: *Oceanography of the Bering
    Sea*, edited by D. N. Hood and E. J. Kelley, 257-277, Institute
    of Marine Sciences, University of Alaska, Fairbanks.

Armstrong, F. A. J., C. R. Stearns, and J. D. H. Strickland, 1967.
The measurement of upwelling and subsequent biological
processes by means of the Technicon AutoAnalyzer and
associated equipment, *Deep-Sea Res. 14*: 381-389.

Ashmole, N. P., 1971. Seabird ecology and the marine environment,
In: *Avian Biology,* edited by D. S. Farner and J. R. King,
1: 233-286, Academic Press.

Bakkala, R. G. and G. B. Smith, 1978. Demersal fish resources of
the eastern Bering Sea: Spring, 1976, Northwest and Alaska
Fisheries Center Processed Report, 234 pp., U.S. Department of
Commerce, National Marine Fisheries Service, Seattle,
Washington.

Beamish, P. C., 1969. *Quantitative Measurement of Acoustical
Scattering from zooplanktonic Organisms,* Ph.D. Dissertation,
University of British Columbia.

Beamish, P. C., 1971. Quantitative measurements of acoustical
scattering from zooplanktonic organisms, *Deep-Sea Res. 18*:
811-822.

Bendschneider, K. and R. J. Robinson, 1952. A new spectrophotometric determination of nitrite in seawater, *J. Mar. Res. 2*:
87-96.

Clarke, M. E., 1978. *Some Aspects of the Feeding Biology of Larval
Walleye Pollock,* Theragra chalcogramma *(Pallas), in the
Southeast Bering Sea,* M.S. Thesis, University of Alaska.

Coachman, L. K. and R. L. Charnell, 1977. Finestructure in outer
Bristol Bay, Alaska, *Deep-Sea Res. 24*: 869-889.

Coachman, L. K. and R. L. Charnell, 1979. On lateral water mass
interaction -- A case study, Bristol Bay, Alaska, *J. Physical
Ocean. 9*: 278-297.

Cooney, R. T. and C. Geist, 1978. Studies of zooplankton and
micronekton in the southeast Bering Sea, In: *Environmental
Assessment of the Southeast Bering Sea: Final Report,*
edited by R. T. Cooney, 238 pp., University of Alaska,
Fairbanks.

Cushing, D. H., 1971. Upwelling and the production of fish, *Adv.
Mar. Biol. 9*: 255-334.

Fournier, R. O., 1978. Biological aspects of the Nova Scotian
    shelf break fronts, In: *Oceanic Fronts in Ocastal Processes*,
    edited by M. J. Bowman and W. E. Esaias, 69-77, Springer
    Verlag, New York.

Fournier, R. O., J. Marra, R. Bohrer, and M. Van Det, 1977.
    Plankton dynamics and nutrient enrichment of the Scotian
    shelf, *J. Fish. Res. Board Can. 34*: 1004-1018.

Frost, B. W. and L. E. McCrone, 1974. Vertical distribution of
    zooplankton and myctophic fish at Canadian weathership P, with
    a description of a new multiple net trawl, Proc. Int. Conf.
    on Engineering in the Oceanographic Environment, Halifax,
    Nova Scotia, *1*: 159-165.

Guillard, R. R. L. and P. Kilham, 1977. The ecology of marine
    planktonic diatoms, In: *The Biology of Diatoms*, edited by
    D. Werner, 372-469, University of California Press, Berkeley.

Halflinger, K., 1978. *A Numerical Analysis of the Benthic Infauna
    of the Southeastern Bering Sea Shelf*, M.S. Thesis, University
    of Alaska.

Herman, A. W. and K. L. Denman. Vertical mixing at the shelf-slope
    water front south of Nova Scotia as measured with "Batfish"
    (in manuscript).

Hunt, Jr., G. L., 1979. Reproductive ecology, foods and foraging
    areas of seabirds nesting on the Pribilof Islands, Environ-
    mental Assessment of the Alaskan Continental Shelf (in press).

Iverson, R. L., H. C. Curl, Jr., H. B. O'Connors, Jr., D. Kirk,
    and K. Zakar, 1974. Summer phytoplankton blooms in Auke
    Bay, Alaska driven by wind mixing of the water column, *Limnol.
    Oceanogr. 19*: 271-278.

Iverson, R. L., T. E. Whitledge, and J. J. Goering. Fine structure
    of chlorophyll and nitrate in the southeastern Bering Sea
    shelf break front (in manuscript).

Joyce, T. M., 1977. A note on the lateral mixing of water masses,
    J. Physical Ocean. 7: 626-629.

Kinder, T. H. and L. K. Coachman, 1978. The front overlaying the
    continental slope in the eastern Bering Sea, *J. Geophys. Res.
    83*: 4551-4559.

King, J. E. and R. L. Pyle, 1957. Observations on seabirds in the
    tropical Pacific, *Condor 59*: 27-39.

Macaulay, M. C., 1978. *Quantitative Acoustic Assessment of Zooplankton Standing Stock*, Ph.D. Dissertation, University of Washington.

McRoy, C. P. and J. J. Goering, 1974. The influence of ice on the primary productivity of the Bering Sea, In: *Oceanography of the Bering Sea*, edited by D. W. hood and E. J. Kelley, 403-421, Institute of Marine Science, University of Alaska, Fairbanks.

McRoy, C. P. and J. J. Goering, 1976. Primary production budget for the Bering Sea, *Marine Sci. Comm. 2*: 255-267.

Murphy, J. and J. P. Riley, 1962. A modified single solution method for the determination of phosphate in natural waters, *Anal. Chim. Acta 27*: 31-36.

Nasu, T., 1974. Movement of baleen whales in relation to hydrographic conditions in the northern part of the North Pacific Ocean and the Bering Sea, In: *Oceanography of the Bering Sea*, edited by D. W. Hood and E. J. Kelley, 345-361, Institute of Marine Science, University of Alaska, Fairbanks.

Patton, C. J. and S. R. Crouch, 1977. Spectrophotometric and kinetic investigation of the Berthelot reaction for the determination of ammonia, *Anal. Chem. 49*: 464-469.

Pingree, R. D., P. R. Pugh, P. M. Holligan, and G. R. Forster, 1975. Summer phytoplankton blooms and red tides along tidal fronts in the approaches to the English Channel, *Nature 258*: 672-677.

Pingree, R. D., P. M. Holligan and R. N. Head, 1977. Survival of dinoflagellate blooms in the western English Channel, *Nature 265*: 266-269.

Savidge, G., 1976. A preliminary study of the distribution of chlorophyll a in the vicinity of fronts in the Celtic and Western Irish Seas, *Estuarine and Coastal Mar. Sci. 4*: 617-625.

Schumacher, J. D., T. H. Kinder, D. J. Pashinski, and R. L. Charnell, 1979. A structural front over the continental shelf of the eastern Bering Sea, *J. Physical Ocean. 9*: 79-87.

Slawyk, G. and J. J. MacIsaac, 1972. Comparision of two automated ammonium methods in a region of coastal upwelling, *Deep-Sea Res. 19*: 521-524.

Strickland, J. D. H. and T. R. Parsons, 1972. A practical handbook
    of seawater analysis, *Fish. Res. Board Can. Bulletin 167*
    (2nd ed.), 310 pp.

Sverdrup, H. U., 1953. On conditions for the vernal blooming of
    phytoplankton, *J. du Conseil 18*: 287-295.

Takahashi, Y. and H. Yamaguchi, 1972. Stock of the Alaska pollock
    in the eastern Bering Sea, *Bull. Jap. Soc. Sci. Fish. 38*:
    418-419.

Urick, R. J., 1967. *Principles of Underwater Sound for Engineers*,
    342 PP., McGraw-Hill Co., Inc., New York.

Whitledge, T. E., S. C. Malloy, and C. J. Patton (in press).
    Automated nutrient analyses in seawater, Oceanographic
    Sciences Division, Brookhaven National Laboratory technical
    report.

Wilimovsky, N. J., 1974. Fishes of the Bering Sea: the state of
    existing knowledge and requirements for future effective
    effort, In: *Oceanography of the Bering Sea*, edited by
    D. W. Hood and E. J. Kelley, 243-256, Institute of Marine
    Science, University of Alaska, Fairbanks.

VI. Estuarine/Shelf Interactions

ANAEROBIC BENTHIC MICROBIAL PROCESSES: CHANGES FROM THE ESTUARY

TO THE CONTINENTAL SHELF[1]

W. J. Wiebe[2]

University of Georgia

ABSTRACT

*Anaerobic microbial processes produce a unique set of conditions for plants and animals in salt marsh estuaries. These processes are most active within the estuary and very nearshore environments. The change in their activity as one **proceeds** seaward affects both inorganic and organic nutrient regeneration and animal activities.*

INTRODUCTION

In recent years marine scientists have begun to recognize the wide variety of roles that non-photosynthetic microorganisms play in estuarine and ocean systems. For example, the microorganisms can facilitate transformations of organic and inorganic material, serve as a food source for animals, cause disease and restrict animal and plant distribution and growth. For a variety of reasons, most attention has been paid to aerobic microbial activities and growth. However, in estuarine sediments where organic matter is abundant, anaerobic activities have the potential to influence greatly the chemical and biological nature of the habitat.

---

[1]Contribution No. 392 of the University of Georgia Marine Institute

[2]Present address: CSIRO, Division of Fisheries and Oceanography, P.O. Box 20, North Beach, Western Australia, 6020.

Indeed, *Baas-Becking and Wood* (1955) suggested that the highly
active, anaerobic, microbial conversion of plant material in
sediment was responsible for the sustained biotic richness of
estuaries. In this paper I wish to discuss the major anaerobic
microbial activities and some consequences of these activities in
the Sapelo Island, Georgia, salt-marsh-estuarine ecosystem and
the surrounding nearshore and coastal waters.

THE CYCLES

The interrelationships between the anaerobic processes are
diagrammed in Fig. 1. In fermentation, the oxidation of one
organic compound is coupled with the reduction of another. Energy
is derived from the oxidation step while the reductive compound
serves as the terminal electron acceptor. Some common end
products of fermentation are acetate, lactate, formate and
dihydrogen.

Methane production results from the anaerobic oxidation of H_2
and reduction of CO_2, which serves as the terminal electron
acceptor. The result is methane. In addition to dihydrogen, a
few organic compounds such as methanol, formate and acetate have
been reported to serve as energy sources for methanogens (see
Zeikus, 1977).

Dissimilatory nitrogenous oxide reduction, or denitrification,
is an obligately anaerobic process and involves the reduction of
nitrate to dinitrogen through a series of steps. The nitrogenous
oxides, NO_3^-, NO_2^-, NO and N_2O, serve as terminal electron acceptors
during the oxidation of organic compounds. Respiratory nitrate
reduction to ammonia via nitrite has been reported for some
clostridia and, in fact, may represent a much more important route
of dissimilatory reduction than has been previously thought.
This process is called anaerobic respiration, since the nitrogenous oxides substitute for oxygen in the terminal steps of
cytochrome-mediated electron flow. One reason anaerobic

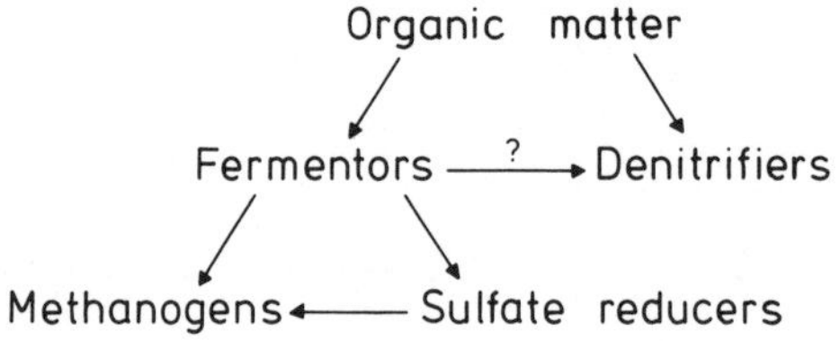

Fig. 1. Major flow of substrate through anaerobes. + =
stimulation, - = competition.

respiration has interested biologists is that, in this case, the anaerobic oxidation of carbon compounds proceeds almost as efficiently as during aerobic oxidation.

The position of denitrification in the nitrogen cycle can be seen in Fig. 2. It is the only process in the cycle that must proceed exclusively under anaerobic conditions. Interestingly, nitrification, the oxidation of ammonia to nitrate via nitrite, is an obligately aerobic process; thus organisms producing nitrate must be spatially separated from those that can dissimilate it. All other processes in the nitrogen cycle can occur under both aerobic and anaerobic conditions. Nitrogen fixation, denitrification and nitrification are (with the exception of two fungi) obligately procaryotic processes.

Dissimilatory sulfate reduction, the reduction of sulfate to hydrogen sulfide via several intermediates, is also an obligately anaerobic process and another case of anaerobic respiration.

THE NITROGEN CYCLE

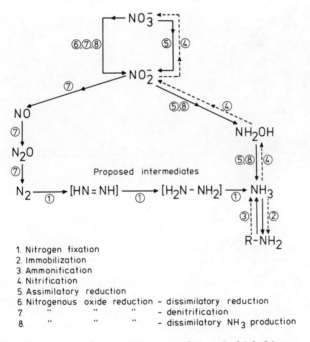

1. Nitrogen fixation
2. Immobilization
3. Ammonification
4. Nitrification
5. Assimilatory reduction
6. Nitrogenous oxide reduction - dissimilatory reduction
7.   "   "   "   - denitrification
8.   "   "   "   - dissimilatory NH_3 production

Fig. 2. Diagram of nitrogen cycle. Solid lines = reductive steps, dashed lines = oxidative steps, combined = neither oxidation nor reduction.

The organic substrates that can be used by the sulfate reducers are very few; in contrast, dissimilatory nitrogenous oxide reducers can metabolize nearly as wide a range of substrates as aerobes. The sulfate reducers use only a few low molecular weight compounds, such as those produced during fermentation. Most leave some terminal organic product, commonly acetate.

The sulfur cycle, as shown in Fig. 3, is, like the nitrogen cycle, a complex cycle involving many possible steps. The actual sequence, the enzymes, and the controls on different steps are less well known than those for nitrogen. Hydrogen sulfide can be produced from amino acids as well as by sulfate reduction; here too the process is obligately anaerobic. Sulfide oxidation proceeds under aerobic conditions and anaerobically in the case of some nitrogenous oxide reducers. In addition to biological transformation, strictly chemical transformation also takes place.

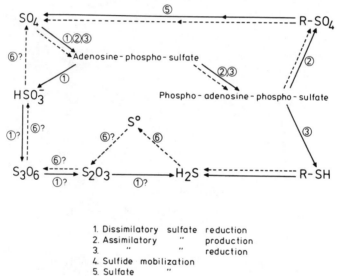

1. Dissimilatory sulfate reduction
2. Assimilatory   "   production
3.   "   "   reduction
4. Sulfide mobilization
5. Sulfate   "
6. Sulfide oxidation

Fig 3. Diagram of sulfur cycle. Solid lines = reductive steps, dashed lines = oxidative steps, combined = neither oxidation nor reduction, ? = hypothesized intermediate. (The intermediates in both oxidation of H_2S and reduction of $SO_4^=$ are poorly known and other pathways and short circuits undoubtedly exist.)

In this section I have briefly introduced the four major
anaerobic processes and their positions within the respective
cycles. Let us now examine:

1)   what environmental factors control their function *in situ*,

2)   how these functions affect aerobic activities, and

3)   the magnitude and consequences of these activities in
     one ecosystem.

CONTROLS ON ANAEROBIC PROCESSES

Whether a particular process occurs in nature and the magnitude to which it occurs depend not only on the presence of
the appropriate organism but also on the milieu. Microorganisms
have ubiquitous distribution but can be active only under specific
conditions. In this section I want to consider the types of
external factors that can exert control on the anaerobic processes.
The list of factors to be considered is not all inclusive, and of
course the factors do not act unilaterally. When we examine a
system for some property, for example methane production, the
result represents the summation of external and intrinsic controls.

Production and maintenance of anaerobic conditions depend
upon the availability of sufficient reduced compounds to permit
the removal of all oxygen and to prevent its penetration into the
habitat. The required amount of reducing compound depends upon
the other features discussed below. In salt-marsh soils which
contain rooted plants, excess reducing agents are available.
However, because plant roots provide passage for oxygen into the
sediments, the aerobic and anaerobic zones are not vertically
segregated; microzones of different conditions can exist side by
(see *Jorgensen*, 1977, for a discussion of sediment microzonation).

Water flow across the surface and within the sediment affects
oxygen availability. In addition, it affects the availability of
inorganic ions. The hydrological properties appear of paramount
importance in controlling sediment anaerobic processes. The
nature of the substrate (e.g. clay, sand, shell), the surface
slope and the rate of sedimentation interact to establish the
physical framework for the fluid dynamic regime. The activities
of animals within sediment may also affect water movement and, in
addition, the exchange of anaerobic products to the aerobic
sediments and the overlying water. Bioturbation occurs on a
microscale, mediated by meiofauna, as well as on a macroscale
(e.g., the burrowing of fiddler crabs). Such activities not only
can increase the oxygen exchange with the sediment but they also
contribute to the redistribution of reduced substrates found at

the surface and provide routes for the venting of gaseous
end products such as methane, hydrogen sulfide and dimethyl
sulfide.

The metabolic capacity of a system is linked to the *in situ*
temperature. Where temperatures vary seasonally, activities may
follow suit. While sequential succession of psychrophilic and
mesophilic populations has been documented in estuarine waters
(*Sieburth*, 1965), *Nedwell and Floodgate* (1972) found that
dissimilatory sulfate reduction virtually ceased in winter and
King and Wiebe (1978a) observed that methanogenesis decreased
dramatically in the winter months in the intertidal saltmarsh
sediments at Sapelo Island, Georgia.

Finally, sporadic events, such as storms, appear to play an
important role in setting environmental conditions (see *Peterson*,
1979). Livingston and Duncan (1979) have pointed out the long
term effects that meterological events can exert in the control of
estuarine populations. *Imberger et al.* (ms. submitted) have
described how heavy rains can facilitate redistribution of surface and subsurface sediments.

INTERACTION OF ANAEROBIC AND AEROBIC PROCESSES

Anaerobic activities affect aerobic environments and
organisms in both positive and negative ways. We have briefly
reviewed the N and S cycles and have observed that several key
steps are obligately anaerobic and that anaerobic metabolism often
yields organic end products.

Fermentation and sulfate reduction yield dissolved organic
carbon (DOC) end products, while a special group of procaryotes
produces methane. These end products diffuse, or are transported
by bioturbation, water flow, etc., into aerobic sediments where
they are either immediately metabolized or escape into the water
or the air. *Baas-Becking and Wood* (1955) stated that this
constant anaerobic transformation and release of DOC was the key
to the productive nature of salt marshes because it releases
heterotrophic feeding from direct temporal dependence on primary
production. Organic matter capable of being metabolized by aerobes
is continually released from the anaerobic sediments even if
primary production is highly seasonal. *Rich and Wetzel* (1978)
have discussed the importance of anaerobic release of DOC in
lakes and suggest that it represents an important source of
nutrition for water column heterotrophs. Methane, while a gaseous
product, has been demonstrated to be efficiently metabolized in
lake water columns with up to 90% of the total amount released
refixed (*Strayer and Tiedje*, 1978). *Atkinson and Richards* (1967)
demonstrated a similar high ratio of capture in anoxic marine

basins. However, in intertidal marshes of Sapelo Island, methane metabolism in the aerobic zone was extremely low; blockage of aerobic metabolism had almost no effect upon the rate of release (*King and Wiebe*, 1978a).

Denitrification results in the loss of fixed nitrogen into the atmosphere as N_2. *Ryther and Dunstan* (1971) speculated that marsh and coastal systems were limited by available nitrogen and, thus, denitrification could serve to control primary production. However, while estuaries might be considered ideal sites for denitrification, *Sherr and Payne* (1978) reported little *in situ* activity in salt marsh soil.

Sulfate reduction results in both negative and positive effects. For example, volatile sulfide has been shown to inhibit plant growth (*Hollis et al.*, 1975), while *Patrick and Khalid* (1974) have shown that sulfate reduction may be necessary for the mobilization of soil phosphate from bound iron phosphate to free phosphate ion. Under aerobic conditions phosphate complexes with iron to form $FePO_4$, but under anaerobic conditions free sulfide preferentially binds iron to form FeS, releasing the inorganic phosphate. *Jorgensen and Fenchel* (1974) discussed this and other aspects of the sediment sulfide production and concluded that its production was an important factor in the movement of inorganic nutrients from the sediment into aerobic zones.

Anaerobic microbial activities reduce the redox potential of sediments; in some cases potentially toxic compounds, such as H_2S, are produced. These compounds can affect the distribution patterns of plants and animals directly (*Theede et al.*, 1969). There is increasing evidence that sulfide can affect plant distribution. It is not sufficient that halophytes are just salt tolerant; in some zones they must be sulfide tolerant as well, or they must develop mechanisms to detoxify their immediate environment. One example of this latter strategy is provided by *Joshi and Hollis* (1977). They demonstrated the synergistic relationship between the sulfide oxidizer *Beggiatoa* and rice plant roots. The *Beggiatoa* oxidizes hydrogen sulfide to sulfate, while the plant roots provide carbon compounds, oxygen in small amounts and catalase. Neither organism can exist in this environment alone.

Microbiologists and geochemists have speculated on the long term geochemical consequences of anaerobic metabolism in nature. *Trudinger et al.* (1972) listed ten sulfide minerals (*e.g.* CuS, PbS, ZnS, FeS, FeS_2) whose biogenic origin had been demonstrated in the laboratory. They did not suggest that large, natural, mineral deposits were necessarily of biogenic origin but rather that such origin was possible. *Trudinger et al.* (1972) also discussed the energy requirements for such activities;

they concluded that for some mineral deposits, such as the Roan
Antelope and Kupferochiefer, present rates of sulfate reduction
and carbon fixation were sufficient to account for mineral
deposition. Estuaries, because of their generally high rates of
primary production and extensive anoxic areas, are considered
historically likely sites for major ore genesis. There is a great
deal of interest at present in using present estuarine activities
to develop models of biorganic ore formation (*e.g. Skyring and
Chambers*, in press). It appears that anaerobic processes are
essential for some steps in formation and deposition of several
metals and that ancient estuaries may have been major sites of
mineral deposition.

Finally, the anaerobic microorganisms serve as food for
animals. In this regard they do not differ from the aerobes, but
they mobilize nutrients and organic matter in regions where the
aerobes cannot function. The role of microorganisms as a major
food source for estuarine animals has been demonstrated recently
by *Wetzel* (1977), *Lopez et al.* (1977) and *Levinton and Lopez*
(1977), among others. These authors showed that some animals
do not derive adequate nutrition directly from marine detritus
but rather that they consume the attached and free microflora.
Often, animals do not possess sufficient enzymatic diversity to
metabolize many of the major compounds produced by plants; they
must rely upon procaryotic organisms to accomplish the initial
breakdown and then consume the microorganisms and their products.
In this sense the anaerobic microflora of saltmarshes and estuaries
perform functions similar to those of the ruminant microflora.

In summary, the anaerobic and aerobic processes in estuaries
and salt marshes interact in a variety of ways. It is in large
part these interactions which produce the unique, rich biotic
environments collectively called salt marsh estuaries.

THE SAPELO ISLAND SALTMARSH

In previous sections, the general consequences of anaerobic
metabolism have been discussed. Now I wish to examine the roles
of these processes in one specific ecosystem, the marshes and
coastal waters around Sapelo Island, Georgia. This region can be
divided into three zones: the saltmarsh estuary, nearshore and
coastal zones.

The Duplin River saltmarsh-estuarine watershed at Sapelo
Island consists of about 1,100 ha with intertidal marshes covering
85% of the total area. There are two high and low tides each day,
with a two- to three-meter tidal amplitude; most of the marsh is
inundated on each tide. The dominant halophyte, *Spartina
alterniflora*, grows in the spring and summer, flowers in fall

and virtually ceases growth in winter. Roots extend into the
sediment and provide an aerobic rhizosphere deep within the soil;
roots are also responsible for the release of DOC into the sediments. Bioturbation by fiddler crabs, polychaetes and meiofauna
facilitates sediment reworking. In addition storms contribute
to the turnover of the soil.

In this system we find two distinct types of plant growth:
tall *Spartina* (TS) plants along the creek banks, and short *Spartina*
forms (SS) over the marsh flats. Plant growth, root distribution
and the microbiology differ greatly in these two zones (see
Christian et al., 1975). The TS zone plants reach a height of about
two meters and roots penetrate the sediment over one meter, while
SS zone plants are less than one meter in height and roots form a
compact mat 15-20 cm in depth. The major differences in anaerobic
microbial activities in the two zones are summarized below.

Denitrification has been measured in TS and SS zones; the
activity appears to be low in both (*Sherr and Payne*, 1978). In
TS soil denitrification is organic substrate limited, while in
the SS soil there is no evidence of substrate limitation. *Hanson*
(1977) measured sediment nitrogen fixation in the same cores used
for denitrification and found activity equal to or slightly less
than denitrification. In this system denitrification does not
appear to play a significant quantitative role in either nitrogen
dynamics or carbon energetics.

Fermentation was measured using ^{14}C glucose heterotrophy
techniques developed by *Christian and Wiebe* (1978). The TS zone
activity was highly substrate-limited compared to activity in the
SS zone. In a series of perturbation experiments *Christian et al.*
(1978) found that removing the above-ground portion of *Spartina* in
the SS zone did not affect fermentation rates for over one year.
They concluded that the organic carbon used for fermentation is
derived from particulate organic carbon in sediment and not
directly from the dead or living roots or root exudates. Thus,
the supply of organic matter which drives fermentation, and subsequently sulfate reduction and methanogenesis, is not affected
by short term fluctuations of halophyte primary production. It
seems most likely that interstitial water diffusion within the
SS zone limits activity as we shall discuss later.

Sulfate reduction activity measured by the conversion of ^{35}SO$_4$
to H$_2$^{35}S was much greater in TS soil and occurred to a greater
depth in the sediment than in the SS soil (*Skyring et al.*, in
press). Sulfide may be partitioned into free and bound forms:
that is, bound as insoluble iron sulfide or free as soluble S$^=$,
HS$^-$ or H$_2$S. The TS zone contained about four times as much total
sulfide as SS soils and virtually all of it was found as bound
sulfide; by comparison, one-third to one-half of the total sulfide

in SS soils occurred as free sulfide (*Oshrain and Wiebe*, ms.
submitted). Recently *Howarth* (1979) examined sulfate reduction in
a Massachusetts saltmarsh and found that much of the recovered ^{35}S
was associated with pyrite. *Skyring et al.* (in press), however,
could find little evidence of rapid pyrite formation is Sapelo
Island soils, so it may be that, within different *Spartina* saltmarshes, processes differ greatly. *Hollis et al.* (1975)
speculated that the SS zone growth form results from the toxicity
of free sulfide to *Spartina* roots, and while the data from Sapelo
Island could be interpreted as substantiating this hypothesis, to
date no specific experiments have been performed to examine this
possibility. Recently, *Nedwell and Abram* (1978) found similar
short and tall plant zones in an English saltmarsh, where *Spartina*
was only a minor constituent. Sulfate and sulfide concentrations
differed in the two zones in a fashion similar to that at Sapelo
Island.

At Sapelo Island, *Oshrain and Wiebe* (ms. submitted) found
that sulfate concentrations in TS soils were constant down to 40
cm, the deepest zone investigated, while SS soil sulfate concentrations decreased below about 3 cm. When the relationship between
salinity and sulfate concentration was examined, it was found that
TS soils contained about the same or slightly less sulfate than
would be predicted by the salinity:sulfate ratio; SS soil ratios
showed a sulfate depletion below 1-3 cm of up to 40%.

The total sulfate concentration (free plus bound) for entire
cores within one zone (that is, the total sulfide found by adding
all of the individually analyzed core sections) varied little from
core to core. However, when specific depths in different cores
were compared (*e.g.* the 3-5 cm horizon), great differences in
sulfide concentration were seen. *Oshrain and Wiebe* (ms. submitted)
interpreted these results as meaning that a fixed proportion of
the primary production was involved in sulfate reduction, but
that the specific depth at which sulfate reduction took place
depended upon local events such as root growth and bioturbation.
While specific events can redistribute this material vertically,
the system is committed to a constant amount of sulfate reduction.
This hypothesis, while still not completely validated, suggests
that there may be a great deal of homeostasis in activities in
these soils.

The differences between the two systems is best seen in the
methane production data. *King and Wiebe* (1978a) found that 0.4
g CH_4-C/m^2/yr were released in the TS soil and about 60 g CH_4-C/m^2/
yr in the SS soil. Production was highly seasonal, virtually all
occurring between April and October. *King and Wiebe* (1978b)
suggested that competition for substrate, between sulfate reducers
and methane producers, may be responsible for the differences in

rates. *Oremland and Taylor* (1978) have experimental evidence for this type of competition in sediments of tropical seagrass beds.

In subsequent experiments, *King and Wiebe* (1978b) demonstrated that the differences in anaerobic metabolism in the two zones were largely the result of differences in the interstitial water flow. *Nestler* (1977) showed that considerable water flow occurs in the TS zone soil but that such flow through the SS zone was greatly restricted. Residence time for water in TS soil is about two days, while water moves only centimeters per month within SS soils.

The difference in water flow between the two zones could explain the differences in sulfate depth profiles and free and bound sulfide concentrations. In TS soils sulfate is continually added; thus, even in the presence of active sulfate reduction, little sulfate depletion would be expected. Iron is also imported with the water, permitting most of the sulfide that is formed to be quickly bound. In SS soils, on the other hand, interstitial water exchange with tidal water is very slow; sulfate appears not to be replaced rapidly enough to prevent local depletion and free iron also is in short supply. To examine the possibility that interstitial water flow was the dominant factor controlling sulfate reduction and methanogenesis, *King and Wiebe* (1978b) placed 7.5 cm diameter cores in the TS marsh soil; some cores were left open, others capped at the top or top and bottom. After two months, control sites and uncapped cores gave identical, low rates of methanogenesis, while cores capped at the top or top and bottom yielded quantities of methane equivalent to those of SS soil. Furthermore, in capped cores, sulfate was depleted by up to 40%. It appears that, of the external factors controlling anaerobic metabolism discussed previously, interstitial water flow is the dominant feature in the Duplin River salt marsh.

The anaerobic processes appear to play important roles in the ecology of the Sapelo Island saltmarsh soil. They may limit plant production in some zones, mobilize inorganic nutrients (*e.g.* phosphate), provide a continuing source of available organic matter to aerobic organisms and restrict the distribution of both micro- and macroorganisms.

THE NEARSHORE ZONE

The nearshore zone is defined here as the region in which the sediments are subject to wave and tidal action. The outer limit of this zone varies with conditions but it extends no further than 5-10 km offshore and to a depth of 10-15 m. This zone represents a transition in many biological characteristics. In the ocean waters near Sapelo Island, transparency (*Campbell,*

pers. comm.) and adenosine triphosophate (ATP) concentrations
(*Bancroft*, 1977) decrease dramatically within a few kilometers
of shore. *Thomas* (1966) measured primary production in transects
across this region and found that carbon fixation decreased from
an average of 285 g C fixed/m^2/yr in the inner 10 km to 130 g C
fixed/m^2/yr seaward of this point. *Hanson and Wiebe* (1977)
reported that heterotrophic activity decreased as one went seaward in this same region, and that the effect of tidal activity
in resuspending microorgamisms and sediment could be detected
several miles to sea.

There has been little work done on the nearshore anaerobic
processes either at Sapelo Island or elsewhere. *Wiebe, Oshrain
and King* (unpublished results) measured sediment methane and
total sulfide concentrations on a track from Doboy Sound, which
is the sea connection with the Duplin River watershed, to about
15 km offshore. Nearshore, the surface layers of muddy sediments
(15-20 cm depths) contained significant methane and sulfide concentrations but values were generally less than 10% of those
found in the saltmarsh soils. Concentrations decreased rapidly
offshore and became virtually undetectable even inshore when
sandy sediments were encountered. Sulfate reduction and
methanogenesis in the upper 20 cm of sediment decreased rapidly
within the nearshore zone and became undetectable often within
sight of the beach. While it is known that sulfide and methane
are present tens of meters below the sediment surface, little is
known about their rates of production. **Production is most likely**
very slow and would have little quantitative effect on either
animal distribution or nutrient regeneration.

THE COASTAL ZONE

In the coastal zone beyond the nearshore waters the transition
from saltmarsh to open ocean sediment is completed. Methane and
sulfide concentration in the upper 10-20 cm are barely detectable
and the sediment redox potential of this zone increases to
positive values. The surface sediments become largely aerobic
environments. As this change occurs, the mode of nutrient
regeneration must necessarily change also. It is no longer
dominated by anaerobic processes but rather by animals feeding on
the benthic microflora. The benthos becomes progressively unlinked
with surface activity (see *Peterson*, 1979, for a discussion of
this effect on animal distribution). The consequence of this
change to nutrient regeneration, for example, is that sediment
nutrients can no longer be mobilized by anaerobic activities.
For example, phosphate probably resides mainly within aerobic
bacteria and can be released only by grazing animals (see *Johannes*,
1964, 1965, for a discussion of this phenomenon). Coastal
sediment organisms must rely on direct supply of organic carbon
to sustain them, and they are not provided with the buffer of

continually produced anaerobic organic compounds. Large seasonal
changes in the types and abundance of benthic animals have been
documented in these waters off Sapelo Island (*Leiper*, 1973) and
this may well be a result of transient food availability.

CONCLUSIONS

The anaerobic microflora within the benthos of salt marsh
and very nearshore sediments appears to play a major role in
organic and inorganic nutrient supply to the aerobic sediment and
water column organisms. Some of the products (*e.g.* sulfide)
can restrict the distribution of plants and aminals. The
dynamics of these processes and the quantitative linking of
sediment with water column activities have only recently been
investigated. Nevertheless, it is evident that activities in
the two zones can be highly interactive. Because of the potential
interactions, a much clearer picture of the causes and effects
of one anaerobic process emerges when it is studied simultaneously
with the other processes and the physical environmental conditions.
Baas-Becking and Wood (1955) predicted, and present work is confirming, that the major factor which causes estuaries to be such
biotically rich environments is the presence of an active anaerobic
microflora that controls inorganic nutrient and organic carbon
flux to the aerobic environments.

ACKNOWLEDGEMENTS

This work was supported by NSF Grant No. DES75-20845 and by
Grants from the Sapelo Island Research Foundation.

REFERENCES

Atkinson, L. P. and F. A. Richards, 1967. The occurrence and
distribution of methane in the marine environment, *Deep-sea
Res. 14*: 673-684.

Baas-Becking, L. G. M. and E. J. F. Wood, 1955. Biological
processes in the estuarine environment. I. Ecology of the
sulfur cycle, *K. Akad. van Wetenshappen, Amsterdam, Proc.
Phys. Sci. Series B. 58*: 160-181.

Bancroft, K., 1977. *The Use of the Adenylate Energy Charge to
Measure Growth State in Microbial Ecosystem*, Ph.D.
dissertation, University of Georgia.

Christian, R. R., K. Bancroft, and W. J. Wiebe, 1975. Distribution
    of microbial adenosine triphosphate in salt marsh sediments
    at Sapelo Island, Georgia, *Soil Sci. 119*: 89-97.

Christian, R. R., K. Bancroft, and W. J. Wiebe, 1978. Resistance
    of the microbial community within salt marsh soils to
    selected perturbations, *Ecology 59*: 1200-1210.

Christian, R. R. and W. J. Wiebe, 1978. Anaerobic microbial
    community metabolism in *Spartina alterniflora* soils, *Limnol.
    Oceanogr. 23*: 328-336.

Hanson, R. B., 1977. Comparison of nitrogen fixation activity in
    tall and short *Spartina alterniflora* salt marsh soils, *Appl.
    Environ. Microbiol. 33*: 569-602.

Hanson, R. B. and W. J. Wiebe, 1977. Heterotrophic activity
    associated with particulate size fractions in a *Spartina
    alterniflora* salt-marsh estuary. Sapelo Island, Georgia, USA,
    and the continental shelf waters, *Mar. Biol. 42*: 321-330.

Hollis, J. P., A. I. Allan, G. Pitts, M. M. Joshi, and I. K. A.
    Ibrahim, 1975. Sulfide diseases of rice on iron-excess soils,
    Acta Phytopate. Academ. Scient. Hungr. 10: 329-341.

Howarth, R. W., 1979. Pyrite: Its rapid formation in a salt-marsh
    and its importance to ecosystem metabolism, *Science 203*:
    49-51.

Imberger, J., T. Berman, R. R. Christian, E. G. Haines, R. B.
    Hanson, L. R. Pomeroy, D. Whitney, W. J. Wiebe, and R. G.
    Wiegert, The influence of water motion on spatial and temporal
    variability of chemical and biological substances in a salt
    marsh estuary, submitted to Limnol. Oceanogr.

Johannes, R. E., 1964. Phosphorus excretion and body size in
    marine animals: microzooplankton and nutrient regeneration,
    Science 146: 923-924.

Johannes, R. E., 1965. Influence of marine protozoa in nutrient
    regeneration, *Limnol. Oceanogr. 10*: 434-442.

Jorgensen, B. B., 1977. Bacterial sulfate reduction within
    reduced microniches of oxidized marine sediments, *Mar. Biol.
    41*: 7-17.

Jorgensen, B. B. and T. Fenchel, 1974. The sulfur cycle of a
    marine sediment model system, *Mar. Biol. 24*: 189-201.

Joshi, M. M. and J. P. Hollis, 1977. Interaction of *Beggiatoa* and rice plant: detoxification of hydgrogen sulfide in the rice rhizoshpere, *Science 195*: 179-180.

King, G. M. and W. J. Wiebe, 1978a. Methanogenesis in a Georgia salt marsh and some factors controlling its production, *Geochim. et Cosmochim. Acta 42*: 343-348.

King, G. M. and W. J. Wiebe, 1978b. Effects of sulfate on methanogenesis in a Georgia salt marsh, p. 88, Abstracts of American Society for Microbiology Annual Meeting.

Leiper, A. S., 1973. *Seasonal Variations in the Structure of Three Shallow Water Benthic Communities Off Sapelo Island, Georgia,* Ph.D. Dissertation, University of Georgia.

Levinton, J. S. and G. R. Lopez, 1977. A model of renewable resources and limitation of deposit-feeding benthic populations, *Oecologia 31*: 177-190.

Livingston, R. J. and J. L. Duncan, 1979. Climatic control of a north Florida coastal system and impact due to upland forestry management, In: *Ecological Processes in Coastal and Marine Systems,* edited by R. J. Livingston, Plenum Press, N. Y. (this volume).

Lopez, G. R., J. S. Levinton, and L. B. Slobodkin, 1977. The effect of grazing by the detritivore *Orchestia grillus* on *Spartina* litter and its associated microbial community, *Oecologia. 30*: 111-127.

Nedwell, D. B. and J. W. Abram, 1978. Bacterial sulfate reduction in relation to sulfur geochemistry in two contrasting areas of salt marsh sediment, *Estuarine and Coastal Mar. Sci. 6*: 341-251.

Nedwell, D. B. and G. D. Floodgate, 1972. Temperature-induced changes in the formation of sulfide in a marine sediment, *Mar. Biol. 14*: 18-24.

Nestler, J., 1977. Interstitial salinity as a cause of ecophenic variation in *Spartina alterniflora, Estuarine and Coastal Mar. Sci. 5*: 707-714.

Oremland, R. S. and B. F. Taylor, 1978. Sulfate reduction and methanogenesis in marine sediments, *Geochim. et Cosmochim. Acta 42*: 209-214.

Oshrain, R. L. and W. J. Wiebe, Sulfate and sulfide in salt marsh soils (submitted).

Patrick, W. H., Jr., and R. A. Khalid, 1974. Phosphate release and absorption by soils and sediments: effect of aerobic and anaerobic conditions, *Science 186*: 53–55.

Peterson, C. H., 1979. Predation, competitive exclusion, and diversity in the soft-sediment benthic communities of estuaries and lagoons, In: *Ecological Processes in Coastal and Marine Systems*, edited by R. J. Livingston, Plenum Press, N. Y. (this volume).

Rich, P. H. and R. G. Wetzel, 1978. Detritus in the lake ecosystem, *Amer. Natur. 112*: 57–71.

Ryther, J. H. and W. M. Dunstan, 1971. Nitrogen, phosphorus, and eutophication in the coastal marine environment, *Science 171*: 1008–1013.

Sherr, B. F. and W. J. Payne, 1978. Effect of the *Spartina alterniflora* root-rhizome system on salt marsh soil denitrifying bacteria, *Appl. Environ. Microbiol. 35*: 724–729.

Sieburth, J. McN., 1965. Seasonal selection of estuarine bacteria by water temperature, *J. Exper. Mar. Biol. Ecol. 1*: 98–122.

Skyring, G. W. and L. A. Chambers, Sulfate reduction in intertidal sediments, In: *Sulfur in Australia*, edited by J. R. Freeney and A. J. Nicholson, Aust. Acad. Sci. Publ., in press.

Skyring, G. W., R. L. Oshrain, and W. J. Wiebe, Assessment of sulfate reduction rates in Georgia marshland soils, *Geomicrobiology* (in press).

Strayer, R. F. and J. M. Tiedje, 1978. *In situ* methane production in a small hypereutrophic, hard water lake: loss of methane from sediments by diffusion and ebullition, *Limnol. Oceanogr. 23*: 1201–1206.

Theede, H., A. Ponat, K. Hiroka, and C. Schleiper, 1969. Studies on the resistance of marine bottom invertebrates to oxygen-deficiency and hydrogen sulfide, *Mar. Biol. 2*: 325–337.

Thomas, J. P., 1966. *Influence of the Altamaha River on Primary Production Beyond the Mouth of the River*, M.Sc. Thesis, University of Georgia.

Trudinger, P. A., I. B. Lambert, and G. W. Skyring, 1972. Biogenic sulfide ores: a feasibility study, *Econ. Geol. 67*: 1114–1127.

Wetzel, R. L., 1977. Carbon resources of a benthic salt marsh invertebrate *Nassarius absolatus* soy (Mollusca:Nassariidae), *Estuar. Proc.* 2: 293–308.

THE ESTUARY/CONTINENTAL SHELF AS AN INTERACTIVE SYSTEM

Rezneat M. Darnell and Thomas M. Soniat

Texas A&M University

ABSTRACT

The estuary/pass/shelf complex is viewed as an interactive system whose internal dynamics are controlled, in large measure, by driving forces external to the system. Internal exchange reflects the force of water mass movement and mixing processes as they interact within the local context. Transport of chemical materials is related to these factors, but it also reflects the individual reactivities of each chemical species. Some information is now available concerning the estuary/shelf exchange of biologically important elements (N, P, Si, and C), and caloric exchange may be roughly calculated from the organic carbon data. The movement and behavior of trace metals (Fe, Mn, Zn, Cd, Cu, and Hg) are also known in some instances. Large rivers represent a special case, since, lacking estuaries, they transport materials directly into the sea. The Mississippi River outflow is presented as an example.

Exchange of living organisms is related to the physical processes, but it also reflects reproductive and behavioral activities. Although phytoplankton is moved passively by water currents, its rapid reproduction permits maintenance of local populations. Zooplankton, including both holoplankters and larval stages of non-planktonic adults, exhibits behavior patterns which enhance survival of local populations and distributive movement to favorable habitats. Larger mobile species traverse the pass under their own power and display regular seasonal patterns of migration. Some information is available concerning the factors which trigger population movements.

From the standpoint of net gain or loss, there seem to be two basic types of systems: those in which there is a net annual export of organic carbon by estuaries and those in which there is a net annual import. Factors underlying these differences are discussed. Although the literature base is still rather thin, intense recent interest in estuary/shelf relations has provided new research tools for pursuit of problems which can be sketched only in outline form at present. Knowledge derived from holistic studies of the estuary/shelf system is greatly needed as a base for wise management of coastal resources.

INTRODUCTION

The problem of estuary/continental shelf relationships has long been recognized. As early as 1911 *Peterson and Jensen* considered the organic matter of the sea off Denmark to be derived from three sources: that produced in the sea itself, that brought down by rivers, and that brought in by winds. Their studies led them to conclude that the organic matter of the sea bottom off Denmark is derived chiefly from decomposing *Zostera* which grows in the shallow bays and estuaries. During subsequent years a great deal of information has accumulated on the subject of estuary/shelf relations, but until recently most of the information has been largely inferential and subjective. The topic seems not to have been the subject of a major review.

The matter is of more than academic interest. For example, *Lindall and Saloman* (1977) have recently determined that, for the states bordering the Gulf of Mexico, about 90 percent of the marine commercial catch and about 70 percent of the recreational fishery catch are estuarine-dependent. Since the Gulf states annually produce about a third of the nation's commercial and recreational catch, the combined estuarine-dependent fisheries of the Gulf states alone are valued at over $640 million. Considering the present rate of human intrusion into the coastal environments, these resources are very much in jeopardy (*Darnell,* 1971; *Darnell et al.,* 1976; *Lindall and Saloman,* 1977). A great deal of recent attention has also centered around the subject of pollutants, their dynamics and entrainment in estuaries, and their passage to the sea.

Relationships between estuaries and continental shelves are quite complex, and they are often difficult and expensive to study. However, considering the importance of the subject and the rapid rate of deterioration of the coastal ecosystems, we can no longer afford to ignore the matter. Our approach in the present discussion is to conceptualize the problem, factor it into components, review the present status of our knowledge, and indicate areas where further work is needed.

No attempt is made to provide a complete literature review, especially since much of the most important recent work is still in report form and has not yet reached the open scientific literature. Since most of the estuary/shelf relations are mediated through the interconnecting passes, reliance will be placed on a related paper dealing with the ecology of passes (see *Darnell*, this volume).

CONCEPTUALIZATION OF THE PROBLEM

Aquatic systems, in general, do not have finite functional boundaries, and details of their operations are often strongly influenced by factors external to the systems under study. This openness and external dynamic programming of internal events leads to conceptual and analytical difficulties for the investigator. In a practical vein one must define the scope of study, recognize external influences and their modes of action, establish import/export pathways, and then proceed with the internal analysis. In other words, it is absurd nowadays to begin any significant ecological study of an aquatic system without first constructing a conceptual model involving the above components. Mathematical and computer models are available for some processes and some subsystems, but the conceptual model is the *sine qua non* for the basic organization of existing knowledge prior to technical analysis. A first-step conceptual model of the estuary/pass/ shelf system is given in Fig. 1.

The estuary and adjacent continental shelf, together with the intervening pass, form a recognizable ecological system. This system is divisible into subsystems including the estuary, pass, and shelf, or it may be divided as plankton, nekton, and benthos, for example. The subsystems and their components relate in various and often predictable ways. Involved are transport and mixing processes, migratory movements, ontogenetic change in modes of life (*e.g.*, benthonic egg to planktonic young to nektonic adult), and trophodynamic exchange of chemicals and energy, to name, a few. Related external systems include the atmosphere, the adjacent land, the river which enters the estuary, and the ocean beyond the adjacent shelf.

EXTERNAL INFLUENCES

The estuary/pass/shelf complex is subject to influence by physical factors from four sources: atmosphere, the land, upstream fresh waters, and open ocean. Each of these will be examined briefly. Atmospheric influence is mediated through light, heat, wind, and precipitation. Light supports photosynthesis of attached algae and vascular plants, benthic diatoms, and phytoplankton.

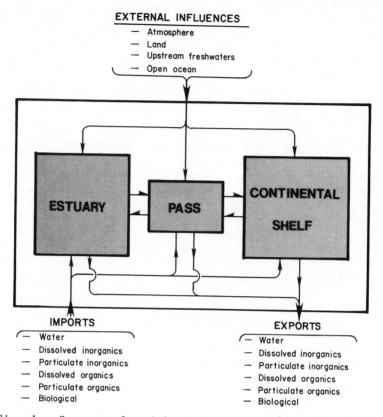

Fig. 1. Conceptual model of the estuary/pass/shelf system and its internal relations.

Heating of the surface layers (by radiant energy and atmospheric heat exchange) may lead to or reinforce density stratification of the water column and to evaporative water loss. If this loss is not balanced by freshwater input, hypersaline conditions may result, although hypersalinity is more characteristic of coastal lagoons than of estuaries, since the latter generally receive some river flow. As a rule, atmospheric temperature tends to be less variable in coastal environments than in inland areas, but sudden and severe shifts in coastal atmospheric temperature are known, and these may result in rapid changes in coastal water temperature with catastrophic biological results (*Gunter*, 1941). In addition to modifying water temperature, the wind is of particular importance through induction of water currents. Wind action may hasten or retard the flow through passes, and it may accentuate water level extremes throughout the system.

Wind-induced currents promote water mass mixing and bottom
roiling which can lead to suspension and redistribution of
inorganic sediments and particulate organic matter. Such mixing
and suspension promote air/water gaseous exchange as well as a
number of chemical processes. Precipitation increases the amount
of freshwater present. Runoff may bring in significant amounts of
dissolved and particulate matter, and the additional freshwater
may induce or intensify circulation patterns and promote transport
of materials from one part of the system to another. Since
coastal precipitation is often quite heavy, these effects may be
quite significant.

Most influence from the land occurs through the upstream
freshwater drainage, but some local influence may also be present.
Dissolved and particulate matter may be introduced through wind
action and surface runoff, and land animals often forage and
scavenge around the water's edge.

Lying between the river and the open sea, the estuary/pass/
shelf complex receives and ultimately transmits to the sea the
river water and much of the material dissolved and suspended
therein. River-borne material includes the water itself, together
with dissolved and suspended inorganic and organic materials, as
well as some living organisms (plankton and some larger forms).
Some of these materials (including coarse sediments and certain
chemical species) remain in the system for extended periods,
whereas others (including dissolved and finely particulate matter)
may pass quickly through. Most materials which enter the system
become physically or chemically altered, but others (such as
silicates) may sometimes pass through with no appreciable change.
The downstream river-related influence is shown in Fig. 2.

The open ocean influences the system in several ways.
Oceanic currents induce eddies and countercurrents in shelf waters,
and oceanic water may transmit various types of surface and
internal wave motion from the deeper basin. Oceanic water is more
constant in temperature and chemical (including salinity)
characteristics, but the movement of oceanic water onto the shelf
may increase the range of variability of the shelf water. In some
cases where there is upwelling or the entrainment of deep ocean
water on the shelf there may be a significant cooling and
introduction of nutrients. In general, however, mixing of surface
oceanic water with shelf water tends to increase the salinity
while reducing the level of nutrients. Numerous species of oceanic
plant and animal life may enter shelf waters, and although the
effects are temporary, some may be locally significant. Especially
important are the predatory oceanic species such as tunas, bill-
fishes, mackerels, some whales, etc. which move in and out
seasonally.

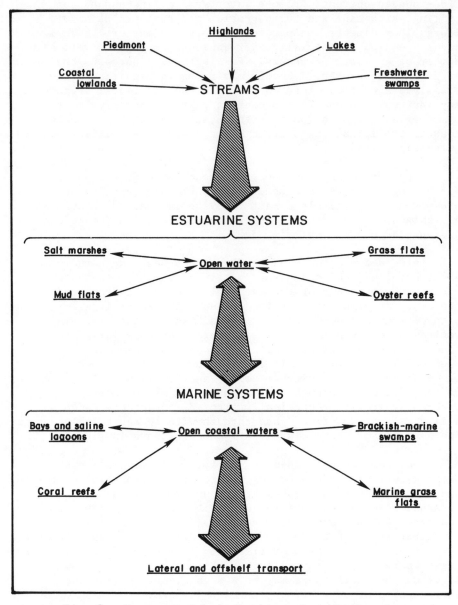

Fig. 2.   Downstream hydrologic series showing inter-
relations of major and minor components.

Most of the above influences are reasonably well understood in a general way, but hard data for a given estuary/pass/shelf system are, more often than not, lacking. Whereas the relationships may readily be expressed in conceptual models, the magnitude of each influence often cannot be realistically estimated for the local situation. Since external influences are often the controlling factors in local system dynamics, it is quite clear that we will not really understand the internal system operations until the external influences are reasonably well in hand. Actually, for best results the internal and external factors should be studied simultaneously. To do so would require the extended effort of an interdisciplinary team, and despite the managerial and other difficulties involved, this type of operation must be carried out on a selected series of our coastal waters in the very near future. It is the only way to obtain some of the most basic information about system function, and it is the best way to understand the complex adventures and ecological effects of some of the chemical pollutants now entering our coastal systems.

IMPORT/EXPORT PHENOMENA

Although import and export have been touched on above, it is important to focus special attention upon these relationships. A generalized import/export equation is given below:

$$K_i + (K_p - K_l) = K_o \tag{I}$$

This equation (modified from *Windom*, 1975) states simply that the rate of input into a system (modified by local production and loss) equals the rate of output of the system. For sophisticated theoretical studies, differential equations would be appropriate in expressing the above relationships, and in many cases these are now available. However, the simple equation given above is adequate for the general purpose of expressing flow-through budgets, and this is an important element of the present analysis. An example of such a budget has been provided by *Wolfe* (1975) (Fig. 3). Here the element manganese is traced from watershed input through the estuary to the estuarine outputs. An internal storage compartment is recognized (the sediments), and the budget is not balanced. Either additional export mechanisms are available, or the system is not in steady state condition.

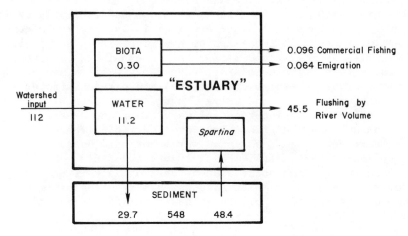

Fig. 3. Annual inputs and exports of the element manganese for the Newport River estuary. (From *Wolfe*, 1975).

In dealing with estuary/pass/shelf systems one must be interested in flow-through budgets. However, superimposed upon and interacting with the flow-through materials are all the dynamics of the internal system itself. It is obvious that the general equation given above for the import/export relations of the entire system may, as well, be applied to internal relations between subsystems, *i.e.*, relations between the estuary and the pass and between the pass and the shelf.

THE ESTUARY/PASS/SHELF SYSTEM

At the outset we will define the three subsystems under consideration. For present purposes, the estuary is the expanded lower reach of the river which is normally characterized by both tidal influence and salinity elevated above that of the river. Fully estuarine conditions do not extend upstream beyond that point. Downstream the estuary is bounded by the pass. The distinction between estuaries and coastal lagoons is not a sharp one, and much of what is said about estuaries will apply to coastal lagoons as well. Estuaries themselves are made up of a series of interrelated subsystems which may include salt marshes, mud flats, grass flats, oyster reefs, and open water.

If hard substrates are present, extensive biofouling communities
generally develop. Distinctions are often made between fjord-type
and soft-sediment-type estuaries. The former occur at high
latitudes where glacial ice has scoured a deep V-shaped basin in
granitic or other hard rock. The latter are found in lower
latitudes where the basin is molded of sedimentary materials by
the action of water currents. Intermediate types are known.

The pass is the constricted connection between the estuary
and the continental shelf. As discussed in a related paper in the
present volume (*Darnell*, 1979), the pass is under strong influence
of the adjacent water systems. Passes tend to be characterized
by deep channels and hydrographic conditions with great time-
dependent variability. Of prime importance are the swift
currents which regularly sweep through the system. Subsystems
include those associated with the water column (plankton and
nekton) and those associated with the substrate (benthos and
lateral wall).

The continental shelf includes the marine waters outside
the pass on out to the shelf break. As noted by *Emery* (1969),
this may be anywhere in the depth range of less than 65 to more
than 200 fathoms. Conventionally the outer edge of the shelf is
taken as about 100 fathoms (or 200 meters). Actually, it is
unrealistic to include the entire shelf in the present system
since the primary estuary/pass influence is limited to the inner
shelf zone, and this influence decreases seaward. Sediment
transported from the estuary tends to be deposited near the mouth
of the pass, and both this sediment and the water which leaves
the pass tend to be swept along the inner shelf by longshore
currents. Furthermore, *Chittenden and McEachran* (1976) have
shown that, for the northwestern Gulf of Mexico, on the inner
shelf the fishes have a strong affinity for the estuaries whereas
further out the fish fauna tends to be relatively independent of
estuaries. They have found the break to occur at about the 22 m
(12 fm) contour, although faunal overlap occurred between 18 and
36 m (10-20 fm). *Defenbaugh* (1976) studied the distribution of
benthic macroinvertebrates of the continental shelf of the
northern Gulf of Mexico, and he also detected a distinct
bathymetric zonation in the faunal distribution. He defined the
inner shelf area as extending out to the 30 m (16.5 fm) isobath.
The inner shelf invertebrate assemblage of *Defenbaugh* shows
estuarine affinities, at least among the more mobile species.
For present purposes, the continental shelf included in the
estuarine/pass/shelf system will be limited to the inner shelf,
and this is tentatively defined as the 36 m (20 fm) contour or
the outer limit of estuary-related ichthyofauna as found by
Chittenden and McEachran (1976). Actually, the outer limit of
estuary/pass influence probably varies on a geographic basis,
and it should be redefined for the different coasts.

Inner shelf subsystems include minimally the water column (with
neuston, plankton, and nekton) and the benthos. The latter may
be further subdivided on the basis of bottom coarseness, presence
of hard substrate for the attachment of sedentary species,
presence of seagrass beds, presence of coral or algal reefs, etc.

HYDROGRAPHIC RELATIONSHIPS

The dominant physical feature of most estuary/pass/shelf
systems is the passage of fresh water from the river to the sea.
Regardless of salt content, in a steady state system with no
seepage or evaporative water loss, the rate of freshwater input
into the upper estuary should equal the rate of freshwater export
to the shelf and open sea $(K_i = K_o)$. Seepage and evaporative
loss are probably negligible in most temperate zone estuaries,
but they may be quite significant in the tropics and subtropics
(including the northern Gulf of Mexico) where there is low
inflow of surface water, large estuarine surface area, and the
temperature is high. Steady state could be achieved only if the
inputs and various influencing factors remained constant, and a
finite time lag would be required for internal equilibration.
The primary factors which induce variability in K_i include
irregularity in river flow, rainfall, and reverse flow from the
sea. External influencing factors also include changes in water
level (due to tides and winds), and differences in temperature
and salt content of the different water masses.

The dynamics of water mass movement and interaction within
estuaries has received a great deal of attention (see, for
example, *Blumberg*, 1976; *Bowden*, 1967; *Pritchard*, 1952, 1967;
Rattray, 1967, and *Ward*, 1976). These studies have involved field
measurements and experiments, laboratory flume studies, mathematical treatment, and computer simulation analysis. The topic
is extremely complex and is still not fully understood. For
the non-specialist, the best general introduction **is** provided by
Pritchard (1952). Space will not permit a full discussion of
estuarine circulation here. Suffice it to say that most temperate and tropical estuaries are stratified and display a
seaward-flowing fresher layer on the surface, a landward-flowing
saline layer on the bottom, and a zone of mixed water in between.
Considerable variability in detail is noted from one estuary to
another and from time to time within a given estuary.

Hydrographic relationships play a central role in estuary/
shelf transport phenomena and in the determination of living conditions throughout the system. To the extent that we understand
mixing and dissolved salt transport we can speak intelligently
about the mixing and transport of other dissolved (and some of
the finer particulate) materials. However, we are still a long

way from understanding just how hydrographic conditions translate
into the transport of coarser sediments. Even further away is
detailed knowledge of how hydrographic conditions bring about
transport of most biological materials. Yet these are goals which
are now recognizable and toward which we may strive in our efforts
to understand the dynamics of coastal systems.

TRANSPORT OF DISSOLVED AND NON-LIVING PARTICULATE MATERIALS

The three systems under consideration represent a linear
array in which the export from one represents the input of the
next. Thus:

System A		*System B*		*System C*
$K_i \rightarrow K_o$	$=$	$K_i \rightarrow K_o$	$=$	$K_i \rightarrow K_o$

Since the pass influence is generally negligible in terms of the
total quantity of transported material, it is often desirable to
ignore the pass influence and deal directly with relations
between the estuary and the shelf. In fact, most of the
literature treats the subject in this manner. However, the pass
influence may not always be negligible, especially during periods
of low flow and where the pass is quite long. As engineering
activities stabilize the passes, line them with solid substrates
where none existed before, and create extensive breakwaters and
jetties which substantially lengthen water passage time, the pass
must assume a more prominent role in the total complex.

Transport takes place in both directions, *i.e.*, from the
estuary to the shelf and from the shelf to the estuary. In fact,
in a stratified system the exchange takes place in both directions
simultaneously, seaward in the surface layer and landward in the
bottom layer. Dissolved substances and fine particulates are
transported freely by both currents, but as might be expected,
particulate materials which are larger, heavier, and denser are
transported largely by the bottom currents, except during periods
of unusual mixing and accelerated currents.

Transport rates show considerable time variability. The rate
of flux reflects the normal variation of current flow associated
with tidal cycle and seasonal stream discharge, as modified by
local weather conditions. Expecially important are wind speed and
direction as well as the degree of violence of the weather. Not
only is there a greater volume of water movement during windy
weather, but roiling of the shallow coastal water places more
materials in the water column to be transported. Thus, one severe
storm or hurricane may transport more particulate material than

all the accumulated tidal flows during the year. Material
transport is also strongly influenced by upstream factors such as
rainfall in the basin, engineering projects (especially damming
and stream shortening) (*Darnell et al.*, 1976) and discharge
from municipal, industrial, and agricultural areas. Mathematical
and computer models are available which simulate material transport through the estuary.

An interesting aspect of the transport problem is the fact
that many materials change state as they pass through a system.
For example, nitrogen brought to an estuary by a river as
dissolved inorganic nitrate may exit from the estuary largely as
particulate nitrogen associated with the phytoplankton. Unfortunately, there is a dearth of quantitative information on the
subject of estuary/shelf transport, and there is even less published information correlating transport with the major physical
forcing functions. Therefore, portions of the following
discussion will necessarily be subjective.

Seaward Transport

River water which passes through the estuary to the shelf
always carries with it a load of dissolved and fine suspended
material. Thus, nutrients, heavy metals, dissolved organics,
industrial and municipal chemicals, as well as particulates,
brought down by river flow will, after some loss to the estuarine
system, be passed on the the shelf. To the extent that mixing
takes place, the seaward flowing current will also transport back
to the shelf a portion of the salt and other solutes which
entered the estuary in the inflowing bottom water. In addition,
any materials which enter the bottom water during its inward
passage may subsequently become entrained in the surface layer
and passed seaward (*Pritchard*, 1969). Of especial importance in
this connection are municipal effluents discharged into deeper
waters of the estuary (*Garside et al.*, 1976). Because of specific
peculiarities and local estuarine circumstances, each chemical
species will undergo its own individual adventures and state
transitions prior to being transported to the shelf. Illustrative
data for some of the more important substances are presented
below.

Nitrogen - The element nitrogen is of special interest
because of its role as a phytoplankton nutrient. In rivers and
estuaries it may exist as organic and inorganic particulates as
well as dissolved organic, ammonium, nitrite, and nitrate. At
least some of the dissolved organics are available to phytoplankters (*Wheeler et al.*, 1977), and the dissolved inorganics
certainly are. Among the inorganics, the phytoplankters
apparently take up the reduced forms preferentially before the

more oxidized forms, *i.e.*, ammonium first, then nitrite, then nitrate. A great deal of recent effort has gone into the elucidation of the transitions and fates of nitrogen within the estuary (see, for example, *McCarthy et al.*, 1977). Of concern here is its transport to the shelf. *Hobbie et al.* (1975) studied nitrogen input, cycling, and output in the Pamlico River estuary of North Carolina. They concluded that about 1,425 tons of nitrogen as nitrate and 744 tons of nitrogen as ammonium were exported annually by this estuary to the neighboring shelf. These amounts represented about half the nitrate and over ninety percent of the ammonium received by the estuary from the entering streams. Working in the Georgia Bight, *Haines* (1975) calculated that the coastal streams annually transported to the shelf about 0.42×10^{10}g of inorganic nitrogen (nitrate plus ammonia). However, *Dunstan and Atkinson* (1976) noted the potential importance of stream discharge of organic nitrogen as well. In coastal streams dissolved organic nitrogen often runs an order of magnitude higher than the dissolved inorganic forms, and whether or not the phytoplankton can pick it up immediately, the dissolved organic nitrogen ultimately becomes available following microbial activity. Also working in the Georgia Bight, they calculated that the coastal streams discharge enough new nitrogen to account for about $3\mu g$-at N liter^{-1} yr^{-1} for the entire water volume of the Georgia Bight area (660×10^{12} liters). Both *Haines* (1975) and *Dunstan and Atkinson* (1976) concluded that whereas the nitrogen derived from stream flow may be important in the nearshore shelf (*i.e.*, in the first 15 miles, or so), from the standpoint of phytoplankton production of the shelf as a whole, the stream contribution must be negligible. Neither study addressed the contribution of particulate organic and inorganic nitrogen.

Some information is becoming available concerning the nitrogen enrichment of shelf waters resulting from municipal sewage discharge. *Bowman* (1977), using published information on effluent discharge rates, nutrient levels in river water, and mean river discharge rates, determined that about 400×10^6 gal day^{-1} of sewage effluent is transported from the East River to Long Island Sound. This material is quite rich in various forms of nitrogen, and *Bowman* discussed the impact of the dissolved inorganic nitrogen compounds on the ecology of Long Island Sound. *Garside et al.* (1976) evaluated the influence of sewage-derived nitrogen on the Hudson River estuary and New York Bight. During most of the year the Hudson estuary discharges about 160 metric tons N day^{-1} into the New York Bight. The highest value of 185 tons N day^{-1} occurred during the spring, and the lowest input of around 120 tons N day^{-1} was seen during the summer months. This study is of special interest because it evaluates the upstream transport of sewage enriched bottom waters and their eventual mixing and downstream transport in the surface layer. The

authors found that this phenomenon varied seasonally depending
upon freshwater outflow.

Phosphorus. As in the case of nitrogen, the element
phosphorus is of special interest in relation to phytoplankton
nutrition. In rivers and estuaries it exists as organic and
inorganic particulates, as well as dissolved organic and inorganic
phosphate. Its cycling within estuaries and coastal lagoons has
been studied by a number of investigators (see, especially,
McRoy et al., 1972, and Pomeroy et al., 1969). Hobbie et al.
(1975) studied phosphate movement through the Pamlico River
estuary of North Carolina. They found that about forty-two percent
of the phosphorus brought to the estuary by rivers and a local
phosphate mine was discharged into the sea (459 tonnes yr^{-1}
export/1081 tonnes yr^{-1} import). Working in the Izembek Lagoon
off the Bering Sea of Alaska, McRoy et al. (1972) found a substantial transfer of phosphorus from bottom sediments into seawater
by the action of Zostera beds. This transfer, in turn, led to a
significant export of reactive phosphorus from the lagoon to the
Bering sea. This export amounted to 1.47 x 10^3 kg per tidal cycle
or about 3 x 10^3 kg P day^{-1} during the growing season. They also
noted that an additional substantial but unmeasured amount of
phosphorus is transported from the lagoon as detached floating
leaves of the eelgrass. Apparently there does not exist a complete phosphorus export budget for any of the nation's estuaries.

Silicon. Dissolved silicon is of interest primarily because
of its importance in supporting growth of diatom populations.
Peterson et al. (1975) studied the distribution and dynamics of
dissolved silicon in San Francisco Bay. The primary souce of
dissolved silicon is the rivers which enter the bay, although
marine intrusive bottom waters supply significant amounts during
the season of low river flow. The level of dissolved silicon in
the bay waters showed great seasonal and areal variation. Considerable quantities of dissolved silicon enter from stream flow,
and when that flow is great in comparison with the rate of diatom
utilization, there is a linear downstream gradient of dissolved
silicon which reflects the degree of mixing of fresh and salt
water. At such times the silicon content is negatively correlated
with salinity. However, during the summer months when stream
input is at a minimum and when phytoplankton populations are high,
the bay acts as a silicon sink. This pronounced phytoplankton
effect is apparently due to two factors, greatly increased phytoplankton density and increased residence time of the fresher
waters within the bay. Although no estimates were given of the
quantity of silicon transported to the ocean, data in the graphs
show that at the most seaward stations (i.e., those closest to the
Golden Gate) reactive silicate silicon concentrations ranged
mostly between 30 and 60 μg-at liter^{-1}.

Organic carbon - general. Organic carbon is associated with and is derived from all living systems. In its various forms it may function as a nutrient or energy source, growth stimulator, growth inhibitor, chemical cue, etc. Thus, organic carbon may be of general interest simply as a source of nutrients or energy, or it may be of special interest as a result of its specific biological activity. Its role, of course, depends upon its chemical and biological state. In the present section we are concerned only with the general movement of the mass of organic carbon from the estuary to the shelf.

Despite the obvious importance of the topic, very few quantitative data are available. Of especial interest is the work of W.E. Odum (1969) and Heald (1969) and of Odum and Heald (1975). These workers reported on the production of particulate organic detritus by the marsh and mangrove swamp of a south Florida estuary, as well as the processes of decomposition, transport to the shelf, and details of consumer nutrition. They found that the estuarine complex exports to the adjacent shelf about 6×10^3 metric tons (dry weight) of suspended particulate detritus annually. This amount is equivalent to an estuarine production of about 1.5 metric tons of particulate detritus acre^{-1} year^{-1}. Most of the material was contributed by decomposing leaves of the red mangrove (Rhizophora mangle), marsh grasses together accounting for less than 15 percent of the total particulate detritus flux. Although the authors did not determine the quantity of organic carbon exported, it may be calculated roughly. From ashing studies the investigators estimated the organic content of detritus to be about 62 percent of the dry weight. If one assumes that 50 percent of the organic loss at ashing is due to organic carbon, it may be concluded that about 31 percent of the weight of the average detritus particle is organic carbon. Therefore, the particulate organic carbon export is on the order of 2×10^3 metric tons C yr^{-1}. Dissolved organic carbon export cannot be calculated, but the few measurements indicated high concentrations of dissolved organic material (10-15 mg C liter^{-1}).

Happ et al. (1977) analyzed the distribution and dynamics of organic carbon in Caminada and Barataria Bays of southern Louisiana. They also provided three estimates of the net flux of total (i.e., dissolved and particulate) organic carbon from Barataria Bay to the Gulf of Mexico. These were based upon three different estimates of flushing rate. Their estimates of the total net flux of organic carbon from Barataria Bay to the Gulf ranged from 43 to 280×10^6 kg C yr^{-1}. For a number of reasons the highest figure seems the most likely, but as the authors state, even this figure must be an underestimation. Field measurements were made in calm weather, and there is reason to believe that extensive export occurs during storms and hurricanes which impinge upon the area with some frequency. It is of interest to note that DOC:POC ratios averaged 6:1, a relatively high value, suggesting that in this area of low tidal energy the detritus removed from the marshes must be reduced to very fine particles and DOC before being exported from marshes to bays and from bays to the shelf. This study is of special interest because of the thoroughness of the analysis.

Organic carbon - specific. As noted above, organic carbon may be of biological importance because of its specific chemical state. Very little solid information is available on the subject of export of biologically active chemicals, but there is enough to indicate a significant, if uninvestigated, process.

Vitamin B_{12} is a very potent growth stimulating substance. Studies by *Burkholder and Burkholder* (1956) and *Starr* (1956) demonstrate the abundance of this substance in various coastal environments of Georgia and its association with bottom muds and suspended particulate matter. In a transect from the marshes through the Duplin River and Doboy Sound to the open shelf, it was determined that the quantity of particulate detritus decreased downstream, but that the vitamin B_{12} content of the detritus increased seaward. Their results are consistent with the hypothesis that marsh-derived particulate substrate is enriched through bacterial action as it is transported to the shelf. This has not been absolutely proven, however, and no quantitative transport data are available.

Chlorophyll is an important photopigment produced by green plants, and analysis of chlorophyll concentration of water columns is often employed as an estimate of phytoplankton standing crop. However, chlorophyll degrades slowly, and in coastal waters it may also be associated with dead phytoplankton and with bits of decomposing thallic algae and vascular plants. *Mann* (1975) studied the exchange of chlorophyll between Bedford Basin and Halifax Harbour, Nova Scotia over a 25-hour period. He concluded that 58 percent of the day's production was exported from the Basin to the Harbour during the period of study. Although quantitative data were not provided, Mann's study does show the import and export of chlorophyll in relation to the two tidal cycles.

Many species which spend a portion of their lives in estuaries and a portion in the sea apparently depend upon chemical cues to aid them in underwater navigation. *Odum* (1970) has summarized much of the available information on this subject. Although the literature is not extensive, it seems conclusive. A number of migratory species are known which react positively to estuarine water, even after it has been filtered. Removal of dissolved organics by charcoal eliminates the response. Further studies cited by *Hasler* (1966) point to the ability of aquatic species to distinguish particular odors from among a complex mixture of

dissolved chemicals. Quantitative data concerning the outflow
of specific organic compounds are lacking, but considering the
remarkable ability of sea-run salmon to home in on their native
streams and the dependence of salmon on olfactory cues, it seems
clear that the water leaving each estuary must possess a unique
chemical "signature."

   Energy. Organically-bound energy, expressed as its heat
equivalent, or calories, is available to support the growth and
metabolism of heterotrophic organisms and, to some extent, of the
phytoplankters. The flow of calories within and between estuarine
subsystems has been investigated by a number of workers. For
example, *Teal* (1962) demonstrated that 45 percent of the net production of a Georgia *Spartina* marsh was available for export to
the estuary proper. *Thayer et al.* (1975) measured the flow of
calories through various compartments of a *Zostera* system in a
North Carolina estuary.

   No information has been found concerning the export of
calories from estuaries to the continental shelf, but such export
can be roughly calculated from data on organic matter or organic
carbon export. *Cummins and Wuycheck* (1971) provide caloric
equivalents for hundreds of types of biological materials. In
summary, they note that producers average about 4.685 Cal g^{-1},
microconsumers 4.958 Cal g^{-1}, and macroconsumers 5.821 Cal g^{-1}
ash-free dry wt. They further state that "the values for detritus,
which include both the organic substrates and the microflora (and,
undoubtedly, certain microfaunal elements also), do not differ
very much from the primary producer values." Actually, the
detritus values range very close to 5.000 Cal g^{-1}, and if this
value is used, the estimate is likely not biased more than 5-10
percent one way or the other. Applying this figure to the data
of *Odum* (1969) and *Heald* (1969) for particulate organic detritus
export from the south Florida estuary, we get an outflow of
18.6 x 10^9 Cal yr^{-1}. The data of *Happ et al.* (1977) for total
organic export from a Louisiana estuary may also provide a rough
estimate of caloric outflow if we can assume that carbon makes up
about 50 percent of the ash-free dry weight of the organic material
exported. Here, the estimate comes out to 2,888 x 10^9 Cal yr^{-1}
or about 150 times as much as from the Florida estuary. The
Bartaria Bay system is considerably larger than the south Florida
estuary (about 200 times as large if a freshwater swamp is
included), and the Louisiana study included total rather than
simply particulate organic carbon. When normalized to the same
unit of area the exports seem to be of about the same order of
magnitude, and both of these subtropical estuaries are clearly of
great importance in supporting production on the adjacent shelves.

Trace metals. Because of their importance in relation to
problems of water pollution, trace metals have been the subject of
much recent investigation. Of interest here is work dealing with
the passage of trace metals from rivers to estuaries and from
estuaries to the continental shelf. Much of the recent work has
been summarized by *Windom* (1975) and *Wolfe* (1975). *Windom* (1975)
studied the flux of trace metals through the estuaries of nine
coastal rivers of South Carolina, Georgia, and Northern Florida.
He distinguished between the dissolved and particulate concentrations of the metals in river water entering the estuaries
(Table 1). He found that none of the iron, a third of the manganese, and over three fourths of the cadmium, copper, and mercury
delivered by the rivers to the estuary were passed on to the
continental shelf. In relation to specific metals he reached the
following conclusions.

- Practically all the iron (both dissolved and particulate)
is lost as the river water passes through the estuary.

- At least part of the dissolved manganese precipitates in
the estuary because the amount delivered to the shelf is less
than that received in dissolved form by the estuary.

- In the case of cadmium and copper the amount delivered to
the shelf is about the same as that entering the estuary in
dissolved form.

- For mercury the loss to the shelf is greater than the
dissolved input, suggesting desorption from the particulate matter.

Lindberg et al. (1975) determined the mercury discharge from
Mobile Bay to the Gulf of Mexico to be 4.8×10^3 kg yr^{-1}. Of this
amount 41.7% was in the dissolved state and 58.3% was associated
with suspended particulates. *Wolfe* (1975) studied the flux of
three trace metals through the Newport River estuary of North
Carolina, and expressed his results in terms of a unit area of
estuarine bottom (*i.e.*, mg m^{-2} yr^{-1}). He found that 29.6% of the
iron (510/1720), 41.4% of the manganese (45.5/112), and 94% of the
zinc (3.4/3.6) entering the estuary from the watershed were
flushed out to the shelf by the river water. In addition, very
small amounts (less than one percent) were lost through biological
emigration and commercial fishing. The discrepancy between the
data of *Windom* and *Wolfe* for the element iron is unexplained. It
has been noted by several authors that organic detritus derived
from the decomposition of *Spartina* and mangrove leaves tends to
become enriched with mercury as it ages, probably as a result of
microbial uptake. Thus, the transport of mercury from the estuary
to the shelf is tied, in some degree, to the transport of particulate organic carbon.

TABLE 1. The flux of certain trace metals through the estuaries of nine rivers of the South Carolina, Georgia, and northern Florida coasts. The data for Fe are in terms of 10^6 kg yr^{-1}. The data for the other metals are in terms of 10^3 kg yr^{-1}. (After *Windom*, 1975).

		Trace Metal Concentration				
		Fe	Mn	Cd	Cu	Hg
River input (K_i)	Dissolved	9.5	909	41.0	222	3.09
	Particulate	52.7	957	11.0	82	0.51
	Total	62.2	1866	52	304	3.6
Sedimentation loss (K_l)		210	1200	9	66	0.4
Estuarine discharge (K_o)		0	666	43	238	3.2
% Flux ($\frac{K_o}{K_i} \times 100$)		0	35.7	82.6	78.3	89.0

Landward Transport

The landward or up-estuary transport of particulate material
by saline bottom currents has long been recognized. For example,
Bumpus (1965), using bottom drifters, showed that for the mid-
Atlantic coast there is a strong tendency for bottom waters of
the inner half of the shelf to move shoreward and to penetrate the
estuaries. Both particulate and dissolved materials enter
estuaries on these bottom currents. *Postma* (1967) provided a
penetrating analysis of the complex processes involved in the
movement of particulate materials from the mouth of the estuary
toward the flats, marshes, and river channel, and he also noted
that some of the particulate material being transported seaward
in the surface flow may settle into the bottom layer and be transported back up toward the river channel. *Postma* (1961) provided
specific data on transport rate as a function of particle size
and tidal velocity in the Dutch Wadden Sea, and he showed that
the finer fractions tend to accumulate differentially as one proceeds further up the estuary from the sea. If the adjacent shelf
is rich in particulate organic material, as in the case of the
North Sea, particulate organics may constitute a significant
fraction of the material brought in. For the Grevelingen Estuary
of the Netherlands, *Wolfe* (1977) calculated that about 31-45 kg C
is received from the North Sea at each tidal cycle, an amount
equivalent to 155-225 g C m^{-2} yr^{-1}). This is the largest organic
carbon input into the estuary, even exceeding that contributed by
local phytoplankton production (130 g C m^{-2} yr^{-1}).

Dissolved materials also pass into the estuary from the shelf
at each tidal cycle. These may be carried by the bottom saline
layer or throughout the water column if the water is well mixed.
In a situation where the water column is stratified, some of the
material brought into the estuary in the bottom layer may become
entrained and exit in the surface layer. For a conservative
substance such as chloride the flux may be calculated from data
on flow rates and concentrations in each layer throughout the
tidal cycle. For non-conservative substances such as oxygen,
which are biologically more interesting, the calculations are
considerably more complex. The total seawater flux can be calculated for chloride, as noted above. Knowledge of the ratio of
dissolved oxygen to chloride in the entering water would give a
figure for the quantity of oxygen taken into the estuary by this
means. However, as noted by *Biggs* (1967) and others, oxygen may
become involved in various chemical and biological processes and
be tied up as sulfate, carbon dioxide, nitrate, etc. before
leaving the estuary. Furthermore, oxygen may enter the estuary
by other routes (river flow, atmospheric exchange, and photosynthesis). Only through a thorough knowledge of the various
processes involved could a full accounting be made of the oxygen
brought in from the sea, and the same applies to other

non-conservative substances. No information has been found in the
literature which provides a budget for the dissolved materials
contributed by shelf waters. However, it may be assumed that the
shelf contribution of dissolved materials such as oxygen and
nutrient salts is significant in all estuaries at some seasons and
for some estuaries (such as those of the warm arid environments
of the south Texas coast) most of the time.

The Budget for Net Flux

Ultimately it becomes important to evaluate the total flux of
materials between the estuary and the shelf to ascertain whether
there is a net gain or loss by the estuary. This process involves
measurement of material flows in both directions through the pass
during complete tidal cycles over some extended period of time.
An example is provided by *Woodwell et al.* (1977). These investigators studied the exchange of dissolved and particulate carbon
between a tidal marsh and Long Island Sound. The basic sampling
unit was one complete tidal cycle, and sampling was carried out
four times a month for 16 months. Taking into account carbon
dioxide as well as dissolved and particulate organic carbon, they
were able to obtain values for the total annual flux of carbon
through the pass.

From this study *Woodwell et al.* (1977) reached the following
conclusions. Net flux in the inorganic carbonate system was
insignificant. There was a significant net inward flux (*i.e.*,
from the Sound) of particulate organic carbon of about 61 g C m^{-2}
yr^{-1}. The dissolved organic carbon showed a possible slight
annual loss of up to 8 g C m^{-2} yr^{-1}. There was a net annual gain
of chlorophyll by the marsh system. Fish populations export a
small amount of organic carbon (about 20 g m^{-2} yr^{-1}) in the fall.
Thus, if there is any regular annual net flow of carbon through
the channel apart from fish, it appears to be inward and small,
and the marsh is apparently a sink for fixed carbon.

The system studied by *Woodwell et al.* is not a typical
estuary since the coastal pond-marsh system is not the expanded
mouth of a stream. Yet the study is of great interest because it
establishes the basic methodology for analyzing net annual flux
and because it challenges the long-held concept of estuaries as
net exporters of fixed carbon.

Very Large Rivers - A Special Case

Very large rivers generally lack significant estuaries, and
in any event, the flow rates are generally so great that normal
estuarine processes are far overshadowed by the force of river

water flowing through. For most very large streams, the river
literally flows into the sea. During flood stage enormous
quantities of particulate and dissolved materials are transported
to the shelf where sedimentation and mixing take place. During
periods of low flow a salt wedge may creep upstream along the
bottom, and some mixing will occur through the interface which
may extend for many miles above the river's mouth. Solute concentrations of river water differ markedly from those of the sea
which it enters. Large rivers are generally richer in nutrients
than is the water of the adjacent shelf, but details reflect
characteristics of the respective drainage basins and, thus, vary
with the individual stream. For example, the Mississippi and Nile
Rivers are rich in nitrates, phosphates, and silicates, while the
Columbia River is low in phosphates, and the Amazon River is
depleted in both nitrates and phosphates relative to the marine
waters into which they flow. Space does not permit full examination of the relations of large rivers with adjacent shelves, and
the Mississippi River will be used as a brief example. Information
on some other large rivers may be found in the following works:
Stefansson and Richards (1963) (Columbia), *Gibbs* (1967), *Hulburt
and Corwin* (1969), *Ryther, Menzel and Corwin* (1967), *Williams*
(1968) (Amazon); *Halim* (1960, 1972), *Halim et al.* (1976), *Sharaf El
Din* (1977) (Nile).

    The Mississippi River annually contributes to the Gulf of
Mexico about 6.8×10^{11} m^3 of water yr^{-1}. With it go about 4.5
$\times 10^8$ tons of sediment (*Shepard,* 1960). *Hoffman* (1974) showed
that about 0.19×10^9 kg of particulate organic carbon (POC) and
1.8×10^9 kg of dissolved organic carbon (DOC) are brought to the
Gulf annually. *Hoffman* (1974) and *Rankin* (1974) provided information on various fractions of this organic material. Building
blocks of humic and fulvic acids were prominent. *Ho and Barrett*
(1975) provided information on various nutrients contributed by
the Mississippi River during a seven month study (Table 2).
Since their samples were not taken directly at the river mouth,
it is possible that their samples were influenced to some extent
by mixing with nearshore Gulf waters. The concentrations of
dissolved and particulate metals in Mississippi River water (taken
near the mouth) are provided by *Trefry* (1977) (Table 3).

    The significance of these values may be grasped by comparing
them with figures given earlier in this paper concerning the contribution of estuaries to the shelf. For example, *Happ et al.*
(1977) estimated that 43 to 280×10^6 kg organic C is supplied
to the Louisiana shelf annually by Barataria Bay, whereas *Hoffman*
(1974) estimated that the Mississippi River annually supplies about
2.99×10^9 kg organic C, or from 10 to 70 times as much. Since
Hoffman's data refer only to the water column (and ignore the
bed load), they clearly represent minimum values for the total
contribution of the Mississippi River to the shelf. The effect

TABLE 2. Quantity of selected nutrients discharged into the Gulf of Mexico by the Mississippi River estimated for the seven month period January–July, 1973. (After *Ho and Barrett*, 1975).

N (as NO_3^- + NO_2^-)	.89 x 10^9 kg/7 months
P (as PO_4^{-3})	.04 x 10^9 kg/7 months
SiO_2	1.3 x 10^9 kg/7 months
Organic – N	.82 x 10^9 kg/7 months
Organic – C	9.9 x 10^9 kg/7 months

of the Mississippi River contributed nutrients in stimulating productivity on the shelf is dramatic (*Riley*, 1937; *Thomas and Simmons*, 1960).

MOVEMENT OF LIVING ORGANISMS

It is a matter of common knowledge that there is a two-way exchange of living organisms between the estuary and the shelf, but quantitative data on exchange rates are scarce, and even information on seasonal programming and transfer mechanisms is, for most species, rather imprecise. Both hydrographic and biological factors are generally involved. From those organisms which are carried passively by water currents (phytoplankton and

TABLE 3. Dissolved and particulate metal concentrations of Mississippi River water. Dissolved metal concentrations are given in mg m^{-3}. Particulate metal concentrations are given in mg kg^{-1}. (After *Trefry*, 1977).

	Fe	Mn	Zn	Pb	Cu	Ni	Cr	Cd	Hg
Dissolved	10	10	10	<1	2	2	0.5	0.1	<0.1
Particulate	——	——	193	46	45	——	——	1.3	——

some bacteria) to those which are capable of moving more or less
independently of currents (larger fishes, porpoises, *etc.*), there
is a broad spectrum of species in which biological activities
must be meshed with hydrographic realities to provide for
successful completion of the life history.

Phytoplankton

The phytoplankton of estuaries varies from species with
freshwater requirements (at the head of the estuary) to those
which require marine salinities (near the estuary mouth). In
between and often overlapping with these two groups are a series
of euryhaline phytoplankters of which species of *Coscinodiscus*
are generally prominent. Depending upon the nature of the
individual estuary and season of the year, the species composition will shift with changes in salinity, temperature, and
nutrient conditions (*Patrick*, 1967; *Riley*, 1967). *Ketchum* (1954)
has shown that for a passive plankton population to maintain
itself in an estuary it must have a reproductive rate sufficient
to offset the portion of the population which is lost by downstream drift. He concluded, "In general it may be said that
local phytoplankton populations may be expected to maintain
themselves within estuaries where the exchange ratios are about
0.5 or less. Greater exchanges would tend to deplete the population more rapidly than it can reproduce itself." The above
calculation is based on flushing during a complete tidal cycle
of 12.4 hrs. In a study of processes contributing to the downstream
decrease of coliform bacteria in a tidal estuary, *Ketchum et al.*
(1952) concluded that bacteriocidal action is the most important
process, followed by predation and dilution and other unevaluated
factors. Their bacterial densities varied from 1×10^5 MPN ml^{-1}
nine miles upstream to 5×10^1 at the mouth of the Raritan
estuary (low tide values). As noted earlier, *Mann* (1975) concluded that 58 percent of the day's production of chlorophyll
was exported from Bedford Basin to Halifax Harbour, Nova Scotia,
over a 25-hour period (*i.e.*, over two tidal cycles). No information has been found on the numbers or biomass of phytoplankters
imported or exported by estuaries, although data on phytoplankton
densities and flushing rates could be employed to give rough
estimates of export rates.

Zooplankton

Scheltema (1975) has recently reviewed the available
literature concerning larval dispersal and gene flow of benthic
invertebrates in estuaries and along coastal regions. Much of
his discussion is pertinent to the present issue, and only a
few salient points will be given here. Whereas most phytoplankton

organisms might be considered to be passive drifters, the same
assumption cannot be made for zooplankters. *Wood and Hargis*
(1971) showed that fine coal particles and oyster larvae do not
exhibit the same patterns of distribution in estuarine water.
The former are transported passively, whereas the latter engage
in selective swimming behavior which enhances their upriver
movement in the estuary. *Bousfield* (1955) showed that nauplii
of estuarine barnacles increase their depth distribution with
each molt so that young nauplii are transported toward the
mouth of the estuary in surface currents, and older nauplii
are transported back toward the head of the estuary on bottom
currents. *Hughes* (1969) has shown that postlarval pink shrimp
become active in the water column when salinity is increased and
reduce or cease swimming in the water column when the salinity
is decreased. If the same behavior applies in the field it
would result in their transport into the estuary on bottom
currents of successive flood tides. These and related studies
show the importance of behavior in determining the distribution
of zooplankton species, and they provide mechanisms whereby life
history patterns may take advantage of hydrologic systems for
transport and perpetuation of the local species populations.
Barlow (1952), however, has shown the importance of reproduction
in maintaining an estuarine population of the copepod, *Acartia*.
In the most landward part of the population reproduction was too
slow to counteract downstream drift, and this segment of the
population was dependent upon recruitment from deeper layers.
In the central area reproduction was sufficient to offset
seaward transport. In the most seaward area loss was so great
that the population was maintained only by recruitment from the
central area. Thus, a truly planktonic species would suffer
some transport from the estuary into the sea if it remained
in the surface waters, and such transport apparently occurs
in *Acartia* and other estuarine zooplankton species. Shelf
species may be swept into the estuaries on bottom currents,
and estuarine species can maintain position by riding bottom
currents during part of the life history. Although estuarine
zooplankton species are sometimes encountered on the shelf, and
shelf species do enter the estuary, quantitative data are
apparently lacking. *Bousfield et al.* (1975) have shown that
a given estuary may simultaneously be inhabited by three rather
distinct zooplankton faunas: an upstream freshwater holoplanktonic group; an estuarine-endemic holoplanktonic group in the
middle sector, mostly at the surface; and a coastal marine
group which penetrates landward in the cold, high-salinity
bottom waters.

It was earlier mentioned that suspended matter may
temporarily settle out in the pass, and certainly some of
the plankton must suffer this fate. If solid substrate is
present populations of filter feeders must remove much plankton

as it traverses the pass, and this effect may be pronounced if long jetties are present (*i.e.*, if the water has a long residence time in the pass). On the other hand, the pass inhabitants may themselves produce planktonic larvae so that the composition of the plankton would change from one end of the pass to the other. This phenomenon apparently has not been investigated.

Nekton and Mobile Benthic Species

The nekton and mobile benthic fauna include those forms which exercise considerable control over their own locomotion and are, thus, largely independent of water movement. Included are adults of squids, shrimp, crabs, fishes, and marine mammals. Although these forms exhibit a wide spectrum of relations with estuaries, two categories are recognized for discussion purposes here, estuarine-related and estuarine-dependent. Estuarine-related species are coastal marine forms which inhabit the estuary with some regularity but which do not require this habitat. Included are such species as sand sharks, bluefish, certain flounders, marine catfish, sea robins, jackfish, and porpoises which regularly forage in estuaries, as well as occasional crabs, shrimp, and fishes which move in during periods of high salinity.

Estuarine-dependent species normally require the estuarine habitat for some stage of the life history. The typical life history for penaeid shrimp, blue crab, and estuarine-dependent fishes involves offshore spawning, movement into the estuary of planktonic or benthonic young, maturation in the estuary, and emigration to the shelf by subadults or adults. Movement into the estuaries generally takes place during spring and early summer months, and movement out of the estuaries occurs during late summer and fall. Details vary with species and location, and they also vary locally depending upon a number of factors, only a few of which are really understood. Since most of the movements and other life history events are seasonal, they are often correlated with temperature *per se* or with change in temperature. This factor may act through its general effect on metabolic processes, as a limiting factor determining the range of an activity, or as a cue or triggering device. Salinity is also important. As noted by *Gunter* (1950) and others, most estuarine-dependent coastal species can tolerate high salinities, but low salinity tolerance is more species specific. The tide (or some tide-related factor) seems to be important in initiation or facilitation of the emigration of brown and pink shrimp from south Texas estuaries (*Copeland*, 1965), and tidal influence may be far more widespread than is currently realized.

These and other factors do not act in isolation, and it seems likely that biological events are really controlled by

correlated factor combinations. Spring floods which rivers bring
to the estuaries may be cold, turbid, nutrient-rich, and swift.
The upper and lower water masses in a stratified estuary possess
many different physical and chemical properties. Tidal flow at
full moon is swifter and of greater amplitude than at other
phases, and quantities of suspended and particulate materials
should be quite different. For these reasons broader correlations
are possible. For example, it has been shown that total fisheries
production in Texas bays and estuaries is positively correlated
with years of heavier rainfall and river runoff (*Chapman*, 1966).
White shrimp production in Texas is favored by fresher estuaries,
and brown shrimp production is favored by less saline conditions
(*Gunter and Edwards*, 1969). In Apalachicola Bay (Fla.), high production of shrimp and blue crabs occurs during years of heavy
precipitation in the upstream river basin, whereas high oyster
production occurs when the precipitation is low (*Livingston et al.*,
1978). Mass migration of species from shallow Texas bays is
triggered by sudden temperature drops associated with the passage
of severe cold fronts during the fall months (*Copeland*, 1965).
However, *Tabb et al.* (1962) reported that young pink shrimp failed
to migrate from estuarine nursery areas of the Everglades during
a year of exceptionally high salinity. Thus, there is some
information concerning the mechanisms which determine migration
and local success of estuarine populations, but all of these merit
more thorough analysis.

Quantitative sampling of larger mobile organisms moving into
and out of estuaries has been carried out in only a few instances
despite the fact that it is not difficult to accomplish, and the
information gained from it is of great theoretical and practical
value. *Simmons and Hoese* (1959) studied the movement of fishes
and invertebrates through a natural pass between a south Texas
barrier island bay and the Gulf of Mexico. The sampling program
lasted 15 months and, despite numerous problems with equipment,
the study provides at least a semiquantitative picture of the
seasonal patterns of movement in both directions through the pass.
Considerably more specimens were taken moving toward the Gulf
than toward the bay. Pronounced seasonal migrations were noted,
with peaks (in both directions) occurring in June and July.
Individual species showed different seasonal patterns of movement
with Atlantic croakers moving Gulfward primarily in May-June,
spotted seatrout in June-August, and southern flounder in
September-December. Gulfward movement of adult blue crabs showed
a large peak in May-June, whereas bayward movement took place
primarily in July-September. Movements of the various species
were correlated with prevailing temperature and salinity conditions
and with life history events of the species. *Copeland* (1965)
carried out a similar but more quantitative study of the movement of fishes and invertebrates through Aransas Pass Inlet in
south Texas. Regular sampling took place three days per week at

maximum flood and ebb tides for a period of a year. The total catch was greatest in May-June and in October, and these peaks were correlated with changes in water level of the bays and changes in temperature. As in the previous study, catches were greatest on ebb tides. Data on average biomass captured, current speeds, and cross-sectional area sampled permitted calculation of total biomass passing seaward through the inlet. On an annual basis it amounted to about 3.19×10^5 kg day^{-1} or 11.65×10^7 kg yr^{-1}. From this information it was computed that the bays supplying the pass produced 233 kg acre^{-1} (576 kg hectare^{-1}) of larger fishes and invertebrates per year. The annual export of penaeid shrimp alone amounted to 3.9×10^6 kg yr^{-1}. *Woodwell et al.* (1977) briefly mentioned that the Long Island marsh exported about 20 g C m^{-2} yr^{-1} as fishes. Assuming that organic carbon makes up half the dry weight and that dry weight is a third of the wet weight of a fish, *Woodwell's* pond would be producing 1,200 kg live fish hectare^{-1} yr^{-1}, or twice that reported by *Copeland* for Texas bays. This result is quite surprising since coastal fishery production per unit area of marshland tends to increase toward the equator (*Turner*, 1977).

DISCUSSION

Conceptually, the estuary-pass-shelf complex may be viewed from two perspectives. On the one hand, the estuary and shelf may be thought of as two different ecosystems, and the relationships between the two would simply represent a matter of ecosytem coupling. Most of the available literature was derived from this approach. Emphasis has generally been upon the estuarine ecosystem, with the pass and shelf being considered simply a source or sink. An alternate view would be to consider the estuary, pass, and at least the inner shelf as a single system consisting of interrelated parts. Interest would focus on all the parts as well as relationships of the parts to each other and to the whole. The difference in these views is not simply a matter of semantics, because the latter perspective embraces the problem which must be faced and solved by all the estuarine-dependent coastal species which must complete portions of their life histories on both sides of the pass.

Although quantitative information on net flux of nutrients and organic matter through passes is still rather scarce, enough is now known to suggest the existence of two basic types of systems, export and import. As stated by *Wolff* (1977),"..... a picture of two different types of estuaries arises, one deriving detritus from salt marshes, mangroves or eelgrass beds and fertilizing the adjacent sea with detritus, the other deriving detritus from the sea and not exporting detritus to any other system." Wolff cites five factors leading to these differences:

export systems should be characterized by 1) high production of
plant material in the intertidal zone (esp. of *Spartina
alterniflora*), 2) high concentrations of organic material floating
at the surface, 3) strong offshore winds, 4) low tidal amplitude
(hence, weak bottom currents), and 5) estuaries which are far
richer in nutrients and phytoplankton than the adjacent shelves.
Import systems would show the reverse characteristics. Additional
factors are surely involved. One is the amount of rainfall and,
hence, surface runoff which characterizes the drainage basin.
High rainfall would bring nutrients down-river to the estuary,
stimulating estuarine production and providing a large volume of
outflow of surface water through the pass. Another factor might
be the frequency of major storms which, in addition to increasing
rainfall, would elevate surface water levels in the estuary,
enhance wave erosion of marshes and swamps, reduce particle size
of detritus, and promote thorough mixing of estuarine waters prior
to and during greatly increased surface water export through the
pass. Undoubtedly, there are additional factors of importance in
determining whether net export or import prevails. These factors
will surely relate to meteorology, hydrography, basin extent and
morphology, sediment types, and relative organic production of
the estuary and inner shelf. In most systems net import-export
relations must change on a seasonal basis. At present it seems
that Barataria Bay of Louisiana (*Happ et al.*, 1977) and the Florida
mangrove swamp (*Odum and Heald*, 1975) are clearly export systems,
whereas the Long Island pond of *Woodwell et al.* (1977) and the
Grevelingen Estuary of the Netherlands (*Wolff*, 1977) are of the
import type. Izembek Lagoon of Alaska (*McRoy et al.*, 1972)
exports phosphorus, and Pamlico Sound of North Carolina (*Hobbie
et al.*, 1975) seems to export both nitrogen and phosphorus, but
not enough is known to classify either of these waters with
certainty. The coastal marshes and estuaries of Georgia and other
Atlantic states also represent an enigma. Earlier studies of
Starr (1956), *Teal* (1962), *Odum and de la Cruz* (1967), and *Odum*
(1968) suggested that high levels of marsh production led to
significant offshore transport and consumer utilization of organic
detritus. However, recent studies by *Haines* (1975, 1977) and
Haines and Dunstan (1975) have questioned the importance of
Spartina-derived detritus in the estuarine and shelf systems, and
have also questioned the importance of estuarine and river exported
nutrients (particularly nitrogen) in supporting shelf populations.
Until more extensive net transport budgets are available, it seems
wise not to attempt to classify the south Atlantic systems.

From the weathering of upstream rocks and the activities of
civilization, streams become enriched with metals and trace
elements. All such materials become entrained to some extent in
estuaries, but each molecular species moves through its own
unique chemical, geological, and biological pathways. Some, such
as iron, are held almost totally within the estuary. Others, such

Bousfield, E. L., 1955. Ecological control of the occurrence of barnacles in the Miramichi estuary, *Bull. Nat. Mus. Canada* *137*: 1-69.

Bousfield, E. L., G. Fiteau, M. O'Neill, and P. Gentes, 1975. Population dynamics of zooplankton in the middle St. Lawrence Estuary, In: *Estuarine Research, V. 1,* edited by L. E. Cronin, 325-351, Academic Press, N. Y.

Bowden, K. F., 1967. Circulation and diffusion, In: *Estuaries,* edited by G. H. Lauff, 15-36, A.A.A.S. Publ. no. 83, Washington, D. C.

Bowman, M. J., 1977. Nutrient distributions and transport in Long Island Sound, *Est. and Coastal Mar. Sci. 5*: 531-548.

Bumpus, D. F., 1965. Residual drift along the bottom on the continental shelf in the middle Atlantic Bight area, *Limnol. Oceanogr. 10* (supplement): R 50-53.

Burkholder, P. R. and L. M. Burkholder, 1956. Vitamin B_{12} in suspended solids and marsh muds collected along the coast of Georgia, *Limnol. Oceanogr. 1*: 202-208.

Chapman, C. R., 1966. The Texas Basins Project, In: *A Symposium on Estuarine Fisheries,* edited by R. F. Smith, A. H. Schwartz, and W. H. Massman, 83-92, Amer. Fish. Soc., Spec. Publ. 3. Suppl. to *Trans. Amer. Fish. Soc. 95(4)*.

Chittenden, M. E., Jr., and J. D. McEachran, 1976. Composition, ecology and dynamics of demersal fish communities on the Northwestern Gulf of Mexico Continental Shelf, with a similar synopsis for the entire Gulf, Texas A&M, Sea Grant Publ. 76. 104 p. TAMU-SG-76-208.

Copeland, B. J., 1965. Fauna of the Aransas Pass Inlet, Texas. I. Emigration as shown by tide trap collections, *Publ. Inst. Mar. Sci., Univ. Texas 10*: 9-21.

Cummings, K. W. and J. C. Wuycheck, 1971. Caloric equivalents for investigations in ecological energetics, *Mitt. int. Verh. Limnol. 18*: 1-158.

Darnell, R. M., 1971. The world estuaries--ecosystems in jeopardy *Intecol. Bull. 3*: 3-20.

Darnell, R. M., 1979. The pass as a physically dominated, open ecological system, In: *Ecological Processes in Coastal and Marine Systems,* edited by R. J. Livingston, Plenum (this volume).

Darnell, R. M., W. E. Pequegnat, B. M. James, F. J. Bensen, and R. E. Defenbaugh, 1976. Impacts of construction activities in wetlands of the United States, U.S. Environmental Protection Agency, Ecol. Res. Series, 600/3-76-045.

Defenbaugh, R. E., 1976. *A Study of the Benthic Macroinvertebrates of the Continental Shelf of the Northern Gulf of Mexico*, Ph.D. Dissertation, Texas A&M University.

Dunstan, W. M. and L. P. Atkinson, 1976. Sources of new nitrogen for the South Atlantic Bight, In: *Estuarine Processes, V. 1*, edited by M. Wiley, 69-78, Academic Press, N. Y.

Emery, K. O., 1969. The continental shelves, *Sci. Amer. 221*: 106-122.

Garside, C., T. C. Malone, O. A. Roels, and B. A. Sharfstein, 1976. An evaluation of sewage-derived nutrients and their influence on the Hudson Estuary and New York Bight, *Est. and Coastal Mar. Sci. 4*: 281-289.

Gibbs, R. J., 1967. Amazon River: environmental factors that control its dissolved and suspended load, *Science 156*: 1734-1737.

Gunter, G., 1941. Death of fishes due to cold on the Texas coast, January, 1940, *Ecol. 22*: 203-208.

Gunter, G., 1950. Seasonal population changes and distributions as related to salinity, of certain invertebrates of the Texas Coast, including the commercial shrimp, *Publ. Inst. Mar. Sci., Univ. Texas 1*: 7-51.

Gunter, G. and J. C. Edwards, 1969. The relation of rainfall and freshwater drainage to the production of penaeid shrimps (*Penaeus fluviatilis* Say and *Penaeus aztecus* Ives) in Texas and Louisiana waters, *FAO Fish. Rep. (57). Vol. 3*: 875-892.

Haines, E. B., 1975. Nutrient inputs to the coastal zone, In: *Estuarine Research, V. 1*, edited by L. E. Cronin, 303-324, Academic Press, N. Y.

Haines, E. B., 1977. The origins of detritus in Georgia salt marsh estuaries, *Oikos 29*: 254.

Haines, E. B. and W. M. Dunstan, 1975. The distribution and relation of particulate organic material and primary productivity in the Georgia Bight, 1973-1974, *Est. and Coastal Mar. Sci. 3*: 431-441.

Halim, Y., 1960. Observations on the Nile bloom of phytoplankton in the Mediterranean, *J. Cons., Cons. Int. Explor. Mer. 26*: 57-67.

Halim, Y., 1972. The Nile and the East Levantine Sea, past and present, In: *Recent Researches in Estuarine Biology*, edited by R. Natarajan, 76-84, Hindustan.

Halim, Y., A. Samaan, and F. A. Zaghoul, 1976. Estuarine plankton of the Nile and the effect of freshwater phytoplankton, In: *Freshwater on the Sea*, edited by S. Skreslet, R. Leinbo, J. B. L. Matthews, and E. Sakshaug, 158-164, Geilo, Norway.

Happ, G., J. G. Gosselink, and J. W. Day. Jr., 1977. The seasonal distribution of organic carbon in a Louisiana estuary, *Est. and Coastal Mar. Sci. 5*: 695-705.

Hasler, A. D., 1966. *Underwater Guideposts*, 155 pp., Univ. of Wisconsin Press, Madison.

Heald, E. J., 1969. *The Production of Organic Detritus in a South Florida Estuary*, Ph.D. Dissertation, Univ. of Miami.

Ho, C. L. and B. B. Barrett, 1975. Distribution of nutrients in Louisiana's coastal waters influenced by the Mississippi River, La. Wildl. and Fish. Comm., Tech. Bull. No. 17.

Hobbie, J. E., B. J. Copeland and W. G. Harrison, 1975. Sources and fates of nutrients of the Pamlico River estuary North Carolina, In: *Estuarine Research, v. 1*, edited by L. E. Cronin 265-286, Academic Press, N. Y.

Hoffman, H. J., 1974. *A Comparison of Organic Matter in River and Sea Water*, M.S. Thesis, Texas A&M University.

Hughes, D. A., 1969. Responses to salinity change as a tidal transport mechanism of pink shrimp, *Penaeus duorarum, Biol. Bull. 136*: 43-53.

Hulburt, E. M. and N. Corwin, 1969. Influence of the Amazon River outflow on the ecology of the western tropical Atlantic. III. The plankton flora between the Amazon River and the Windward Islands, *J. Mar. Res. 27*: 55-72.

Ketchum, B. H., 1954. Relation between circulation and planktonic populations in estuaries, *Ecology 35*: 191-200.

Ketchum, B. H., J. C. Ayers, and R. Vaccaro, 1952. Processes contributing to the decrease of coliform bacteria in a tidal estuary, *Ecol. 33*: 247-258.

Lindall, W. N., Jr., and C. H. Saloman, 1977. Alteration and destruction of estuaries affecting fishery resources of the Gulf of Mexico, *Marine Fisheries Review 39*: 1-7.

Lindberg, S. E., A. W. Andren, and R. C. Harriss, 1975. Geochemistry of mercury in the estuarine environment, In: *Estuarine Research, v. 1*, edited by L. E. Cronin, 64-107, Academic Press, N. Y.

Livingston, R. J., F. G. Lewis, III, and G. G. Kobylinski, 1978. Long-term changes in epibenthic fisheries and invertebrates, In: *Short- and Long-term Effects of Forestry Operations on Water Quality and the Biota of the Apalachicola Estuary (North Florida, U.S.A.),* edited by R. J. Livingston, 356-409, Fla. Sea Grant, Tech. Paper No. 5.

Mann, K. H., 1975. Relationship between morphometry and biological functioning in three coastal inlets of Nova Scotia, In: *Estuarine Research, v. 1*, edited by L. E. Cronin, 634-644, Academic Press, N. Y.

McCarthy, J. J., W. R. Taylor, and J. L. Taft, 1977. Nitrogenous nutrition of the plankton in Chesapeake Bay. I. Nutrient availability and phytoplankton preferences, *Limnol. Oceanogr. 22*: 996-1011.

McHugh, J. L., 1976. Estuarine fisheries: Are they doomed?, In: *Estuarine Processes, v. 1*, edited by M. Wiley, 15-27, Academic Press, N. Y.

McRoy, C. P., R. J. Barsdate, and M. Nebert, 1972. Phosphorus cycling in an eelgrass (*Zostera marina* L.) ecosystem, *Limnol. Oceanogr. 17*: 58-67.

Odum, E. P., 1968. A research challenge: Evaluating the productivity of coastal and estuarine water, 63-64, 2nd Sea Grant Conf., Grad. School Oceanography, U. of Rhode Island, Newport.

Odum, E. P. and A. A. de la Cruz, 1967. Particulate organic detritus in a Georgia salt marsh-estuarine ecosystem, In: *Estuaries,* edited by G. H. Lauff, 383-388, A.A.A.S. Publ. no. 83, Washington, D. C.

Odum, W. E., 1969. *The Structure of Detritus Based Food Chains in a South Florida Mangrove System*, Ph.D. Dissertation, University of Miami.

Odum, W. E., 1970. Insidious alteration of the estuarine environment, *Trans. Amer. Fish. Soc. 99*: 836-847.

Odum, W. E. and E. J. Heald, 1975. The detritus-based food web of an estuarine mangrove community, In: *Estuarine Research v. 1*, edited by L. E. Cronin, 265-286, Academic Press, N. Y.

Patrick, R., 1967. Diatom communities in estuaries, In: *Estuaries*, edited by G. H. Lauff, 311-315, A.A.A.S. Publ. no. 83, Washington, D. C.

Peterson, C. G. J. and P. B. Jensen, 1911. Valuation of the sea. I. Animal life of the sea bottom, its food and quantity, *Rept. Danish Biol. Sta. 20*: 1-78.

Peterson, D. H., T. J. Conomos, W. W. Broenkow, and E. P. Scrivani, 1975. Processes controlling the dissolved silica distribution in San Francisco Bay, In: *Estuarine Research, v. 1*, edited by L. E. Cronin, 153-187, Academic Press, N. Y.

Pomeroy, L. R., R. E. Johannes, E. P. Odum, and B. Roffman, 1969. The phosphorus and zinc cycles and productivity of a salt marsh, In: *Symp. Radioecol.*, edited by D. J. Nelson and F. C. Evans, 412-419, Proc. 2nd Nat. Symp., Ann Arbor Mich.

Postma, H., 1961. Transport and accumulation of suspended matter in the Dutch Wadden Sea, *Netherlands Journal of Sea Research 1*: 148-190.

Postma, H., 1967. Sediment transport and sedimentation in the estuarine environment, In: *Estuaries*, edited by G. H. Lauff, 158-179, A.A.A.S. Publ. no. 83, Washington, D. C.

Pritchard, D. W., 1952. Estuarine hydrography, *Advances in Geophysics 1*: 243-280.

Pritchard, D. W., 1967. Observations of circulation in coastal plain estuaries, In: *Estuaries*, edited by G. H. Lauff, 37-44, A.A.A.S. Publ. no. 83, Washington, D. C.

Pritchard, D. W., 1969. Dispersion and flushing of pollutants in estuaries, *J. Hydraul. Div., Am. Soc. Civil Engs. 95*: 115-124.

Rankin, J. G., 1974. *Chemical and Physical Characteristics of Dissolved Organic Matter Isolated from the Mississippi River Delta and the Gulf of Mexico,* Ph.D. Dissertation, Texas A&M University.

Rattray, M., Jr., 1967. Some aspects of the dynamics of circulation in fjords, In: *Estuaries,* edited by G. H. Lauff, 52-62, A.A.A.S. Publ. no. 83, Washington, D. C.

Riley, G. A., 1937. The significance of the Mississippi River drainage for biological conditions in the northern Gulf of Mexico, *J. Mar. Res. 1*: 60-74.

Riley, G. A., 1967. The plankton of estuaries, In: *Estuaries,* edited by G. H. Lauff, 316-326, A.A.A.S. Publ. no. 83, Washington, D. C.

Ryther, J. H., D. W. Menzel, and N. Corwin, 1967. Influence of the Amazon River outflow on the ecology of the western tropical Atlantic. I. Hydrography and nutrient chemistry, *J. Mar. Res. 25*: 69-83.

Scheltema, R. S., 1975. Relationship of larval dispersal gene-flow and natural selection to geographic variation of benthic invertebrates in estuaries and along coastal regions, In: *Estuarine Research, v. 1,* edited by L. E. Cronin, 372-391, Academic Press, N. Y.

Sharaf El Din, S. H., 1977. Effect of the Aswan Dam on the Nile flood and on the estuarine and coastal circulation pattern along the Mediterranean Egyptian coast, *Limnol. Oceanogr. 22*: 194-207.

Shepard, F. P., 1960. Mississippi delta: marginal environments, sediments, and growth, In: *Recent Sediments of the Northwest Gulf of Mexico,* edited by F. P. Shepard, F. B. Phleger and T. H. van Andel, 56-81, Amer. Assoc. of Petrol. Geol, Tulsa Oklahoma.

Simmons, E. G. and H. D. Hoese, 1959. Studies on the hydrography and fish migrations of Cedar Bayou, a natural tidal inlet on the central Texas coast, *Publ. Inst. Mar. Sci., Univ. Texas 6*: 56-80.

Starr, T. J., 1956. Relative amounts of vitamin B_{12} in detritus from oceanic and estuarine environments near Sapelo Island, Georgia, *Ecology, 37*: 658-664.

Stefansson, U. and F. A. Richards, 1963. Processes contributing to the nutrient distribution off the Columbia River and Straight of Juan de Fuca, *Limnol. Oceanogr.* 8: 394-410.

Tabb, D. C., D. L. Dubrow, and A. E. Jones, 1962. Studies on the biology of the pink shrimp, *Penaeus duorarum*, in Everglades Park, Florida, State Bd. Cons., Univ. Miami Marine Lab, Tech. Ser., 37: 1-30.

Teal, J. M., 1962. Energy flow in the salt marsh ecosystem of Georgia, *Ecology 43*: 614-624.

Thayer, G. W., S. M. Adams, and M. W. LaCroix, 1975. Structural and functional aspects of a recently established *Zostera marina* community, In: *Estuarine Research, v. 1,* edited by L. E. Cronin, 518-540, Academic Press, N. Y.

Thomas, W. H. and E. G. Simmons, 1960. Phytoplankton production in the Mississippi delta, In: *Recent Sediments of the Northwestern Gulf of Mexico,* edited by F. P. Shepard, F. B. Phleger, and T. H. van Andel, 103-116, Amer. Assoc. of Petrol. Geol., Tulsa, Oklahoma.

Trefry, J. H., 1977. *The Transport of Heavy Metals by the Mississippi River and Their Fate in the Gulf of Mexico,* Ph.D. Dissertation, Texas A&M University.

Turner, R. E., 1977. Intertidal vegetation and commercial yields of penaeid shrimp, *Trans. Amer. Fish. Soc. 106*: 411-416.

Ward, G. H., Jr., 1976. Formulation and closure of a model of tidal-mean circulation in a stratified estuary, In: *Estuarine Processes, v. 2,* edited by M. Wiley, 365-378, Academic Press, N. Y.

Wheeler, P., B. North, M. Littler, and G. Stephens, 1977. Uptake of glycine by natural phytoplankton communities, *Limnol. Oceanogr. 22*: 900-910.

Williams, P. M., 1968. Organic and inorganic constituents of the Amazon River, *Nature (Lond.) 218*: 937-938.

Windom, H. L., 1975. Heavy metal fluxes through salt-marsh estuaries, In: *Estuarine Research, v. 1,* edited by L. E. Cronin, 137-152, Academic Press, N. Y.

Wolfe, D. A., 1975. The estuarine ecosystem(s) at Beaufort, North Carolina, In: *Estuarine Research, v. 1,* edited by L. E. Cronin, 645-671, Academic Press, N. Y.

Wolff, W. J., 1977. A benthic food budget for the Grevelingen
    estuary, the Netherlands, and a consideration of the
    mechanisms causing high benthic secondary production in
    estuaries, In: *Ecology of Marine Benthos*, edited by
    B. C. Coull, 267-280, University of South Carolina Press,
    Columbia.

Wood, L., and W. J. Hargis, 1971. Transport of bivalve larvae
    in a tidal estuary, In: *Fourth European Marine Biology
    Symposium*, 29-44, Cambridge University Press.

Woodwell, G. M., D. E. Whitney, C. A. S. Hall, and R. A. Houghton,
    1977. The Flax Pond ecosystem study: exchanges of carbon
    in water between a salt marsh and Long Island Sound,
    Limnol. Oceanogr. 22: 833-838.

LAWRENCE G. ABELE, Department of Biological Science, Florida State University, Tallahassee, Florida 32306

LARRY P. ATKINSON, Skidaway Institute of Oceanography, Skidaway Island, P. O. Box 13687, Savannah, Georgia 31406

SUSAN S. BELL, Belle W. Baruch Institute for Marine Biology and Coastal Research, University of South Carolina, Columbia, South Carolina 29208

JACKSON O. BLANTON, Skidaway Institute of Oceanography, Skidaway Island, P. O. Box 13687, Savannah, Georgia 31406

RONALD J. BOBBIE, Department of Biological Science, Florida State University, Tallahassee, Florida 32306

DONALD F. BOESCH, Virginia Institute of Marine Science, Gloucester Point, Virginia 23062

L. K. COACHMAN, Department of Oceanography, University of Washington, Seattle, Washington 98195

ALBERT W. COLLIER, Department of Biological Science, Florida State University, Tallahassee, Florida 32306

R. T. COONEY, Institute of Marine Science, University of Alaska, Fairbanks, Alaska 99701

BRUCE C. COULL, Belle W. Baruch Institute for Marine Biology and Coastal Research, University of South Carolina, Columbia, South Carolina 29208

REZNEAT M. DARNELL, Department of Oceanography, Texas A & M University, College Station, Texas 77843

JOHN W. DAY, JR., Coastal Ecology Laboratory, Center for Wetland Resources, Louisiana State University, Baton Rouge, Louisiana 70803

PAUL K. DAYTON, Scripps Institution of Oceanography, La Jolla, California 92093

D. A. DIETERLE, Oceanographic Sciences Division, Brookhaven National Laboratory, Upton, New York 11973

RICHARD C. DUGDALE, Bigelow Laboratory for Ocean Services, West Booth Bay Harbour, Maine 04575

THOMAS W. DUKE, Gulf Breeze Environmental Research Laboratory, Sabine Island, Gulf Breeze, Florida 32561

JAMES L. DUNCAN, Cordis Corporation, P. O. Box 370428, Miami, Florida 33137

PAMELLA ERICKSON, National Cancer Institute, National Institutes of Health, Bethesda, Maryland 20014

T. S. ENGLISH, Department of Oceanography, University of Washington, Seattle, Washington 98195

527

JOHN S. FISHER, Department of Environmental Science, University of
Virginia, Charlottesville, Virginia   22903

J. J. GOERING, Institute of Marine Science, University of Alaska,
Fairbanks, Alaska   99701

EVELYN B. HAINES, University of Georgia Marine Institute, Sapelo
Island, Georgia   31327

CHARLES S. HOPKINSON, Coastal Ecology Laboratory, Center for Wetland Resources, Louisiana State University, Baton Rouge,
Louisiana   70803

G. L. HUNT, JR., Department of Ecology and Evolutionary Biology,
University of California, Irvine, Irvine, California

RICHARD L. IVERSON, Department of Oceanography, Florida State
University, Tallahassee, Florida   32306

PAUL A. LAROCK, Department of Oceanography, Florida State University, Tallahassee, Florida   32306

REUBEN LASKER, NOAA/NMFS Southwest Fisheries Center, La Jolla,
California   92038

JEFFREY S. LEVINTON, Department of Ecology and Evolution, State
University of New York at Stony Brook, Stony Brook, New
York   11794

ROBERT J. LIVINGSTON, Department of Biological Science, Florida
State Unviersity, Tallahassee, Florida   32306

M. C. MACAULEY, Department of Oceanography, University of Washington, Seattle, Washington   98195

C. P. MCROY, Institute of Marine Science, University of Alaska,
Fairbanks, Alaska   99701

DUANE A. MEETER, Department of Statistics, Florida State University,
Tallahassee, Florida   32306

JANET S. NICKELS, Department of Biological Science, Florida State
University, Tallahassee, Florida   32306

WILLIAM E. ODUM, Department of Environmental Science, University
of Virginia, Charlottesville, Virginia   22903

CHARLES H. PETERSON, Institute of Marine Sciences, University of
North Carolina, Morehead City, North Carolina   28557

JAMES C. PICKRAL, Department of Environmental Science, University
of Virginia, Charlottesville, Virginia   22903

LAWRENCE G. POMEROY, Institute of Ecology, University of Georgia,
Athens, Georgia   30602

ROBERT W. POOLE, Division of Biology and Medicine, Brown University,
Providence, Rhode Island   02912

W. S. REEBURG, Institute of Marine Science, University of Alaska,
Fairbanks, Alaska   99701

BEN W. RIBELIN, Department of Biological Science, Florida State
University, Tallahassee, Florida   32306

MICHAEL B. ROBBLEE, Department of Environmental Sciences, University of Virginia, Charlottesville, Virginia   22903

PETER F. SHERIDAN, U. S. Environmental Protection Agency, Bears
Bluff Field Station, P. O. Box 368, John's Island, South
Carolina   29455

JOESPH L. SIMON, Department of Biology, University of South
     Florida, Tampa, Florida 33620
SAMUEL C. SNEDAKER, Rosenstiel School of Marine and Atmospheric
     Science, University of Maimi, Rickenbacker Causeway, Miami,
     Florida 33149
THOMAS M. SONIAT, Department of Oceanography, Texas A & M University,
     College Station, Texas 77843
GORDON THAYER, Southeast Fisheries Center, Beaufort Laboratory,
     Beaufort, North Carolina 28516
DAVID THISTLE, Department of Oceanography, Florida State University,
     Tallahassee, Florida 32306
ARTHUR G. TINGLE, Oceanographic Sciences Division, Brookhaven
     National Laboratory, Upton, New York 11973
J. J. WALSH, Oceanographic Sciences Division, Brookhaven National
     Laboratory, Upton, New York 11973
DAVID C. WHITE, Department of Biological Science, Florida State
     University, Tallahassee, Florida 32306
T. E. WHITLEDGE, Division of Oceanographic Sciences, Brookhaven
     National Laboratory, Upton, New York 11973
WILLIAM J. WIEBE, Department of Microbiology, University of
     Georgia, Athens, Georgia 30602
GLENN C. WOODSUM, Department of Biological Science, Florida State
     University, Tallahassee, Florida 32306
JOSEPH C. ZIEMAN, Department of Environmental Sciences, University
     of Virginia, Charlottesville, Virginia 22903
RITA T. ZIEMAN, Department of Environmental Sciences, University of
     Virginia, Charlottesville, Virginia 22903
JOSEPH ZINDULIS, Rutgers Marine Sciences Center, New Brunswick,
     New Jersey 08903

DONALD F. BOESCH, Virginia Institute of Marine Science, Gloucester
    Point, Virginia   23062
THOMAS L. BOTT, Stroud Water Research Center of the Academy of
    Natural Sciences of Philadelphia, R. D. 1, Box 512, Avondale,
    Pennsylvania   19311
IVER M. BROOK, Rosenstiel School of Marine and Atmospheric Science,
    University of Miami, Rickenbacker Causeway, Miami, Florida
    33149
PETER CASTRO, Biological Sciences Department, California State
    Polytechnic University, Pomona, California   91768
BRUCE C. COULL, Belle W. Baruch Institute for Marine Biology and
    Coastal Research, University of South Carolina, Columbia,
    South Carolina   29208
L. EUGENE CRONIN, Chesapeake Research Consortium, 1419 Forest
    Drive, Suite 207, Annapolis, Maryland   21403
ARMANDO A. DE LA CRUZ, Department of Biological Sciences, P. O.
    Drawer GY, Mississippi State, Mississippi   39762
REZNEAT M. DARNELL, Department of Oceanography, Texas A & M Univer-
    sity, College Station, Texas   77843
GARY E. DAVIS, U. S. National Park Service, South Florida Research
    Center, Everglades National Park, Homestead, Florida   33030
PAUL K. DAYTON, Scripps Institution of Oceanography, A-001, La
    Jolla, California   92093
ALAN R. EMERY, Royal Ontario Museum, 100 Queen's Park, Toronto,
    Ontario, Canada   M5S 2C6
CHARLES R. FUTCH, Bureau of Marine Science and Technology, Florida
    Department of Natural Resources, Tallahassee, Florida   32304
ROBERT H. GORE, Smithsonian Institution, Ft. Pierce Bureau, Box
    196 RFD 1, Fort Pierce, Florida   33450
JAMES G. GOSSELINK, Coastal Ecology Laboratory, Center for Wetland
    Resources, Louisiana State University, Baton Rouge, Louisiana
    70803
COURTNEY T. HACKNEY, Department of Biology, University of South-
    western Louisiana, Lafayette, Louisiana   70505
EVELYN B. HAINES, The University of Georgia Marine Institute,
    Sapelo Island, Georgia   31327
ERIC HEALD, Tropical BioIndustries Development Company, 9000 South-
    west 87th Court, Suite 104, Miami, Florida   33176
JOHN E. HOBBIE, Marine Biological Laboratory, Woods Hole, Massachu-
    setts,   02543
RICHARD L. IVERSON, Department of Oceanography, Florida State
    University, Tallahassee, Florida   32306

EDWIN A. JOYCE, JR., Bureau of Marine Science and Technology,
    Florida Department of Natural Resources, Tallahassee, Florida
    32304
PETER A. JUMARS, Department of Oceanography WB-10, University of
    Washington, Seattle, Washington 98195
WILLIAM KRUCZYNSKI, U. S. Environmental Protection Agency Region 4,
    345 Courtland Street NE, Atlanta, Georgia 30308
ROBERT J. LIVINGSTON, Department of Biological Science, Florida
    State University, Tallahassee, Florida 32306
D. B. NEDWELL, Department of Biology, University of Essex, Wivenhoe
    Park, Colchester CO4 3SQ, England
SCOTT W. NIXON, Graduate School of Oceanography, Narragansett Bay
    Campus, University of Rhode Island, Kingston, Rhode Island
    02881
JAMES J. O'BRIEN, Department of Meteorology, Florida State Univer-
    sity, Tallahassee, Florida 32306
WILLIAM E. ODUM, Department of Environmental Sciences, University
    of Virginia, Charlottesville, Virginia 22903
RONALD S. OREMLAND, U. S. Geological Survey, Water Resources Divi-
    sion MS21, 345 Middlefield Road, Menlo Park, California
    94025
ROBERT W. POOLE, Division of Biology and Medicine, Brown University,
    Providence, Rhode Island 02912
LAWRENCE R. POMEROY, Institute of Ecology, University of Georgia,
    Athens, Georgia 30602
DANIEL S. SIMBERLOFF, Department of Biological Science, Florida
    State University, Tallahassee, Florida 32306
DAVID W. SMITH, Department of Biological Sciences, University of
    Delaware, Newark, Delaware 19711
HARRIS B. STEWART, JR., U. S. Department of Commerce, NOAA,
    901 South Miami Avenue, Miami, Florida 33130
DONALD R. STRONG, JR., Department of Biological Science, Florida
    State University, Tallahassee, Florida 32306
RICHARD SWARTZ, Environmental Protection Agency Marine Science
    Center, Newport, Oregon 97365
KENNETH TENORE, Skidaway Institute of Oceanography, Skidaway
    Island, Savannah, Georgia 31406
GORDON W. THAYER, NOAA National Marine Fisheries Service, Beaufort
    Laboratory, Beaufort, North Carolina 28516
JOHN H. TIETJEN, Department of Biology, City College of New York,
    New York New York 10031
DAVID THISTLE, Department of Oceanography, Florida State Univer-
    sity, Tallahassee, Florida 32306
ROBERT W. VIRNSTEIN, Harbor Branch Foundation, Route 1, Box 196,
    Fort Pierce, Florida 33450
ROBERT WHITLATCH, Marine Laboratory, University of Connecticut,
    Noank, Connecticut
GEORGE M. WOODWELL, The Ecosystems Center, Marine Biological
    Laboratory, Woods Hole, Massachusetts 02543